Cognition-Based Evolution

Natural Cellular Engineering and the Intelligent Cell

William B. Miller, Jr.

Banner Health Systems
Bioverse Foundation
Paradise Valley, Arizona
USA

CRC Press
Taylor & Francis Group
Boca Raton London New York

CRC Press is an imprint of the
Taylor & Francis Group, an **informa** business

A SCIENCE PUBLISHERS BOOK

Cover credit:
Background image – Image taken from Pixabay License.
Top left image – Image taken from Pixabay License and modified by the author.
Top right image – Image taken from Wiki commons. The file is licensed under the Creative Commons Attribution-Share Alike 3.0 Unported license. Attribution: User:SubtleGuest
Top centre image – Top background from Freepik.com. Bottom foreground content by author, William B. Miller, Jr.

First edition published 2023
by CRC Press
6000 Broken Sound Parkway NW, Suite 300, Boca Raton, FL 33487-2742

and by CRC Press
4 Park Square, Milton Park, Abingdon, Oxon, OX14 4RN

© 2023 William B. Miller, Jr.

CRC Press is an imprint of Taylor & Francis Group, LLC

Library of Congress Cataloging-in-Publication Data (applied for)

ISBN: 978-1-032-26147-8 (hbk)
ISBN: 978-1-032-26148-5 (pbk)
ISBN: 978-1-003-28676-9 (ebk)

DOI: 10.1201/9781003286769

Typeset in Times New Roman
by Radiant Productions

Dedication

To
Linda, Lauren, and Scott

Preface

A major point of emphasis in this book is that our intelligent cells utilize chance events to channel creativity. Chance entered my life through an unanticipated encounter with a 'boy named Sue' on a blustery fall day 15 years ago. 'Sue' was the towering Tyrannosaurus Rex fossilized skeleton that dominates the magnificent rotunda of the Chicago Field Museum. I had visited the museum one late afternoon seeking relaxation after many days of intensive medical meetings. There was no purpose other than a pleasant diversion. Before that moment, my interest in evolution had been superficial. Yet, one glimpse of that titanic creature changed everything. Despite obvious differences in scale, I was startled by unanticipated resemblances between the major bones of the T. Rex and our human ones. My companion that afternoon was an astute physician and colleague who had joined me on this brief excursion. When I expressed my amazement at these anatomic similarities and my further impression that the conventional evolutionary narrative seemed critically lacking, his rejoinder was adamant. I was failing to properly appreciate geological time-spans. It was fanciful to think otherwise. For reasons that still elude me, a gauntlet had been thrown.

My broad medical training as a physician was a singular advantage. I was already struggling to reconcile the patterns of infectious disease that could be observed in my daily medical practice with the then governing perspectives of microbial life that existed several decades back. To my appraisal, their consistent patterns of behavior suggested preferences for specific body locations. If that was the case, they likely had a degree of sensory awareness and intelligence. That thought sparked a preliminary investigation of the intricacies of the microbiome and, rather unexpectedly, led to a second career as an evolutionary biologist focusing on this novel perspective.

In all endeavors, it is always a matter of timing, perseverance, a modicum of innate abilities, and luck. Each of these ingredients contributes to the innumerable connections necessary to succeed in any complex endeavor. Some of these fortuitous relationships merit special mention. Foremost among these was the early endorsement of my unconventional thoughts by my brilliant, now-deceased brother-in-law, Spencer W. Franck, Jr. He discerned a kernel of merit in my arguments when so many others were dismissive. Without his encouragement at a pivotal moment, my efforts would have dwindled, and this volume would never have been written.

I was also blessed to meet a remarkable biologist, John Torday, who became my early invaluable collaborator. The opportunities to exchange ideas and debate his unconventional insights were indispensable to all contained herein. My additional scientific collaborators over the last several years also claim my particular thanks:

Frantisek Baluška, Arthur Reber, Jean Guex, Francisco Enguita, Ana Lúcia Leitão, and Perry Marshall. Although not direct collaborators, a few other scientists deserve personal note due to their overarching influence: Denis Noble, Frank Ryan, Luis Villarreal, Arnold De Loof, Brian J. Ford, James Shapiro, John Lieff, Michael Levin, Chris Fields, Eugene Koonin, Guenther Witzany, Pamela Lyon, Pedro Marijuán, Jaime Cárdenas-García, Loris Zamai, Richard Jacobs, and Tam Hunt. However, it must be emphasized that this book could never have materialized without the hundreds of exceptional scientists cited in this volume. Isaac Newton stated it best in a letter to Robert Hooke in 1675: "If I have seen further it is by standing on the shoulders of Giants". My hope is that this volume will burnish their remarkable achievements and serve as a stepping stone to the next better truth.

These acknowledgments would be woefully incomplete without emphasizing the paramount relationship that sustained this whole enterprise and whose value I regard as immeasurable. None of this would have been possible absent the steadfast and loving support of my inestimable wife, Linda.

Finally, my special thanks goes to Taylor and Francis/CRC Press. Their foresight enables this first volume specifically devoted to Cognition-Based Evolution.

Contents

Chapter 1

Introduction
Evolution Recast

"Sooner or later you must move down an unknown road that leads beyond the range of the imagination, and the only certainty is that the trip has to be made".

Bruce Catton, historian (1899–1978)

In a celebrated 1973 essay, the renowned evolutionary biologist Theodosius Dobzhansky famously exclaimed: "Nothing in biology makes sense except in the light of evolution" (Dobzhansky 2013). Generations of biologists have co-opted this striking phrase in defense of Darwinian evolution by natural selection. The maxim still holds, but decades later, it is now found to be incomplete. Based on definitive research, an illuminating amendment is mandated: "Nothing in biology and evolution makes sense except in the light of cognition".

Scientific investigations warrant that all cells are intelligent at their scale (Ford 2004, 2009, 2017, Lyon 2015, Miller 2016a, Majumdar and Pal 2017, Reber 2019). Consequently, biological and evolutionary development have been upended. Cognition-Based Evolution inaugurates from within this secure, unorthodox base, firmly placing all biology and evolution within a framework of 'the intelligent measuring cell.' (Miller 2016a, 2018, Baluška and Miller 2018, Miller et al. 2019, Torday and Miller 2020, Miller et al. 2020a, Baluška et al. 2021). From this fresh vantage, a contemporary, encompassing evolutionary narrative unfolds that overcomes the accumulating manifest deficiencies within conventional Neodarwinian evolutionary theory.

The initial premises of evolution theory had many forebears but commenced decidedly with the publication of Darwin's seminal, 'On the Origin of Species' in 1859 (Darwin, 2004 [1859]). It was greeted by an immediate onslaught of insistent countervailing attacks on both religious and scientific grounds. Nevertheless, its principle tenets have achieved undeniable primacy, reaching such an extent that Darwinism and evolution are viewed as an identity by the majority of the scientific community and the general public. Any casual web search on evolution populates with nearly limitless sources insisting on the hallmark principles of the modern interpretation of Darwinism: evolution is caused by natural selection acting on random genetic variations that express as gene frequencies in populations. Over decades, natural selection took center stage in our evolutionary narrative. Although there is no question that natural selection plays a significant role in evolutionary development, the 20th century emphasis on that process as a driving force obscured that its consequences must always be post-facto to preceding variations. Consequently, a predominant interpretation of the scope of natural selection has emerged that significantly misinterprets its role. As will be explored, a more productive evolutionary focus centers on the exact manner in which biological variations arise to yield sustainable heritable products. Cognition-Based Evolution (CBE) endorses a robust alternative to the conventional Neodarwinian narrative of random genetic variations. Multicellular evolutionary variation emerges from the collective problem-solving actions of intelligent, measuring cells.

Ironically, Darwin's success has become its inadvertent reciprocal limitation. The beguiling elegance of his theory of natural selection forcefully gripped the imaginations of scientists and the general public with uncommon facility. Its popular reduction to 'survival of the fittest' swiftly became its implicit embodiment. Yet, placing the theory of evolution by natural selection in those terms was a disservice to the meticulous depth of Darwin's thoughts.

Part of its enduring primacy is that once this alluring theory is considered, it seems intuitively obvious. Thomas Henry Huxley (1825–1895), a prominent English biologist and contemporary admirer of Darwin, examined natural selection theory and despondently stated: "How extremely stupid not to have thought of that" (Huxley 1887). Yet, the concept of natural selection was not novel to Darwin. Nor is it the reason that he deserves his place among the pantheon of preeminent scientific intellects. The sharp point of Darwin's brilliance was his deductive observation that creatures inevitably varied over time and in so doing, propelled evolutionary changes. He had no idea how those variations occurred, but he knew that biological flexibility was a consistent product of breeding in animal husbandry and agriculture. He termed these variations 'sports', adopting the common terminology of breeders. With singular insight, he concluded that these variations represented the feedstock of continuous evolutionary change from a common progenitor forward (Miller 2013). He also intuited a further crucial, incumbent linkage. The persistence of these variations depended on their reproductive success as a function of natural selection. In a flash of insight, Darwin was the first to fully grasp the correct biological hierarchy. Variations occur first; selective influences follow. Cognition-Based Evolution (CBE) distinguishes itself from conventional Neodarwinism precisely on these grounds by concentrating on the sources of variation rather than the influence of selection.

Unquestionably, the Darwinian theory of evolution by natural selection has undergone its own form of adaptive evolution for over 100 years. There have been persistent attempts to refresh it by incorporating fresh scientific discoveries, such as the rapidly expanding science of genetics. Consequently, its current embodiment is represented by contemporary Neodarwinism, epitomized as the Modern Synthesis and serving as the current mainstay of evolutionary thought. This modern iteration of Darwinism rests on four cardinal buttresses: a) evolution is due to predominantly random heritable genetic variations, b) the persistence of these variations relates to reproductive success subject to natural selection, c) longitudinal descent through these successive random variations is necessarily gradual, and d) the target of selective influence is primarily centered at the level of the visible macroorganic form (Futuyma 2017, Koonin 2012).

Cognition-Based Evolution (CBE) stands apart from each of these particulars. Most significantly, this entirely different evolutionary explanation inaugurates from a unorthodox base compared to Neodarwinism. All cells are intelligent, and further, that intelligence defines the living state and all subsequent evolutionary development. Beginning from this unprecedented source, CBE proceeds in an entirely distinct manner from Neodarwinism: variations and their deployment as adaptations are the result of intelligent cells collectively assessing and communicating information to problem-solve. This collaborative process is a form of natural engineering, and it is this self-organizing impulse that directly drives multicellularity. Significantly, this constructive co-engineering process requires the co-participation of the three cellular domains (Prokaryota, Archaea, Eukaryota) and the virome. By this means, CBE methodically explains why all multicellular eukaryotes are oligatory holobionts comprised by participants from all of the elemental domains. Multicellular eukaryotes are the unicellular domains' means of continuously exploring and adapting to environmental cues, thereby permitting their permanent planetary perpetuation.

Although not generally recognized, biology has remained an orphan among its brethren sciences. It alone had not uncovered a method of becoming a field of exacting measurement rather than observation. Undoubtedly, sophisticated statistical analysis is applied to population genetics, but it has added less than hoped since it offers no predictive value. While other fields were transforming in the last several decades, especially physics, evolutionary biology remains a largely descriptive enterprise (Moss 2012, Nicholson 2012).

Decades ago, initial genetic discoveries gave tantalizing hope that fundamental causations might be revealed through its deeper scrutiny. Yet, fifty years later, evolutionary biology remained mired in post-facto associations and presumptive correlations. Although our technical abilities to understand genomic patterns have grown explosively, these have yielded no breakthroughs for the 'why' of what we can observe. The genetic building blocks that underpin all living things are being scrupulously identified, but why they are arranged precisely as they are remains as elusive now as decades ago. Biology's principle handicap has been that other than its faith in natural selection, the exact mechanisms that determine biological or evolutionary development have not been illuminated (Scheiner 2010). Despite much research, a coherent central theory of biology beyond Neodarwinian assumptions remained lacking (Torday 2015). Consequently, the long-term major questions about evolution are still currently addressed through compartmentalized theories developed within siloed disciplines (Schreiner 2010).

For example, the issue of the persistence of life is primarily addressed by general biology. On the other hand, the reasons why organisms change and diversify are regarded as the purview of evolutionary biology. While the field of genetics dominates the topic of faithful heritability, the structure and function of cells is the domain of cellular biologists. Further yet, the mystery of how individual organisms maintain their self-integrity and what it means to be an organism is considered appropriate for the philosopher's perch. In contrast, why organisms are distributed as they are is the realm of ecology. Among these varied disciplines, selection has been the single uniting theoretical glue. In the conventional view, natural selection somehow hammers, shapes, and nudges endless iterations of purposeless genetic variations into constructive biological outcomes.

Accordingly, the evolutionary literature has tended to concentrate on issues of the precise targets of selection with the expectation that if these were sufficiently identified, definite answers might ensue. Instead, what is required is a fresh vantage on evolutionary development tasked to resolve the largest issues in biological and evolutionary development, offering coherent answers that had previously eluded Darwinists. Those paramount questions are surprisingly few. Are evolutionary processes random or not? If not, why not? Is there any purpose to the billions of years of evolution that have been documented? If so, what is its deterministic aim? If evolution achieves direction, what force or combination of forces propel it?

Beyond this limited palette, five further issues and their related questions require explanation (Miller 2016a). First, expressly how is heritable variation achieved? Is genetic transmission almost entirely vertical in eukaryotes, or are horizontal genetic exchanges commonplace? Secondly, is selection ultimate causation or just one force among others? Third, is genetic inheritance the authentic controlling process in evolutionary development? Is biology just related to gene frequencies as explicitly taught, or are there other factors to consider? Fourth, is evolution random or not? If evolution is not random, what is the identifiable object of selection? And lastly, why are all multicellular eukaryotes holobionts without exception?

Fortunately, the last decades have provided a wellspring of illuminating research and provocative theorization. These are enabling explorations of biological and evolutionary development along pathways inconceivable to Darwin, his contemporaries, and initial successors. Two specific advances have been singularly important. First among these is the experimental and theoretical research confirming that all cells are cognitive self-referential agents (Baluška and Mancuso 2009, Shapiro 2011, Trewavas and Baluška 2011, Miller 2013, 2016a, 2018, Dodig-Crnkovic 2014, Lyon 2015). Crucially, cognition is present at every scale and defines life (Ford 2004, 2009, 2017, Trewavas and Baluška 2011, Miller 2013, 2016a, Baluška and Levin 2016, Vallverdú et al. 2018, Reber 2019, Baluška et al. 2021, 2022). Critical research substantiates an expansive suite of cellular faculties that were unimagined a mere two decades ago (Ford 2004, 2009, 2017, Lyon et al. 2021). True, cellular awareness is limited. Yet, it is effective and of sufficient magnitude to permit cells and cellular microbes to assess, communicate and deploy information to consistently support their individual homeorhetic preferences. Homeorhesis denotes a dynamical system that tends to return to a particular trajectory of energetic flux as opposed to the earlier concept of homeostatic equilibrium. (Miller 2013, 2016a, Baluška and Levin 2016, Miller et al. 2020a). Consequently, in a cognitive frame, information processing is the core cellular activity and underlies all resulting cell-cell communications that lead to the full array of biological outcomes. Most importantly, basal cognition does not merely enable collaboration, cooperation, co-dependency, and effective competition among cells but specifically impels it (Ben-Jacob and

Levine 2006, Ben-Jacob 2009, Miller 2016a, 2018, Miller and Torday 2018, Miller et al. 2020a,b). The basis of these conjoining actions is the cellular capacity to measure informational inputs. In so doing, each cell can engage in non-random intentional actions through cognitive information processing. Consequently, cognitive cells should be appraised as deliberate measuring instruments (Miller 2016a, Miller et al. 2019, 2020a,b, Baluška and Miller 2018, Baluška et al. 2021, 2022). Further, and most critically, those measurements are directly aimed at sustaining cellular homeorhesis versus the constant flow of environmental stimuli to protect the self-identity of every cell. This process, over time, is adaptive evolution, assuring the continuous cellular assimilation of the environment.

The second requisite shift critical to a productive reappraisal of Neodarwinism is our contemporary understanding of the multicellular eukaryotic form. Metagenomic techniques have revealed that all multicellular eukarytoes are holobionts as intimate entwined, reciprocating partnerships between eukaryotic cells and a vast microbial cohort (Miller 2016b). These vast confederacies are linked cellular ecologies with constituents of mixed cellular species that function together so seamlessly as to act as a single being but is manifestly not. It has only been with the recent advent of advanced genomic and metagenomic analysis that the astounding extent of microbial participation in eukaryotic life could be adequately grasped (Miller 2016b). Pertinently though, the collaborations that vitalize this collective form of life are exactly centered within ubiquitous cognition and its adjunctive cell-cell communication (Schauder and Bassler 2001, Visick and Fuqua 2005, Torday and Rehan 2012). Consequently, Darwinian evolution must be recast as an exploration of why linkages of these specific participants consistently recapitulate as a vast totality of intimately reciprocating ecologies as reproducing holobionts (Miller 2013, 2016a).

In addition to these essentials, other transformative elements are required to compel a new evolutionary narrative. Chief among these is our burgeoning understanding of the complexities and extent of cell-cell communication galvanized by cellular cognition. Crucially, how cognitive cells appraise information and deploy it can now be more effectively analyzed. Critical, too, has been the several decade-long accumulation of knowledge about epigenetic (non-genetic influences on gene expression) impacts and their heritable transmission, effecting both immediate and heritable changes on holobionts (Jablonka et al. 1992, Jablonka and Lamb 1998, Skinner 2015) Skinner 2015). Until just a few years ago, epigenetic influences were considered merely transient and arbitrated through natural selection (Feng et al. 2010, Houri-Zeevi and Rechavi 2017). For decades, biologists remained averse to accepting the crucial role of epigenetics in evolutionary development, tainted by disdain for previously discarded Larmarckianism that had championed the possibility of inheritance of acquired traits. Yet, contemporary research asserts that a surprisingly consequential range of non-random epigenetic influences affect flexible genomes, assuming a substantial role in evolutionary development (Jablonka and Lamb 2008, Laland et al. 2015, Torday and Miller 2020, Miller et al. 2021). Therefore, the prior presumption of strict vertical inheritance is no longer tenable.

The authority of 20th century Neodarwism rested on its confident assumption that evolution is the chronicle of differential fitness of heritable variations resulting from random genetic mutations within central genomes (Bowler 1989, Poelwij et al.

2007). Within that framework, it seemed logical that evolution can be reasonably ascribed to a matter of changing gene frequencies disciplined by natural selection (Koonin 2009). However, 21st century research animates a thorough rethinking of genetic variation and evolutionary development. Consequently, Cognition-Based Evolution (CBE) merely follows the time-honored scientific tradition of periodic revision of any dominating theory based on fresh evidence. It was no different with early 20th century Darwinists. When confronted with the newly evolving science of genetics, they accommodated it by folding its initial findings into the standard architecture of Darwinism, thereby expanding it into an amended Modern Synthesis. However, there is a salient difference in this present revisionary instance compared to any prior. CBE is not just a further appendage to Darwinism but a thoroughly disruptive reconceptualization. Specifically, it argues that evolution is an informational interactome based on self-referential cellular dynamics and not a direct result of natural selection or gene frequencies.

Renaissance artists used compositional chiaroscuro to create an intense, dramatic light in their canvases, heightening their emotional tension. Originated by Leonardo Da Vinci in the late 15th century, these powerful effects were exemplified in the works of Caravaggio with their arresting contrasts between light and shadow. Although superficial viewing of one of these paintings suggests that the details are singularly focused within the brilliantly highlighted features of the figures, close inspection of any of these canvases reveals that these master artists paid meticulous attention to every aspect of the shadows. They instinctively understood that the complexity of the shadows required their intense concentration since they were as consequential to the whole as any intensely illuminated portion. Consequently, they filled the large dark shadows with texture, gradation, and variety. In so doing, they were fulfilling a neuroesthetic impulse to achieve a specific type of symmetry, balance, and complexity (Correa-Herran et al. 2020). Experiments show that if the shadow details are graphically eliminated from the image, it is rendered lifeless. By analogy and due to both instinct and previous technological limitations, our concepts of evolutionary processes have been based on rigorous examination of what had been readily accessible in biological light. Now, well armed with fresh scientific evidence and technical advances, we can turn our methodical scrutiny to the rich details in biology's former shadows.

Indeed, the explicit focal difference between Neodarwinism and Cognition-Based Evolution is the gap between that metaphorical light and shadow, the seen and the unseen, the conspicuous macroorganic reproductive organisms, and the microscopic world of cells and microbes. Until recently, and unlike its brethren 21st century counterpart sciences such as chemistry or physics, biology was based primarily on field observations rather than laboratory experimentation. As an observational and descriptive field, evolutionary biology and evolutionary ecology have largely concentrated on the macroscopic form. Naturally, this scrutiny drove evolutionary theory to accommodate and rationalize those ready, limited observations. Although the microscope had been invented by the early 1600s, few evolutionary biologists in the prior century paid great attention to the microscopic sphere, their means of communication, or their immunological rules. These were of little interest compared to the intriguing world of our dazzling, visible plants and animals. The shadowed microscopic details received much less scrutiny. Now, in our contemporary era, both the microscopic and macroorganic sides can be fully honored. Accordingly, evolutionary biology can be productively reappraised as a complex holobionic entanglement, perpetually and faithfully rooted within its cellular origins across planetary evolution (Miller 2013, 2016a, Miller and Torday 2018, Miller et al. 2020a).

From this new ground state, a cohesive integrated evolutionary path emerges, connecting the primoridal unicellular forms to the complexities of multicellular life. Through this means, some long-contested questions are settled. Does evolution have any identifiable direction? Are there any exact principles of evolution akin to physical laws? If not, are there cellular rules that can be identified that permit useful prediction of biological outcomes? Is creativity exclusive to humankind, or is it evident at all living scales? Is natural selection the true shaping force in evolution as its ultimate causation? If not, what else is driving biological outcomes? Is it true that our evolutionary narrative reduces to gene frequencies as population geneticists contend, or do other factors have paramountcy? This volume with address each of those dilemmas.

As is always the case, fresh questions naturally derive from within those answers. For example, what are the contextual behaviors of the constituent cells within those tissue ecologies that comprise us as holobionts? Why do multicellular eukaryotes engage in an obligatory recapitulation through a unicellular stage? Why is there a continued assumption that the microbial fraction of a holobiont is little more than a consequential appendage instead of an essential participant (Miller 2016a, Corning and Szathmáry 2015, Theis et al. 2016)? Most pertinently, why has the role of universal basal cognition in biological development been previously ignored by most biologists?

Beyond these readily apparent questions, CBE stimulates several unique queries that have been overlooked in prior considerations of evolutionary development. First, 'Who is the observer and who is a participant?' The other, 'Who is serving and who is being served?' (Torday and Miller 2016). These two lead to the most essential question in biology that is only conceivable within a cognitive framework that energizes cooperative engineering between the cellular domains and the virome: 'Whose problems are being solved?'

One of the reasons that these have not been previously explored is that some of the answers do not fit earlier preconceived notions of biological hierarchies. Indeed, as will be explained, we are not the unambivalent aim of evolutionary development on the planet. Instead, evidence indicates that the only continuous direction in evolution is the consistent perpetuation of the four elemental domains, comprised by three cellular domains (Prokaryota, Archaea, Eukaryota), and the virome that functions as its flexible intermediary (Miller and Torday 2018). These primordial forms of life, and only these, have exerted permanent collective dominion over the planet for billions of years. All the stunning forms of visible life that engross us are their aggregates. Necessarily, then, the essential cellular constituents of each of those creatures represent the central action of life on the planet.

In comparison, the macroorganic species that rivet our attention are transient planetary occupants. Mounting evidence supports the disquieting conclusion that the visible organisms that form the light of biology to our keen but superficial appraisal are largely subordinate to our invisible co-partners in the realm of shadows. Disconcerting as it might seem, the unicellular world achieves its continuous protection and permanent planetary place through us. Once this is grasped, an in-depth reappraisal of evolution is justified based on self-referential cellular requisites and the collective, collaborative, and competitive behaviors that are its consequence. The pertinent details lie in their efficient patterns of cell-cell communication and the intricacies of the informational matrix that permits their conjoint action (Torday and Rehan 2012, Miller 2018, Miller et al. 2020b).

Neodarwinism has dominated evolution over the last half-century, melding natural selection with an early, limited grasp of genetics. Random genetic variations, the absolute primacy of natural selection, and the presumption that the macroorganic reproductive unit represents the obvious target of selection are its principle tenets. Yet, each of these is now known to be incorrect. Consequently, evolution and biological development can now undergo a comprehensive re-evaluation based on the specific qualities of cognition that trigger cell-cell signaling of shared information among reciprocating, receptive cells. From within this new, compelling framework, a fresh universal synthesis emerges. Correctly centering evolutionary biology within cellular cognition places it squarely within the realm of measurement. At long last, evolutionary biology is propelled beyond descriptive associations and correlations into the realm of testable predictions subject to refutation. It is only in this way that biology can meet Popper's Falsification Principle—for a theory to be considered scientific, it must be testable so that it might be proved false (Helfenbien and DeSalle 2005).

Achieving this goal rests on scientific advances in microbiology, metagenomics, molecular biology, ecology, and immunology. All of these fields have dramatically progressed in the last decades. Prior to CBE, evolutionary biology had yet to fully incorporate them. Based on accumulated advances across the scientific spectrum, a surprising central principle of life has been uncovered. The universal bond of cellular life is the protection of the self-integrity of the collaborating participants (Miller et al. 2019). Cells have learned that there is a common currency of life. Survival is best attained by being in service to others.

It is part of human nature that what we see is often based on inherent presumptions. As John Lubbock, the British naturalist, keenly observed: "What we see depends mainly on what we look for". Consequently, generations of scholars have attributed evolutionary development to natural selection and presumed that it could shape random genetic variations if given enough time. This is evolution by accident. In its stead, Cognition-Based Evolution offers an unreservedly different perspective. Biological and evolutionary development is bound by the shared cognitive awareness of the three cellular domains and their resulting patterns of co-engineering with their companion virome, enjoined in a continuous, common planetary narrative for over 3.8 billion years. As the physician and scientist Jon Lieff (2020) elegantly concludes in 'The Secret Language of cells', "Although cells are considered the most basic characteristic that defines life, it is actually the conversations among cells, and also the conversations that take place inside them, that determines biological activity and produces the essence of life". That communication has one perpetual aim—to protect the integrity of self-aware cells. To succeed, cells are equipped to sense and assess environmental information to permit their own self-directed and intelligence-based non-random adaptations. With this as the common cellular nomenclature, CBE unmoors evolution from its observational and descriptive roots and refashions biology into the scientific study of the cellular measurement of information and its productive communication. Consequently, the primary focus of biology shifts from cataloging what can be seen and counted with our eyes to the precise measurement of cellular dynamics at the microscopic scale.

CBE answers many confounding questions that have bedeviled evolutionary biology for over 150 years. Yet, undoubtedly, and in requisite scientific symmetry, the resolution of old riddles bindingly equates with the birth of parallel unknowns. And each of these fresh puzzles will stimulate its own dogged and creative inquiries, fathomed through pioneering and unconventional solutions that will serve to reveal the next set of mysteries.

> *"Discovery follows discovery. Each both raising and answering questions, each ending a long search and each providing the new instruments for a new search".*

—Robert Oppenheimer, physicist

References

Baluška, F. and Levin, M. 2016. On Having No Head: Cognition throughout Biological Systems. Frontiers in Psychology 7(902).

Baluška, F. and Mancuso, S. 2009. Deep evolutionary origins of neurobiology: Turning the essence of "neural" upside-down. Communicative & Integrative Biology 2(1): 60–65.

Baluška, F. and Miller, Jr, W.B. 2018. Senomic view of the cell: Senome versus Genome. Communicative & Integrative Biology 11(3): 1–9.

Baluška, F., Miller, W.B. and Reber, A.S. 2021. Biomolecular Basis of Cellular Consciousness via Subcellular Nanobrains. International Journal of Molecular Sciences 22(5): 2545.

Baluška, F., Reber, A.S. and Miller, W.B. 2022. Cellular sentience as the primary source of biological order and evolution. Biosystems 218: 104694.

Ben-Jacob, E. 2009. Learning from Bacteria about Natural Information Processing. Annals of the New York Academy of Sciences 1178: 78–90.

Ben-Jacob, E. and Levine, H. 2006. Self-engineering capabilities of bacteria. Journal of The Royal Society Interface 3(6): 197–214.

Bowler, P.J. 1989. Evolution: The History of an Idea. Berkeley, CA, United States: University of California Press.

Corning, P.A. and Szathmáry, E. 2015. "Synergistic selection": A Darwinian frame for the evolution of complexity. Journal of Theoretical Biology 371: 45–58.

Correa-Herran, I., Aleem, H. and Grzywacz, N. 2020. Evolution of Neuroaesthetic Variables in Portrait Paintings throughout the Renaissance. Entropy: 22(2): 146.

Darwin, C. 2004. On the Origin of Species, 1859. United Kingdom: Routledge.

Dobzhansky, T. 2013. Nothing in biology makes sense except in the light of evolution. The American Biology Teacher, 75(2): 87–91.

Dodig-Crnkovic, G. 2014. Modeling Life as Cognitive Info-Computation. *In*: A. Beckmann (ed.). Computability In Europe. Berlin: Springer.

Feng, S., Jacobsen, S.E. and Reik, W. 2010. Epigenetic Reprogramming in Plant and Animal Development. Science 330(6004): 622–627.

Ford., B.J. 2004. Are Cells Ingenious? Microscope 52(3/4): 135–144.

Ford, B.J. 2009. On Intelligence in Cells: The Case for Whole Cell Biology. Interdisciplinary Science Reviews 34(4): 350–365.

Ford, B.J. 2017. Cellular intelligence: Microphenomenology and the realities of being. Progress in Biophysics and Molecular Biology 131: 273–287.

Futuyma, D.J. 2017. Evolutionary biology today and the call for an extended synthesis. Interface Focus 7(5): 20160145.

Helfenbein, K.G. and DeSalle, R. 2005. Falsifications and corroborations: Karl Popper's influence on systematics. Molecular Phylogenetics and Evolution 35(1): 271–280.

Houri-Zeevi, L. and Rechavi, O. 2017. A Matter of Time: Small RNAs Regulate the Duration of Epigenetic Inheritance. Trends in Genetics 33(1): 46–57.

Huxley, T.H. 1887. On the Reception of the 'Origin of Species'. pp. 179–204. *In*: F. Darwin (ed.). The life and letters of Charles Darwin, including an autobiographical chapter. London: John Murray.

Jablonka, E. and Lamb, M.J. 1998. Epigenetic Inheritance and Evolution: The Lamarckian Dimension. Oxford, UK: Oxford University Press.

Jablonka, Eva, Lachmann, M. and Lamb, M.J. 1992. Evidence, mechanisms and models for the inheritance of acquired characters. Journal of Theoretical Biology 158(2): 245–268.

Jablonka, Eva and Lamb, M.J. 2008. The Epigenome In Evolution: Beyond The Modern Synthesis. Vosgis Herald 12.

Koonin, E. V. 2009. Darwinian evolution in the light of genomics. Nucleic Acids Research 37(4): 1011–1034.

Koonin, Eugene V. 2012. The logic of chance: the nature and origin of biological evolution. Upper Saddle River, NJ: Pearson Education.

Laland, K.N., Uller, T., Feldman, M.W., Sterelny, K., Müller, G.B., Moczek, A., … Odling-Smee, J. 2015. The extended evolutionary synthesis: its structure, assumptions and predictions. In: *Proceedings* of the Royal Society B: Biological Sciences (Vol. 282, p. 20151019).

Lieff, J. 2020. The Secret Language of Cells: What Biological Conversations Tell Us About the Brain-Body Connection, the Future of Medicine, and Life Itself. Dallas, Texas: BenBella Books.

Lyon, P. 2015. The cognitive cell: bacterial behavior reconsidered. Frontiers in Microbiology 6(264).

Lyon, P., Keijzer, F., Arendt, D. and Levin, M. 2021. Reframing cognition: getting down to biological basics. Philosophical Transactions of the Royal Society B: Biological Sciences 376(1820): 20190750.

Majumdar, S. and Pal, S. 2017. Bacterial intelligence: imitation games, time-sharing, and long-range quantum coherence. Journal of Cell Communication and Signaling 11(3): 281–284.

Miller, W.B. 2013. The Microcosm Within: Evolution and Extinction in the Hologenome. Boca Raton, Florida, United States: Universal Publishers.

Miller, W.B. 2016a. Cognition, Information Fields and Hologenomic Entanglement: Evolution in Light and Shadow. Biology 5(2): 21.

Miller, W.B. 2016b. The Eukaryotic Microbiome: Origins and Implications for Fetal and Neonatal Life. Frontiers in Pediatrics 4(96).

Miller, W.B. 2018. Biological information systems: Evolution as cognition-based information management. Progress in Biophysics and Molecular Biology 134: 1–26.

Miller, W.B., Baluška, F. and Torday, J.S. 2020a. Cellular senomic measurements in Cognition-Based Evolution. Progress in Biophysics and Molecular Biology 156: 20–33.

Miller, W.B., Enguita, F.J. and Leitão, A.L. 2021. Non-Random Genome Editing and Natural Cellular Engineering in Cognition-Based Evolution. Cells 10(5): 1125.

Miller, W.B. and Torday, J.S. 2018. Four domains: The fundamental unicell and Post-Darwinian Cognition-Based Evolution. Progress in Biophysics and Molecular Biology 140: 49–73.

Miller, W.B., Torday, J.S. and Baluška, F. 2019. Biological evolution as defense of 'self'. Progress in Biophysics and Molecular Biology 142: 54–74.

Miller, W.B., Torday, J.S. and Baluška, F. 2020b. The N-space Episenome unifies cellular information space-time within cognition-based evolution. Progress in Biophysics and Molecular Biology 150: 112–139.

Moss, L. 2012. Is the philosophy of mechanism philosophy enough? Studies in History and Philosophy of Science Part C: Studies in History and Philosophy of Biological and Biomedical Sciences 43(1): 164–172.

Nicholson, D. J. 2012. The concept of mechanism in biology. Studies in History and Philosophy of Science Part C: Studies in History and Philosophy of Biological and Biomedical Sciences, 43(1): 152–163.

Poelwijk, F.J., Kiviet, D.J., Weinreich, D.M. and Tans, S.J. 2007. Empirical fitness landscapes reveal accessible evolutionary paths. Nature 445(7126): 383–386.

Reber, A.S. 2019. The First Minds: Caterpillars, Karyotes, and Consciousness (Illustrate). New York, NY, USA: Oxford University Press.

Schauder, S. and Bassler, B.L. 2001. The languages of bacteria. Genes & Development 15(12): 1468–1480.

Scheiner, S.M. 2010. Toward a Conceptual Framework for Biology. The Quarterly Review of Biology 85(3): 293–318.

Shapiro, J.A. 2011. Evolution: A View from the 21st Century. Upper Saddle River, NJ: FT Press.

Skinner, M.K. 2015. Environmental Epigenetics and a Unified Theory of the Molecular Aspects of Evolution: A Neo-Lamarckian Concept that Facilitates Neo-Darwinian Evolution. Genome Biology and Evolution 7(5): 1296–1302.

Theis, K.R., Dheilly, N.M., Klassen, J.L., Brucker, R.M., Baines, J.F., Bosch, T.C.G., … Bordenstein, S.R. 2016. Getting the Hologenome Concept Right: an Eco-Evolutionary Framework for Hosts and Their Microbiomes. MSystems 1(2).

Torday, J.S. 2012. Evolutionary Biology: Cell-Cell Communication, and Complex Disease (1st ed.). Hoboken, New Jersey: Wiley-Blackwell.

Torday, J.S. 2015. A Central Theory of Biology. Medical Hypotheses 85(1): 49–57.

Torday, J.S. and Miller, W.B. 2016. Biologic relativity: Who is the observer and what is observed? Progress in Biophysics and Molecular Biology, 121(1): 29–34.

Torday, J. and Miller, W. 2020. Cellular-Molecular Mechanisms in Epigenetic Evolutionary Biology. Cham, Switzerland: Springer.

Torday, J.S. and Rehan, V.K. 2012. Evolutionary Biology, Cell-Cell Communication and Complex Disease. Hoboken, New Jersey: Wiley-Blackwell.

Trewavas, A.J. and Baluška, F. 2011. The ubiquity of consciousness. EMBO Reports 12(12): 1221–1225.

Vallverdú, J., Castro, O., Mayne, R., Talanov, M., Levin, M., Baluška, F., … Adamatzky, A. 2018. Slime mould: The fundamental mechanisms of biological cognition. Biosystems 165: 57–70.

Visick, K.L. and Fuqua, C. 2005. Decoding Microbial Chatter: Cell-Cell Communication in Bacteria. Journal of Bacteriology 187(16): 5507–5519.

CHAPTER 2

Neodarwinism
Evolution in the 20th Century

"Mendel and Darwin made the perfect couple. The sad thing is, of course, like most perfect couples they never met".

Steve Jones—Professor of Genetics,
University College of London,
BBC, In Our Time, Genetics, Dec 13, 2001

There is no dispute that the publication of Darwin's *On the Origin of Species* in 1859 and the ensuing *Descent of Man* in 1871 initiated a raucous evolutionary debate within both scientific and public circles that roils to this day (Darwin, 1859, 1871). With singular insight, Darwin declared that evolution was due to continuous small biological variations whose persistence was governed by natural selection. Contrary to prevailing religious doctrine and horrifying many, Darwin insisted that the evolution of all living creatures was the result of descent from common ancestry through a continuous stream of gradual modifications (Bowler, 1983, 2003). Over the following years, evolutionary theory has been continually modified, attempting to remain concordant with fresh scientific evidence. Nonetheless, contentious issues remain involving the most basic processes of evolution, such as how heritable variations occur, the origin of organic complexity, and the mode and tempo of evolutionary change.

When Darwin offered his influential work in 1859, he was unaware of the existence of genes. His theories were based on scrupulous visual observation and skillful, intuitive reasoning. Even though cells had been discovered nearly 200 years before, Darwin could not conceive of their relevance to evolutionary changes. His work led him to offer two linked conclusions that remain essential to modern Neodarwinism: evolution proceeds through the gradual accumulation of inherited variations shaped by natural selection.

Notably, Darwin was not the first to propose evolution by natural modification or even the concept of natural selection. His grandfather Erasmus Darwin, a British physician, had articulated a concept of a 'struggle for existence' based on the works of even earlier classical scholars (Krumbhaar 1931). Separately, Patrick Matthew, a Scottish horticulturist, had proposed a theory of natural selection in an 1831 book 'Naval Timber and Arboriculture'. Darwin was acquainted with his work, though unwilling to credit it (Rampino 2011). In that 1831 book, and in a following one in 1860, Matthew argued that natural selection is fundamental to the origin of species. His ideas differed from Darwin's Origin since he es poused an episodic narrative of evolutionary change with long periods of evolutionary stability intermittently disrupted by catastrophic mass extinctions (Rampino 2011). Unquestionably, Matthew's insights were keen. In an 1831 journal article, he offered an apt general theory of natural selection: "As the field of existence is limited and pre-occupied, it is only the hardier, more robust, better suited to circumstance individuals, who are able to struggle forward to maturity" (Matthew 1831, p. 384).

Darwin's most unsettling moment was inadvertently triggered by another English naturalist, Alfred Russel Wallace. In a provocative, detailed correspondence with Darwin conducted from his research post in the South Seas and prior to the publication of the 'Origin', Wallace outlined a putative mechanism for evolution that highlighted his detailed concept of natural selection. At first, Darwin was devastated. Wallace had articulated exactly those principles he was preparing to announce after 20 years of meticulous groundwork. In desperation, he wrote to Wallace and requested Wallace's permission to use his ideas for his own presentation. Graciously, Wallace consented, acknowledging that Darwin's formidable reputation in the scientific community was necessary to achieve any thoughtful examination of such an explosive theory (Ruse 2008).

Though often overlooked, Darwin was adding to an already lively debate about the origin of living creatures stretching back over two thousand years. In the early 19th century and preceding Darwin's perspectives by over 50 years, French naturalist Jean-Baptiste Lamarck formulated a revolutionary notion of species variation and organic transformation. He argued a theory of acquired characteristics to account for evolutionary changes. Organisms could gain or lose capacities based on their habits. Those traits of greater use were strengthened, and those of little or no advantage dwindled. Over great periods, small changes would accumulate with individuals inheriting features that their ancestors preferentially exercised (Bowler 2003). Common examples of Lamarckianism are increases in the length of giraffes' necks from stretching to reach leaves or blacksmiths utilizing upper body strength for work having stronger children. In this manner, patterns of respective use or disuse governed heritable shifts in response to environmental demands. Although Lamarck was not the first to promulgate a theory of acquired characteristics, he was the first to coherently articulate and defend it, much as Darwin was the first to convincingly argue for a theory of natural selection (Gould 2002). Darwin did not dismiss this theory. On the contrary, he was partially inclined to Larmarck's ideas. In his 1868 book, 'Variation in Plants and Animals under Domestication', Darwin offered a variant of Lamarckism as a secondary mechanism in evolution termed 'Pangenesis'.

Larmarck's theories were well-regarded throughout much of the 19th century, with many scientists advocating that the inheritance of acquired characteristics was an important evolutionary mechanism. It was championed by several influential biologists, including the renowned German biologist Ernst Haeckel, adding his own form of support through his recapitulation theory, "ontogeny recapitulates phylogeny". Offering various illustrations, he hypothesized that the developmental stages from an embryo to an adult organism represented successive stages in the evolution of its prior ancient ancestors (Bowler 2003). Nonetheless, with the discovery of genes and the formulation of Neodarwinism, Lamarckian acquired traits became discredited among academics to the point of ridicule (Grafton et al. 2013). However, in the modern era, this concept has been revived as the scientific field of epigenetics, now regarded as indispensable in any coherent evolutionary narrative. Epigenetic mechanisms are a necessary part of the evolutionary origin of cell differentiation (Stearns and Hoekstra 2000), and transgenerational inheritance of acquired characteristics is known to proceed along multiple pathways (Singer 2009, Rosenberg et al. 2009, Colson and Raoult 2012, Miller et al. 2021).

Even in his own time, adherents of Darwin were aware that he had never offered a precise explanation of how natural selection created new species (Bowler, 1983). George Romanes, an English physiologist and an academic friend of Darwin, tackled the issue by emphasizing variations in reproductive success through physiological selection as a source of new species, coining the term 'neo-Darwinism'. Ultimately, new discoveries in genetics and mutation theory overcame the apparent problems with Darwin's original premise of variation based on blending inheritance, asserting that progeny inherit an average of the characteristics of their parents. Direct observation contradicted it. With the discovery of Mendelian inheritance through genes, Darwin's natural selection could be merged with genetics, leading to the Neodarwinian Modern Synthesis, which remains the currently dominant evolutionary theory (Gould 2002, Miller 2013).

Key to this direction was the pioneering statistical approach to population genetics developed by Fisher, Haldane, and Wright in the 1920s and 1930s. Their work represented a formalized study of genetic variation focusing on the distribution of gene frequencies operating under natural selection. However, even from this vantage, how genetic frequencies produced species remained unaccounted. The German evolutionary biologist Ernst Mayer provided one answer in 1942 in 'Systematics and the Origin of Species', emphasizing a concept of allopatric speciation in which subpopulations within species diverge due to geographic isolation limiting reproductive exchanges. Through unspecified means, this reproductive isolation yields separate species over time, presumably on the basis of a general genetic drift (Larson 2004). Mayr regarded reproductive isolation as an absolute prerequisite for speciation, introducing the largely agreed upon definition of a species as, "groups of actually or potentially interbreeding natural populations, which are reproductively isolated from other such groups" (de Queiroz 2005). This enduring presupposition remains a cornerstone of evolutionary biology as an essential element of the Modern Evolutionary Synthesis.

It is little remembered that Darwin's theories did not gain widespread acceptance until the 1930s and 1940s with the emergence of the Modern Synthesis. That rapprochement between Darwin's foundational ideas and Mendelian genetics wedded evolutionary theory to the belief that genetic variations arise by chance through random genetic mutation as reproductive errors. By the latter part of the 20th century, this general unification of Darwinian natural selection with genetics became codified as the 'Modern Evolutionary Synthesis'. Its core element was a staunch belief that random genetic variations propelled evolutionary changes, and any genetic shifts could only be properly understood within the context of gene frequencies within populations due to selective pressures. Most of these random genetic variances were ascribed to replication coding errors as chance DNA point (nucleotide) mutations. However, other factors were considered to have subsidiary relevance, including random genetic flows secondary to geographic population barriers, genetic drift within a population's gene pool due random events other than natural selection, the origination of new populations due to founder effects in geographically isolated populations, and non-random and assortative mating within populations. Explicitly then, the Modern Synthesis was committed to the primacy of genes, reproductive fitness, selective advantage, and evolution by gradual modification (Provine 2004, Koonin 2009, Shapiro 2011).

During the last century, several challenges to standard Darwinism have been mounted, especially over the controversial issue of gradualism in evolution. In 1936, Richard Goldschmidt, a political immigrant to the United States and former Director of the Kaiser Wilhelm Institute of Biology, forcefully argued against gradualism in evolution. In 'The Material Basis of Evolution' in 1940, he wrote, "the change from species to species is not a change involving more and more additional atomistic changes, but a complete change of the primary pattern or reaction system into a new one, which afterwards may again produce intraspecific variation by micromutation" (Gould 1977). Goldschmidt's attempt to reconcile macromutations with micromutations attracted near universal disapprobation largely derided among evolutionary biologists who strongly preferred their conventional narrative of countless minimal mutational changes (Gould 1982). How these microevolutionary shifts might become macroevolutionary phenotypes was poorly grasped, but presumably, it would simply iron out over time. How microevolutionary changes become biological novelty at the macro level was a similar mystery. Nonetheless, gradualism in evolution by chance errors during reproduction remained conventional theory despite our ready observations that identifiable mutations in plants and animals display nearly universally harmful effects.

In response to this conundrum, the Japanese biologist Motoo Kimura proposed the 'neutral theory' of molecular evolution as one potential solution. In this framework, most mutations are neutral regarding selection pressure and, consequently, do not directly affect fitness (Kimura 1968). Mutations need not be overtly positive since purifying selection would eliminate the most harmful mutation and the residual beneficial biological results could accrue over time through genetic drift (Kimura 1991, Ohta and Gillespie 1996). Unfortunately, further research in population genetics has offered no direct support. As an alternative, in the 1960s, W.D. Hamilton and Maynard Smith advanced a theory of 'kin selection', achieving substantial popularity and becoming a centerpiece of Neodarwinism. Kin selection proposes that genes seek their own preservation, ultimately expressing at the level of gene frequencies in populations. Any acts of altruism among organisms were merely manifestations of 'inclusive fitness' within the larger gene pool. Richard Dawkins (1989) was the most ardent proponent of this approach, insisting that genes are immortal and must be regarded as the primary unit of selection, existing only to preserve themselves. His book, 'The Selfish Gene', which articulates this theory, is frequently listed among the most influential books in history. Discouragingly, that book and several that followed amplifying the same theme have seduced two generations of biologists into becoming incautiously trapped within its confines.

Despite the nearly universal acceptance of Neodarwinism in the 20th century as a Modern Evolutionary Synthesis, many scholars expressed concerns about its deficiencies. Massimo Pigliucci (2007) offered an incisive summary essay about the long journey from Darwin to the Modern Evolutionary Synthesis titled 'Do we need an extended evolutionary synthesis?' (Pigliucci 2007). In Darwin's time, two questions were paramount as the emerging discipline of evolutionary biology struggled to separate itself from religious beliefs: (1) how can the history and diversity of life be explained, and (2) how can the apparent match between form and function in organisms be reconciled? As Pigliucci emphasized, biology had mistakenly become a theory of genes rather than a theory of forms. Others, too, recognized this limitation, and by the end of the 20th century, evolutionary biologists were prepared to reluctantly reconsider the primacy of the Modern Synthesis.

Critical to this re-evaluation were two salient developments. First was the acceptance of an endosymbiosis theory based on innovative research of biologist Lynn Margulis and others, which acknowledged that symbiotic cooperation was crucial to cellular life (Gray 2017). Secondly, the commonality of horizontal gene transfers among the cellular domains and the virome was revealed, thoroughly undercutting the previous implicit presumption within the Modern Synthesis that evolution depended exclusively on Mendelian vertical inheritance (Torday and Miller 2020).

As a result, a willingness slowly gathered to explore an evolutionary narrative extending beyond random genetic occurrences and natural selection. Among many leaders, Lynn Corporale asserted that flexible genomes could respond in a non-random and strategic manner with natural selection acting to enhance the possibility of favorable genetic mutations (Corporale 2003, Corporale and Doyle, 2013). Fodor and Piattelli-Palmarini (2010) argued that natural selection could not explain evolution. Instead, they emphasized the need for creativity in evolution which natural selection could never supply. The emerging science of epigenetics had a conspicuous impact on the debate. Jablonka and Lamb (2008) championed epigenetics as a series of Lamarckian adjustments to environmental impacts (Jablonka 1999, Jablonka and Lamb 2008). Crucially, these were not random genetic adjustments, and some were heritable.

Confirming the importance of these epigenetic mechanisms has led to a more comprehensive attempt at an alternative to the Modern Synthesis as an Extended Evolutionary Synthesis (EES) arguing that phenotype can be flexibly modified by epigenetic inputs yielding developmental and phenotypic plasticity. In this framework, organisms continually adapt, even at the level of mature organisms (Laland et al. 2014, 2015). In striking contrast to standard dogma, heredity extends beyond a fixed genome with the transmission of developmental, metabolic, and behavioral resources from parent to offspring based on environmental encounters.

Further, these epigenetic modifications not only affect somatic cells but can also be transmitted to the germ line. Within this new synthesis, mutually reinforcing interactions should be regarded as 'niche construction' activities between organisms and their environments. By this ongoing feedback mechanism, organisms can shape their own trajectories in response to environmental stresses through consistent reciprocity between the capacities of an organism and the external environment. Consequently, ESS offered a more pluralistic approach to evolutionary development than the gene-centric Modern Synthesis, emphasizing ecological interactions, plasticity, reciprocal constructive processes, and even cultural adaptations to help account for organismal complexity (Müller 2017).

Rose and Oakley (2007) substantially influenced this shift in opinion. They insisted that the prevailing viewpoint within the Modern Synthesis that the genome was a "well-ordered library of genes", each with its distinct function, was unsustainable based on contemporary research. Furthermore, research had also uncovered a plethora of genetic influences that disrupt any simplistic narrative of genetic regimentation, including horizontal genetic transfers, the protean influence of transposable elements, differential meiotic segregation, genetic duplications, and the vast inventory of RNAs that can directly affect phenotype.

The computational evolutionary biologist Eugene Koonin sought to undermine another arm of the Modern Synthesis, asserting that natural selection is not the only force that shapes evolution (Koonin 2009). Even further, it might not even be dominant, insisting that evolutionary development might yet yield "the discovery of simple 'law-like' regularities" with a more significant impact than natural selection (Koonin 2009, p. 1028).

Along with these emerging trends, many biologists foresaw that a systems approach to evolution was imperative. The influential biologist Carl Woese and the physicist Nigel Goldenfeld were among those arguing that a critical integration of evolutionary theory with microbiology and molecular biology was needed to supplant the weary and out-moded Modern Synthesis (Woese and Goldenfeld 2009).

The last several years have seen a rising academic chorus contending that gene-centered evolutionary biology is insufficient to explain many evolutionary mechanisms (Noble 2020). In 2021, eminent British biologist Denis Noble offered a withering critique of the Modern Synthesis and its 80-year reign as the dominant evolutionary narrative (Noble 2021). Recent findings in molecular biology disclose that the Modern Synthesis is based on four illusions. First, natural selection cannot account for evolution since it is not a driving force. Instead, natural selection is passive, and intentionality is necessary to support life and its diversity. Secondly, the previously sacrosanct Weissmann Barrier that invoked the hereditary flow of information from germ line to somatic cells and never in reverse has been vitiated by modern epigenetic studies. Third, Darwin had proposed that there should be a direct means of communication between body cells and heredity germ line cells. He termed those invisible particles 'gemmules'. The Modern Synthesis derided the notion. Yet, Darwin's conjecture has been vindicated with the discovery of extracellular vesicles (exosomes). These lipid-based cell membrane-derived cargo vessels carry various chemicals along with DNA and RNA. Research indicates that these can act as intermediaries between somatic and germ line cells, enabling their communication and the efficient transfer of resources. Lastly, and perhaps the greatest of these illusions is the Central Dogma which insisted that the flow of information was always from DNA to RNA to proteins and never the reverse, now known to be invalid (Shapiro 2009).

James Shapiro (2021) amplified these arguments, noting that the basic concepts of the genome that undergird the Modern Synthesis are mainly incorrect. That breakdown can be sourced to multiple genetic discoveries: the abundance and fluidity of transposable elements, the inherent flexibility of genetic expression from multiple sources of cellular feedback, and the plethora of repetitive non-coding DNAs that can act as mobile elements and produce non-coding RNAs with a substantial functional properties (Shapiro 2021). These findings lead to an overwhelming different view of the genome. What was once an inviolate heritable unit is now acknowledged to be a malleable 'read-write' cell-modifiable DNA database that encodes a suite of RNA and protein sequences (Shapiro 2016).

In 2010, the evolutionary biologist Richard Lewontin argued that the standard narrative of evolution by natural selection could not account for evolved forms of life that can be observed (Lewontin 2010). Consequently, there must be an immense amount of biology that is missing from Neodarwinism. A growing body of evidence supports this conclusion (Pigliucci 2007, Baluška 2009, Witzany 2010, Shapiro 2011, Torday 2015a,b, Miller 2016, 2018, Miller and Torday 2018, Torday and Miller 2020, Miller et al. 2021).

Nonetheless, replacing the old synthesis requires an innovative substitute. Corning (2020) argues for a more inclusive biology, stressing the fundamental role of cooperation. That point is emphasized by quoting biologist Richard Michaud's assertion that "cooperation is now seen as the primary creative force behind ever greater levels of complexity and organization in all of biology" (Corning 2020, p. 4). Corning also notes the opinion of mathematical biologist Martin Nowak who calls cooperation "the master architect of evolution," arguing further that this elemental collaborative impulse among living organisms impels purposiveness as evolution's most fundamental property. This essential dictum contradicts the stances of two well-known Neodarwinists whose perspectives remain steeped in 20th century biology. Evolutionary biologist John H. Campbell has steadfastly championed that "changes in the frequency of alleles by natural selection are evolution" (Corning, 2020, quoting Campbell 1994, p. 68). In his popular 'The Selfish Gene', Richard Dawkins had solemnly intoned, "We are survival machines—robot vehicles blindly programmed to preserve the selfish molecules known as genes" (Dawkins 1989/1976 p. ix). Both are unreservedly incomplete as both regard purpose in biology as anathema.

Given the continuous and robust advances in biology over the last century, evolutionary theory has clearly undergone its own form of descent by gradual modification. Where must we look to make further progress to build on the brilliance of generations of thoughtful and talented scientists and philosophers? One clue comes from physicist Max Delbrück's definition of genetics as a ".....far-reaching, logically closed, strict science. It is quantitative without making use of a physical measurement system" (Baverstock 2019). Delbrück insisted that biology could not match chemistry in rigor since chemistry is underpinned by a stable atom capable of interrogation. On the other hand, the natural unit of biology is genetics which is mutable and therefore independent of consistent measurement. This essential difference accounts for the gap between genotype and phenotype since genetics is 'blind' to its processes. Undoubtedly, this seemed accurate when genes were thought to be in charge of biology. However, when genes are correctly assessed as tools of competent cells, then how cells deploy genes to solve problems becomes a distinctly pertinent measurable.

The lack of focus on the measurement problem in biology has been noted by others (Houle et al. 2011). Measurement is central to all scientific thought as the necessary means of assessing the relationship between theory and reality. Consequently, any 21st century biology must search for means of skillfully using measurements that might have predictive value, just like in other sciences.

Reaching that goal requires an updated and thoroughly reformulated cell theory. This volume proposes that answer, transitioning biology away from last century's dogmas. Accumulated research permits a cohesive alternative. Biology centers within cognition, information management, the self-organizing capacities of self-aware cells, the flexibility of their genomes, and the fundamental cellular dynamics that propel the robust dynamic intercourse among the cellular domains and the virome (Miller 2013, 2016, Baluska and Miller 2018, Miller et al. 2020). Goldenfeld and Woese (2011) specify the cusp: " Self-reference is a specific feature of biological systems and not physical systems...." (p. 9). As the following chapters will detail, all cells are self-aware, self-organizing, problem-solving entities. Based on this cohering bedrock, Neo-Darwinism is supplanted, and a fresh ecobiological construct is launched that embraces the essential reciprocations that propel planetary evolution. Accordingly, biology's infelicitous romance with allele frequencies is permanently severed, as many have urged (Rybnikov et al. 2015). As Marijuán and Navarro (2022) insist, " a new way of thinking capable of a parsimonious synthesis is needed". As will be outlined in detail, that new path is the flow of information among self-referential cells.

References

Baluška, F. 2009. Cell-Cell Channels, Viruses, and Evolution. Annals of the New York Academy of Sciences 1178: 106–119.

Baluška, F. and Miller, Jr, W.B. 2018. Senomic view of the cell: Senome versus Genome. Communicative & Integrative Biology 11(3): 1–9.

Baverstock, K. 2019. Polygenic scores: Are they a public health hazard? Progress in Biophysics and Molecular Biology 149: 4–8.

Bowler, P.J. 1983. The Eclipse of Darwinism: Anti-Darwinian Evolution Theories in the Decades around 1900 (1st ed.). Johns Hopkins University Press.

Bowler, P.J. 2003. Evolution: The History of an Idea (3rd ed.). Berkeley (California): University of California Press.

Caporale, L. 2003. Darwin In the Genome: Molecular Strategies in Biological Evolution. New York City, NY, USA: McGraw-Hill.

Caporale, L.H. and Doyle, J. 2013. In Darwinian evolution, feedback from natural selection leads to biased mutations. Annals of the New York Academy of Sciences 1305: 18–28.

Colson, P. and Raoult, D. 2012. Lamarckian evolution of the giant Mimivirus in allopatric laboratory culture on amoebae. Frontiers in Cellular and Infection Microbiology 2(91).

Corning, P.A. 2020. Beyond the modern synthesis: A framework for a more inclusive biological synthesis. Progress in Biophysics and Molecular Biology 153: 5–12.

Darwin, C.R. 1859. On the origin of species by means of natural selection (1st ed.). London: John Murray.

Darwin, C.R. 1871. The Descent of Man and Selection in Relation to Sex, Volume 2. Cambridge: Cambridge University Press.

Dawkins, R. 1989. The Selfish Gene (2nd ed.). Oxford, UK: Oxford University Press.

de Queiroz, K. 2005. Ernst Mayr and the modern concept of species. Proceedings of the National Academy of Sciences 102(suppl_1): 6600–6607.

Fodor, J. and Piattelli-Palmarini, M. 2010. What Darwin Got Wrong. New York, NY, USA: Farrar, Straus and Giroux.

Goldenfeld, N. and Woese, C. 2011. Life is Physics: Evolution as a Collective Phenomenon Far From Equilibrium. Annual Review of Condensed Matter Physics 2: 375–399.

Gould, Stephen J. 1977. The Return of Hopeful Monsters. Natural History 86: 22–30.

Gould, Stephen J. 1982. The uses of heresy: An introduction to Richard Goldschmidt's The Material Basis of Evolution. New Haven: Yale University Press.

Gould, Stephen Jay. 2002. The Structure of Evolutionary Theory. Harvard: Belknap Press.

Grafton, R.Q., Nelson, H.W., Lambie, N.R. and Wyrwoll, P.R. 2013. A Dictionary of Climate Change and the Environment: Economics, Science, and Policy. Northampton, MA, USA: Edward Elgar Publishing Ltd.

Gray, M.W. 2017. Lynn Margulis and the endosymbiont hypothesis: 50 years later. Molecular Biology of the Cell 28(10): 1285–1287.

Houle, D., Pélabon, C., Wagner, G.P. and Hansen, T.F. 2011. Measurement and Meaning in Biology. The Quarterly Review of Biology 86(1): 3–34.

Jablonka, E. 1999. Epigenetic Inheritance and Evolution: The Lamarckian Dimension. Oxford: Oxford University Press.

Jablonka, E. and Lamb, M.J. 2008. The Epigenome In Evolution: Beyond The Modern Synthesis. Vosgis Herald 12.

Kimura, M. 1968. Evolutionary Rate at the Molecular Level. Nature 217(5129): 624–626.

Kimura, M. 1991. Recent development of the neutral theory viewed from the Wrightian tradition of theoretical population genetics. Proceedings of the National Academy of Sciences 88(14): 5969–5973.

Koonin, E.V. 2009. Darwinian evolution in the light of genomics. Nucleic Acids Research 37(4): 1011–1034.

Krumbhaar, E.B. 1931. Erasmus Darwin and His Relation to the Doctrine of Evolution. Annals of Medical History 3(5): 487–500.

Laland, K., Uller, T., Feldman, M., Sterelny, K., Müller, G.B., Moczek, A., … Strassmann, J.E. 2014. Does evolutionary theory need a rethink? Nature 514(7521): 161–164.

Laland, K.N., Uller, T., Feldman, M.W., Sterelny, K., Müller, G.B., Moczek, A., ... and Odling-Smee, J. 2015. The extended evolutionary synthesis: its structure, assumptions and predictions. Proceedings of the Royal Society B: Biological Sciences 282(1813): 20151019.

Larson, E.J. 2004. Evolution: The Remarkable History of a Scientific Theory. New York City, NY, USA: Modern Library.

Lewontin, R.C. 2010. Not So Natural Selection. New York City, NY, USA: The New York Review of Books.

Marijuán, P.C. and Navarro, J. 2022. The biological information flow: From cell theory to a new evolutionary synthesis. Biosystems 213: 104631.

Matthew, P. 1831. On naval timber and arboriculture: with critical notes on authors who have recently treated the subject of planting. Longman, Rees, Orme, Brown, and Green; London and Adam Black, Edinburgh.

Mayr, E. 1942. Systematics and the Origin of Species from the Viewpoint of a Zoologist. New York City, NY, USA: Columbia University Press.

Miller, W.B. 2013. The Microcosm Within: Evolution and Extinction in the Hologenome. Boca Raton, Florida, United States: Universal Publishers.

Miller, W.B. 2016. Cognition, Information Fields and Hologenomic Entanglement: Evolution in Light and Shadow. Biology 5(2): 21.

Miller, W.B. 2018. Biological information systems: Evolution as cognition-based information management. Progress in Biophysics and Molecular Biology 134: 1–26.

Miller, W.B., Baluška, F. and Torday, J.S. 2020. Cellular senomic measurements in Cognition-Based Evolution. Progress in Biophysics and Molecular Biology 156: 20–33.

Miller, W.B., Enguita, F.J. and Leitão, A.L. 2021. Non-Random Genome Editing and Natural Cellular Engineering in Cognition-Based Evolution. Cells 10(5): 1125.

Miller, W.B. and Torday, J.S. 2018. Four domains: The fundamental unicell and Post-Darwinian Cognition-Based Evolution. Progress in Biophysics and Molecular Biology 140: 49–73.

Müller, G.B. 2017. Why an extended evolutionary synthesis is necessary. Interface Focus 7(5): 20170015.

Noble, D. 2020. Editorial: Charles Darwin, Jean-Baptiste Lamarck, and 21st century arguments on the fundamentals of biology. Progress in Biophysics and Molecular Biology 153: 1–4.

Noble, D. 2021. The Illusions of the Modern Synthesis. Biosemiotics 14: 5–24.

Ohta, T. and Gillespie, J.H. 1996. Development of Neutral and Nearly Neutral Theories. Theoretical Population Biology 49(2): 128–142.

Pigliucci, M. 2007. Do We Need An Extended Evolutionary Synthesis? Evolution 61(12): 2743–2749.

Provine, W.B. 2004. Ernst Mayr. Genetics 167(3): 1041–1046.

Rampino, M.R. 2011. Darwin's error? Patrick Matthew and the catastrophic nature of the geologic record. Historical Biology 23(2–3): 227–230.

Rose, M.R. and Oakley, T.H. 2007. The new biology: beyond the Modern Synthesis. Biology Direct 2: 30.

Rosenberg, E., Sharon, G. and Zilber-Rosenberg, I. 2009. The hologenome theory of evolution contains Lamarckian aspects within a Darwinian framework. Environmental Microbiology 11(12): 2959–2962.

Ruse, M. 2008. 2. Alfred Russel Wallace, the Discovery of Natural Selection, and the Origins of Humankind. pp. 20–36. *In*: O. Harman (ed.). Rebels, Mavericks, and Heretics in Biology. New Haven: Yale University Press.

Rybnikov, S.R., Frenkel, Z.M., Kirzhner, V.M. and Korol, A.B. 2015. Complex Dynamics of Multilocus Genetic Systems: A Revisit of Earlier Findings in Relation to Ecosystem Evolution. Botanica Pacifica.

Shapiro, J.A. 2009. Revisiting the Central Dogma in the 21st Century. Annals of the New York Academy of Sciences 1178: 6–28.

Shapiro, J.A. 2011. Evolution: A View from the 21st Century. Upper Saddle River, NJ: FT Press.

Shapiro, J.A. 2016. The basic concept of the read–write genome: Mini-review on cell-mediated DNA modification. Biosystems 140: 35–37.

Shapiro, J.A. 2021. Response to Denis Noble's Article "The Illusions of the Modern Synthesis", Biosemiotics. Biosemiotics 14: 73–78.

Singer, E. 2009. A Comeback for Lamarckian Evolution? Cambridge, MA, USA: MIT Technology Review.

Stearns, S. and Hoekstra, R. 2000. Evolution: an Introduction. Oxford, UK: Oxford University Press.

Torday, J.S. 2015a. The cell as the mechanistic basis for evolution. WIREs Systems Biology and Medicine 7(5): 275–284.

Torday, J.S. 2015b. A Central Theory of Biology. Medical Hypotheses 85(1): 49–57.

Torday, J. and Miller, W. 2020. Cellular-Molecular Mechanisms in Epigenetic Evolutionary Biology. Cham, Switzerland: Springer.

Witzany, G. 2010. Biocommunication and natural genome editing. World Journal of Biological Chemistry 1(11): 348.

Woese, C.R. and Goldenfeld, N. 2009. How the Microbial World Saved Evolution from the Scylla of Molecular Biology and the Charybdis of the Modern Synthesis. Microbiology and Molecular Biology Reviews 73(1): 14–21.

CHAPTER 3

21st Century Evolution
The Intelligent Measuring Cell

"A goal for the future would be to determine the extent of knowledge the cell has of itself and how it utilizes this knowledge in a "thoughtful" manner when challenged".

Barbara McClintock (1902–1992) American geneticist and 1983
Noble Prize winner in Medicine or Physiology

In 1995, the distinguished biologist Eörs Szathmáry and John Maynard Smith deleting Maynard Smith (1920–2004) offered rewarding insight into biological and evolutionary development: "Developmental biology can be seen as the study of how information in the genome is translated into adult structure, and evolutionary biology of how the information came to be there in the first place" (Szathmáry and Smith 1995). Maynard was substantially remaking evolutionary biology by placing it into the realm of information and meaning (Ogbunugafor 2020). Decades later, a comprehensive grasp of how biological information is received, assessed, and deployed is finally unfolding.

Many of those answers could not have been anticipated by Maynard since the crucial analytic element in the constructive application of information in biology was not rigorously considered during the last century. That missing essential was the deciding realization that self-referential cognition is central to all living processes (Baluška and Mancuso 2009, Shapiro 2011, 2021, Trewavas and Baluška 2011, Miller 2013, 2016, 2018, Dodig-Crnkovic 2014, Lyon 2015, Miller and Torday 2018, Miller et al. 2019). Darwin could not have had any idea of its ubiquity in his era, nor for that matter, could Maynard imagine it. Sophisticated contemporary resources were needed to enable this imperative transition. The evolving disciplines of metagenomics and epigenomics had to be developed, and a contemporary understanding of the holobionic nature of all eukaryotic life was similarly indispensable (Miller 2013). As these fields were in their infancy at the turn of the 20th century, their revelations were unavailable to earlier evolutionary biologists. Fortunately, our present moment offers a privileged insight; only cognitive agents can account for biological and evolutionary development on our planet.

The last two decades have confirmed that the computation of information underscores all life on the planet (Bray 2009). Necessarily, this active information assessment confers meaning, triggering predictive action (Freddolino and Tavazoie 2012). In 2007, James Shapiro, a microbiologist at the University of Chicago, observed, "Forty years' experience as a bacterial geneticist has taught me that bacteria possess many cognitive, computational and evolutionary capabilities unimaginable in the first six decades of the twentieth century" (p. 807). Further intensive research on the remarkable suite of metabolic processes demonstrated by bacteria and all cells, including our individual eukaryotic cells, has confirmed this pioneering assessment. Only a few years later, in 2011, Shapiro made a further crucial assertion: "Life requires cognition at all levels" (p. 7).

In his influential 2020 book 'What is Life?', Nobel prize-winner Sir Paul Nurse focused squarely on a prime deficiency of last century's biological framework and still bedevils it. Biologists continue to underestimate the remarkable capacities of cells. Nurse enumerates four key features of life: a) the cellular form represents life's basic unit, b) genes serve as information content helping to sustain life and reproduction, c) evolution by natural selection is life's adaptative force, and d) chemistry energizes life and permits life's functions. Yet, even this dextrously articulated summary manages to exclude biology's principle, obligatory element. Evidence confirms that every living entity is granted self-referential cognition, representing the underlying factor that stimulates the categorical cellular properties propelling planetary life. It is cellular cognitive awareness that underlies cellular adaptation. From this central fact, it becomes clear that natural selection does not drive evolution. Natural selection is a post-facto filter of forerunning intelligent cellular adaptive solutions to environmental stresses.

It is now evident that cognition is embodied in the cellular form, present throughout the cellular domains, and exhibited by all plants and animals (Miller 2016, Reber and Baluška 2021, Baluška et al. 2021, Fields et al. 2021). This perspective has been effectively codified as the Cellular Basis of Conscious equating self-referential cognition and sentience as demarcating the origin of life (Reber 2019). Unquestionably, this cellular cognitive capacity does not match our human gifts. Nonetheless, self-referential cognition at all scales is sufficient to enable the reception of information, its measuring assessment, and its purposeful deployment and communication to other cognitive agents. All of these actions are tasked toward supporting crucial cellular homeorhesis that maintains cellular equipoise as its preferential state (Miller 2016, 2018, Miller and Torday 2018). Homeorhesis signifies the properties of some dynamical systems to return to a preferred state of flux after any external displacement despite coexisting random noise. Thus, it is better suited to describing cellular biology than homeostasis, which indicates the tendency to remain at a stable equilibrium, a state better applied to non-living dynamic systems (Chuang et al. 2019).

In biological terms, cognition permits a cell to distinguish internal states from the external environment, thereby upholding itself against environmental stresses. This consistent maintenance of cellular homeorhesis within a proscribed living bandwidth represents the essence of cellular self-identity (Miller et al. 2019). Unquestionably, biological development and planetary evolution flow from this fundamental platform of self-referential identity. All living forms, whether microbial biofilms, whales, weasels, fleas, mushrooms, flowering plants, or sequoias, are the further living expression of cellular cognition writ into complex, interwoven biological forms. Through copious cell-cell communication, all biology enacts through cellular collaboration, cooperation, trading of resources, and competition, having been thrust forward from the first cellular instantiation of self-referential awareness (Miller 2016, 2018, Miller et al. 2019).

During the latter part of the 20th century, cells were considered living robots (Dawkins 1976). However, further research discloses that even bacteria have an extensive cognitive toolkit. Bacteria exhibit remarkably capable and diverse behaviors based on a sophisticated sensing apparatus including abundant communication, the active trading of resources, various means of motility, the mutualistic deployment of proxies, functional memory, and a limited type of sociality evidenced by multicellularity swarming behaviors (Freddolino and Tavazoie 2012, Baluška and Miller 2018). All cells, even prokaryotes with the smallest genomes, can communicate and form signal-based symbioses, coordinate for adaptation and predation, and have molecular kin recognition systems that permit self/non-self discrimination to sustain self-identity (Shapiro 2021). Undoubtedly, this range of contingent action is an elemental level of cognitive function as deliberative awareness of status at scale. Importantly, these faculties support the enhanced levels of collective sensing, cooperation, and interdependence that characterize all cells, including bacteria and their complex biofilms. Notably, all of these functions require substantial levels of memory and information processing. From these particulars, the essence of biological cognition can now be readily appraised. Cognition at all scales is directed toward cellular problem-solving (De Loof 2015, Tasoff et al. 2015).

As both eukaryotic and microbial cells exhibit a similar range of faculties, there is little mystery why they might cooperate in holobionts, such as in the human gut. Both microbial cells and eukaryotic cells engage in productive partnerships requiring specializations for the mutual production of resources (Gevers et al. 2012). In this cooperative form, their responses to environmental stresses are not based solely on individual assessments but instead encompass community concerns involving entire cellular networks as local ecologies (D'Souza et al. 2014). This cooperation drive is based on shared information and is sufficiently compelling that it regularly leads to auxotrophs. These highly specialized cells cannot produce certain essential metabolites themselves but are still capable of contributing necessary metabolites to a collective ecological community. Auxotrophs can be free-living but are also prevalent in symbiotic cellular ecologies, driving biosynthetic gene depletion as an environmental adaptation (D'Souza et al. 2014). Auxotrophs stake their self-referential equipoise and their lives upon trading for resources and the cooperative exchange of information. This interchange could only be maintained if such cells have an expectation (prediction) of reciprocity through a high degree of mutualistic entanglement (Roskelley and Bissell 1995). Accordingly, these cells have risked their survival on high levels of collaboration and co-dependency. Clearly, however, these interrelationships could only exist if all participants were cognitive, information-assessing, and communicative agents.

Thus, this mutual reciprocity testifies that cooperative action is the centerpiece of biological activity (Miller 2016, 2018). Necessarily, these active, conditional exchanges and the high levels of collaboration they entail could only exist between cognitive entities that are sufficiently aware of status to permit this degree of entanglement. Therefore, basal cognition underpins all levels of cellular organization and is implicit in evolutionary development (Shapiro 2011, Miller 2016, 2018). Cellular survival depends on mutual reciprocation between biological entities at all scales to permit the successful assimilation of the external environment to maintain cellular equipoise (Miller 2018). Indeed, this overriding impulse is of sufficient magnitude that it is not confined to either the prokaryotic colonial form or complex holobionts. These same actions are similarly displayed by free-living bacterial cells.

Two decades of research now establish that bacterial complexity is far greater than had been understood in the 20th century. Bacteria have an elaborate sensory apparatus supporting their cognitive toolkit. At all cellular levels, cells extract sensory information and communicate widely and deliberatively (Miller 2016). Bacteria communicate so abundantly and effectively that it has been described as 'chatter' (Visick and Fuqua 2005). Bacterial biofilms are a colonial form of multicellular collaboration sustained by sophisticated intracellular and cell-cell communication (Williams et al. 2007, Torday and Miller 2020). To function successfully in all planetary environments, the cellular faculties that enable complex biofilm architecture indisputably indicate a capacity to measure and calculate at scale (Ben-Jacob 2009, Ben-Jacob and Levine 2006, Ford 2009, 2017, Baluška and Levin 2016, Miller 2016, Miller et al. 2020a). Such capacities are not confined to bacteria. Research confirms that not only can cells measure, they must. Maintaining the regularity of eukaryotic cell size despite asymmetric cellular divisions in different tissues in multicellular organisms depends on the cellular measurement of the concentration of a critical cell protein within a tightly balanced feedback regulatory system based on the cellular assessment of its own DNA content (D'ario et al. 2021). Accordingly, all cells have self-referential cognitive awareness at scale that permits the measuring assessment of information. It is this critical faculty that underlies collective sensing and energizes the co-aligned deployment of resources and resulting interdependence that defines our biological system (Baluška and Miller 2018, Miller et al. 2019, 2020a).

At all scales, including unicells, memory and cognitive information processing are purposively deployed towards problem-solving (Ramanathan and Broach 2007, Miller 2016, 2018). Beyond microbial cells, the eukaryotic cells of holobionts exhibit similar faculties, enabling complex tissue ecologies comprised by many types of differentiated cells and an associated microbiome. The human gut exemplifies the depth and intimacy of this interdependent relationship. Human health depends directly upon the metabolites produced by the human gut microbial ecology and the resources that are traded between it and gut lining eukaryotic cells (Gevers et al. 2012). Importantly though, all cells within the tissue ecologies of holobionts are self-referential cognitive agents, and their collaborations permit each cell to sustain its individual state of homeorhetic preference (Miller 2016). Within this collective form, all constituent cells within any tissue ecology can respond to global environmental cues in a coordinated manner (Ingber 2003). There is no longer any doubt that these biological outcomes are the result of cognitive awareness (Miller 2016, 2018, Reber 2019). Indeed, the sophistication of these interrelationships is sufficiently robust that accurate modeling of the interchanges of microbial metabolic products within microbial ecologies is best modeled within a framework that mirrors economic equilibrium theory based on human patterns of behavior (Tasoff et al. 2015).

The origin of self-referential cognitive self-awareness is not known. However, what is well-acknowledged is that cognition is an essential attribute of life at every scale (Trewavas and Baluška 2011, Shapiro 2011, Lyon 2015, Miller 2016, 2018, Baluška and Levin 2016, Ford 2017, Vallverdú et al. 2018, Reber 2019, Miller et al. 2019, Torday and Miller 2020, Reber and Baluška 2021). Every living organism at any scale exhibits self-referential awareness, and conversely, there is no living organism that does not. Consequently, the instantiation of self-referential cognition was the commencement of the living state as phenomenal experientiality (Miller 2016, Miller et al. 2019, Reber 2019). As Baluska and Reber (2021, p. 32) have emphatically stated, "Life and sentience are co-terminous. Subjectivity, feelings, *consciousness* is an inherent feature of all life".

Naturally, theories abound about the mystery of the origin of cognition. Some propose that conscious self-referential awareness represents a sudden orthogonal skew as a phase transition of energy along the thermodynamic scale, arising from basic physical forces (Walker and Davies 2013, Miller 2016, 2018). In that circumstance, cognition should be regarded as a state function (a value that is independent of the number of steps needed to attain it) alongside others, such as entropy or enthalpy (Miller 2016, 2018). Derivative to that instantiation, cognitive self-awareness would have arisen without the need for any selective influence (Miller 2016, 2018).

Others contend that self-referential awareness is a precondition from the Singularity of the Big Bang and is invested in all matter in the universe as panpsychic cosmic consciousness (Torday 2018). This latter perspective is not new and has an ample history in Western metaphysics, with extensive roots in pre-Socratic Greek philosophy, the writings of Renaissance scholars such as Giordano Bruno (Brüntrup 2016), the philosophies of the mathematician-philosopher Alfred North Whitehead, and recent re-assertions by the contemporary philosopher, David Chalmers. Certainly, Eastern thought has readily accommodated a panpsychic stance. For instance, the Buddhist patriarch Zhanran (1711–1782), following an ancient tradition, wrote: "Who, then, is "animate" and who "inanimate"? Within the assembly of the Lotus, all are present without division. In the case of grass, trees and the soil...whether they merely lift their feet or energetically traverse the long path, they will all reach Nirvana" (Groner 2000). Still, others conceive that consciousness was a gradual process. In this framework, consciousness is based on emergent phenomena within complex systems that leads from 'life' to 'nervous systems' to 'special neurobiological features' to 'conscious phenomenology' that finally form the basis of true subjectivity (Feinberg and Mallatt 2020). Consciousness would therefore be hierarchical rather than the result of an instantiation. Yet, others argue that conscious self-reference emerged as a special case of thermodynamic quantum coherence (Rodríguez-Rosario et al. 2013).

Recent research supports that quantum processes are vital to cellular function and can be presumed to contribute to consciousness (Aerts et al. 2013, Al-Khalili and McFadden 2014, Torday 2018). One supposition for the origin of cognitive self-awareness is that it represents a quantum state, rising as a phenomenon of coherence due to quantum reactions within cellular boundaries in which appropriate resonant energies might confederate. These can be considered intracellular points of intersection of relevant energy/information transfers where external inputs and internal processes coapt as foci of biological discrimination, forming preferential internal states. The 'Orch OR' hypothesis of Hammeroff and Penrose (1996, 2014) advocates for this theory of consciousness by arguing that quantum computations in microtubules inside the cells of the brain orchestrate information processing as entangled superpositions of possibilities whose wave functions collapse into our experience of consciousness (Hameroff 2021).

Although the actual mechanism of consciousness is unknown, two essentials requirements can be identified, explaining why it is embodied in the cellular form. Consciousness of the sort that we can reference requires a boundary mechanism separating an outside from and inside. And further, that boundary system must link to retrievable and deployable memory. Absent this specific combination, conscious awareness cannot exist.

Surprisingly, the issue of preference as a critical aspect of cognitive self-awareness receives scant attention. However, any examination of disease processes and how certain microbes target specific tissues as favorable habitats confirms that cellular microbes have preferences as one element of their cognitive state (Miller 2013). Although how preferences arose remains mysterious, it is noteworthy that the nucleotides that form our chromosomes have distinct, exclusive pairings, which might be considered a type of preference at the chemical level. Perhaps this became the basis for its reiteration as a feature of self-referential awareness. Karl Popper, one of the most influential 20th century philosophers, considered the exclusion of curiosity and preferences from Darwinism as a critical limitation, regarding these traits as the active, creative elements of evolution (Neimann 2021).

Yet, as compelling as any of these theories might be, our working vantage is much more direct. CBE accepts cellular cognition as a directly observable, verifiable, and basal living platform. And, further, this self-referential endowment is specifically embodied within the intelligent, measuring cell. Consequently, cognition becomes our ground state for understanding planetary biology and its evolution. All theoretical debates about its origin, although of compelling interest, can be left aside. By choosing this self-evident framework, we can concentrate only on those cellular faculties that can be directly observed and measured and leave any further theorization about the mystery of the origin of cognition to others.

However, it is pertinent to briefly touch on some cellular attributes that might contribute to cognitive self-awareness. It certainly can be considered whether there is an intracellular component of minimal cognition. For instance, it has been demonstrated that mitochondria can demonstrate individualized reactions to cellular environmental stresses (Ford 2017). It is also known that they can meet such stresses with individual flexibility, controlling distinct molecular cascades that contribute to cellular sensing and guide cellular reactions to viral incursions (Galluzzi et al.

2012, Ford 2017). Moreover, mitochondria have extensive interactions with other intracellular organelles, such as the tubular endoplasmic reticulum, which also has critical roles in maintaining cellular homeorhetic balance and is also believed to support self-identity (Marchi et al. 2014).

Most recently, the intriguing concept of molecular nanobrains has been introduced. Specific components of cells may be self-aware in and of themselves, exhibiting cognitive properties that co-align with other cellular structures to produce effective cell-based cognition. Ribosomal protein networks are believed to participate significantly in cellular information processing, analogous to simple nervous systems with which they exhibit shared architectural features (Timsit and Grégoire 2021). The excitable cellular plasma membrane, internal organelle membranes, centrosomes, and microtubules can also be regarded as potentially distinguishable nanobrains capable of analzying sensory fields and producing those bioelectric fields upon which cells depend (Baluška et al. 2021). Indeed, the plasma membrane may represent the primary nanobrain of the highly flexible eukaryotic cell that arose from the endosymbiogenic merger between two ancient and independently conscious cell lines.

How the virome fits into consciousness is also unknown. Certainly, viruses are essential within our evolutionary narrative since all critical cellular functions, such as replication, translation, and repair, are of viral origin (Bannert and Kurth 2004, Koonin 2009, Feschotte and Gilbert 2012). Whether viruses are conscious entities has been actively considered and will be discussed further in Chapter 10.

What no longer remains theoretical is that sophisticated cognitive activity has been confirmed at every living scale. It is now clear that it is merely human conceit that conflates our distinctive form of intelligence as the only type imbued with a level of consciousness invested with genuine self-awareness (Miller et al. 2019, Reber 2019, Baluška et al. 2021). Commenting on prokaryote intelligence, neuroscientist Antonio Demasio (2019, pp. 53–54) notes: "Bacteria are very intelligent creatures; that is the only way of saying it, even if their intelligence is not being guided by a mind with feelings and intentions and a conscious point of view. They can sense the conditions of their environment and react in ways advantageous to the continuation of their lives. Those reactions include elaborate social behaviors. They can communicate among themselves—no words, it is true, but the molecules with which they signal speak volumes. The computations they perform permit them to assess their situation and, accordingly, afford to live independently or gather together if need be. There is no nervous system inside these single-celled organisms and no mind in the sense that we have. Yet they have varieties of perception, memory, communication, and social governance". Notably, these faculties permit sophisticated computations and comparative measurements of information. Moreover, bacteria collaborate to form sophisticated communities deploying sufficient cognitive complexity to enable learning and prediction (Ben-Jacob and Levine 2006, Ford 2009, Pinto and Mascher 2016).

Among unicellular eukaryotes, the unicellular amoeboid plasmodial slime mold *Physarum polcephalum* can compute solutions to the traveling salesman problem, commonly used in computer science to measure computer program functionality by finding the shortest route between multiple separate locations that must be visited

(Zhu et al. 2018). Amoeba can readily form, store, and recall memories to solve complex problems (Kramar and Alim 2021). Incontrovertibly, brains are not required to achieve cognitive prowess. Jellyfish have no centralized brain. Yet, experiments prove they are self-aware and intelligent, exhibiting complex behaviors and advanced problem-solving abilities (Pookottil 2013).

Even those animals that we have traditionally regarded as unintelligent display remarkable abilities. Many animals, including ants and anglerfish, have numerical competence (Reznikova and Ryabko 2011). Honeybees are capable of numerical cognition, including addition and subtraction, and further, they can use colors for symbolic representations (Howard et al. 2019). Plants are also intelligent, forming memories of environmental stresses with the capacity to evaluate large volumes of information, learn, and problem-solve (Karpiński and Szechyńska-Hebda 2010, Gagliano et al. 2014).

In his profoundly influential 1944 book 'What is Life', the revered Noble Prize-winning physicist Erwin Schrödinger (1887–1961) pronounced that biological organization was based on 'negentropy', as a highly ordered state that contravened the 2nd Law of Thermodynamics (Schrödinger 1944). He further reasoned that the distinctive molecular structure of genetic material represents an aperiodic crystal arrangement, accounting for living 'order from disorder' by animating an essentially limitless number of non-repetitive molecular configurations as the basis of hereditary memory (Malkov et al. 2020). This insight has stimulated dogged attempts to account for biology through physics by applying thermodynamic qualities such as the concept of the free energy principle that has been used to model neuronal processes (Ramstead et al. 2018). All biological systems are ordered, sustained over time, and resist decay. The free energy principle asserts that this self-organization is secondary to the active diminution of internal entropy within dynamical biological systems by the minimization of variational free energy. In so doing, and due to natural selection, an organism remains within "a relatively small, bounded set of states—its viability set —within the total set of possible states that it might occupy (i.e., its phase state)", as a low entropic state (Ramstead et al. 2018, p. 3). Notably, though, in this framework, natural selection drives this privileged entropic zone. However, even if the principle is correct, its justification is not. Natural selection is a filter of preceding variations. It has no directional force in and of itself and the impetus for the minimization of variational free energy must come from an energetic process. Certainly, active self-awareness could qualify as that propulsive entity.

Sir Paul Nurse in 'What is Life' (2020) offered an alternative, penetrating assessment of the essence of life compared to Neodarwinism, "Within their insulating membranes, cells can establish order..... they display a sense of purpose". Using the word 'purpose' is a pregnant term in biological circles, especially since purposeful intentionality cannot be achieved by energy flows divorced from a deliberate origin or pathway. In biology, there is only one credible source. Purpose derives from self-referential cognition. The only purpose that ought to be ascribed to any cell is its need to maintain the integrity of its instantiated self-awareness as its required and preferred state of homeorhetic flux. Undoubtedly though, any sense of purpose requires the 'knowing' state that can only be achieved through intelligence and a linked ability to appraise and deploy information that can ultimately yield deliberate, serviceable actions.

Certainly, those probability states that correlate with low entropy are associated with energy efficiency. Seeking life-preserving energy conservation requires narrow boundary states (low entropy) within which cells are constrained to constantly revisit (Ramstead et al. 2018). This phenomenon has been described as an 'attractional' effect even at the level of thermodynamics. Biological entities maintain their integrity by systematically remaining within an 'attractional' bandwidth of states. By doing so, cells gravitate towards a free energy minimum that consistently resists random fluctuations. It follows by definition that the characteristic state an organism chooses to maintain is its most probable state (Miller et al. 2019). Pertinently, this action can be placed within statistical terms as suppression of 'surprisal', permitting an organism to sustain its homeorhetic equipoise (Friston 2010). However, within the framework of cognition, all of these energetic impulses can be effectively reduce. They are all intended to reach and sustain cellular states of preference.

Although the Free Energy Principle can be meaningfully applied to cognitive entities, concomitant deficiencies remain. Most notably, the mechanistic Free Energy Principle does little to explain how 'surprisal' is actively suppressed over the course of evolution since this principle is placed within the rubric of natural selection as the channeling force. The specific problem is that natural selection is not a force. It is a filter. What then might engage living processes to diminish surprisal? In the living state, cognition is the active process that promotes the suppression of surprisal by anticipating external environmental stresses that might situate an organism at or beyond its upper boundaries. Notably, surprisal equates with low predictive value (Friston et al. 2006). Consequently, the preferential equipoise of any living entity is directly related to prediction, as the correlation between information assessment and the deployment of resources to confront environmental stresses successfully. Indeed, the cardinal difference between the inanimate and the living states is the 'knowing' engagement with information sufficient to sustain cellular equipoise. Necessarily, this linkage is dependent on the cellular capacity for prediction (Miller et al. 2019). Since cellular predictions are directly reliant on the discriminating assessment of often-confusing environmental information, it follows that productive information-based cellular prediction must be a measurement (Miller et al. 2020a,b). Accordingly, cellular measurement is an embedded feature of instantiated self-reference as a categorical attribute of the basal cognitive living state (Miller 2018, Miller et al. 2020a,b).

Therefore, cognition is the 'knowing' state that permits the cellular maintenance of its individual state of preference. Yet, this preferred state can only be maintained through the assessment of environmental cues as a stream of information. Necessarily then, the measured appraisal of information as prediction constitutes the crux of basal cognitive awareness, which in turn equates with a minimization of variational free energy. Obviously then, energy flows within living organisms are a function of the cellular measurement of environmental cues.

From this background, and in succeeding chapters, it will be shown that there is a functional, obliged set of interlinked relationships in biology that represent its actual dynamics and evolutionary narrative. Biology can only be properly comprehended if viewed through the lens of cellular cognition. From that cognitive base, homeorhetic equipoise equates with self-referential preferential living flux, dependent on the maximization of information security as the suppression of surprisal. This action directly correlates with the efficiency of cellular prediction. Reciprocally, this heightened predictive value equates with the minimization of variational free energy. Yet, effective prediction among cells, as will be further discussed in the following chapter, depends entirely on the effective transfer of information among cellular partners to enable mutual assessment to improve predictive value. Two imperative derivatives follow. Within this complex narrative, only cognition can exert these confluences. And secondly, the efficient transfer of valid information as improved prediction and its resulting energy efficiencies is best achieved through collective assessment, thereby justifying the multicellular form.(Miller 2016, Miller et al. 2019, Miller et al. 2020a). It is only through this coupling of forces that cells can achieve a 'knowing' self-referential assessment of status that represents their 'place' within the complex architecture of biofilms or holobionts (Miller and Torday 2018, Miller et al. 2020a,b).

One of the major blind-end convictions that swayed biologists during the last century can now be fully and finally expunged. Cells are neither machines nor living automata (Bongard and Levin 2021). Instead, they are purposeful agents intent on problem-solving within their respective contexts. Cognitive biologist Pamela Lyons (2015, p. 174) affirms this: "Biological cognition is the complex of sensory and other information-processing mechanisms an organism has for becoming familiar with, valuing, and [interacting with] its environment in order to meet existential goals, the most basic of which are survival, [growth or thriving], and reproduction". Necessarily, these information-driven goals can only be obtained through the self-referential measuring capacity of cells.

Although the origin of cognitive self-awareness is not known, its centrality to all of life is indisputable. Erwin Schrödinger's masterful statement sums it best: "Consciousness cannot be accounted for in physical terms. For consciousness is absolutely fundamental. It cannot be accounted for in terms of anything else" (Schrödinger 1984, p. 334).

References

Aerts, D., Broekaert, J., Gabora, L. and Sozzo, S. 2013. Quantum structure and human thought. Behavioral and Brain Sciences 36(3): 274–276.

Al-Khalili, J. and McFadden, J. 2014. Life on the Edge: The Coming of Age of Quantum Biology. London: Bantam Press.

Baluška, F. and Levin, M. 2016. On Having No Head: Cognition throughout Biological Systems. Frontiers in Psychology 7.

Baluška, F. and Mancuso, S. 2009. Deep evolutionary origins of neurobiology: Turning the essence of "neural" upside-down. Communicative & Integrative Biology 2(1): 60–65.

Baluška, F. and Miller, Jr, W.B. 2018. Senomic view of the cell: Senome versus Genome. Communicative & Integrative Biology 11(3): 1–9.

Baluška, F., Miller, W.B. and Reber, A.S. 2021. Biomolecular Basis of Cellular Consciousness via Subcellular Nanobrains. International Journal of Molecular Sciences 22(5): 2545.

Baluška, F. and Reber, A.S. 2021. The Biomolecular Basis for Plant and Animal Sentience: Senomic and Ephaptic Principles of Cellular Consciousness. Journal of Consciousness Studies 28(1–2): 31–49.

Bannert, N. and Kurth, R. 2004. Retroelements and the human genome: New perspectives on an old relation. Proceedings of the National Academy of Sciences 101(Suppl_2): 14572–14579.

Ben-Jacob, E. 2009. Learning from Bacteria about Natural Information Processing. Annals of the New York Academy of Sciences 1178: 78–90.

Ben-Jacob, E. and Levine, H. 2006. Self-engineering capabilities of bacteria. Journal of The Royal Society Interface 3(6): 197–214.

Bongard, J. and Levin, M. 2021. Living Things Are Not (20th Century) Machines: Updating Mechanism Metaphors in Light of the Modern Science of Machine Behavior. Frontiers in Ecology and Evolution, 9.

Bray, D. 2009. Wetware: A Computer in Every Living Cell. New Haven: Yale University Press.

Brüntrup, G. 2016. Emergent Panpsychism. pp. 48–72. *In*: Panpsychism. Oxford, UK: Oxford University Press.

Chuang, J.S., Frentz, Z. and Leibler, S. 2019. Homeorhesis and ecological succession quantified in synthetic microbial ecosystems. Proceedings of the National Academy of Sciences 116(30): 14852–14861.

D'Ario, M., Tavares, R., Schiessl, K., Desvoyes, B., Gutierrez, C., Howard, M. and Sablowski, R. 2021. Cell size controlled in plants using DNA content as an internal scale. Science 372(6547): 1176–1181.

D'Souza, G., Waschina, S., Pande, S., Bohl, K., Kaleta, C. and Kost, C. 2014. Less is more: selective advantages can explain the prevalent loss of biosynthetic genes in bacteria. Evolution 68(9): 2559–2570.

Damasio, A.R. 2019. The Strange Order of Things: Life, Feeling, and the Making of Cultures (Reprint). New York, NY: Vintage.

Dawkins, R. 1976. The Selfish Gene. Oxford, UK: Oxford University Press.

Dawkins, R. and Davis, N. 2017. The Selfish Gene. Macat Library.

De Loof, A. 2015. From Darwin's On the Origin of Species by Means of Natural Selection... to The evolution of Life with Communication Activity as its Very Essence and Driving Force (= Mega-Evolution). Life: The Excitement of Biology 3(3): 153–187.

Dodig-Crnkovic, G. 2014. Modeling life as cognitive info-computation. In Computability in Europe. Cham, Switzerland: Springer.

Feinberg, T.E. and Mallatt, J. 2020. Phenomenal Consciousness and Emergence: Eliminating the Explanatory Gap. Frontiers in Psychology 11.

Feschotte, C. and Gilbert, C. 2012. Endogenous viruses: insights into viral evolution and impact on host biology. Nature Reviews Genetics 13(4): 283–296.

Fields, C., Glazebrook, J.F. and Levin, M. 2021. Minimal physicalism as a scale-free substrate for cognition and consciousness. Neuroscience of Consciousness 2021(2).

Ford, B.J. 2009. On Intelligence in Cells: The Case for Whole Cell Biology. Interdisciplinary Science Reviews 34(4): 350–365.

Ford, B.J. 2017. Cellular intelligence: Microphenomenology and the realities of being. Progress in Biophysics and Molecular Biology 131: 273–287.

Freddolino, P.L. and Tavazoie, S. 2012. Beyond Homeostasis: A Predictive-Dynamic Framework for Understanding Cellular Behavior. Annual Review of Cell and Developmental Biology 28: 363–384.

Friston, K. 2010. The free-energy principle: a unified brain theory? Nature Reviews Neuroscience 11(2): 127–138.

Friston, K., Kilner, J. and Harrison, L. 2006. A free energy principle for the brain. Trends in Cognitive Sciences 13(7): 293–301.

Gagliano, M., Renton, M., Depczynski, M. and Mancuso, S. 2014. Experience teaches plants to learn faster and forget slower in environments where it matters. Oecologia 175: 63–72.

Galluzzi, L., Kepp, O. and Kroemer, G. 2012. Mitochondria: master regulators of danger signalling. Nature Reviews Molecular Cell Biology 13(12): 780–788.

Gevers, D., Knight, R., Petrosino, J.F., Huang, K., McGuire, A.L., Birren, B.W., ... Huttenhower, C. 2012. The Human Microbiome Project: A Community Resource for the Healthy Human Microbiome. PLoS Biology 10(8): e1001377.

Groner, P. 2000. Saicho: The Establishment of the Japanese Tendai School. Honolulu, Hawaii: University of Hawaii Press.

Hameroff, S. 2021. 'Orch OR' is the most complete, and most easily falsifiable theory of consciousness. Cognitive Neuroscience 12(2): 74–76.

Hameroff, S. and Penrose, R. 1996. Orchestrated reduction of quantum coherence in brain microtubules: A model for consciousness. Mathematics and Computers in Simulation 40(3–4): 453–480.

Hameroff, S. and Penrose, R. 2014. Consciousness in the universe. Physics of Life Reviews 11(1): 39–78.

Howard, S.R., Avarguès-Weber, A., Garcia, J.E., Greentree, A.D. and Dyer, A.G. 2019. Numerical cognition in honeybees enables addition and subtraction. Science Advances 5(2).

Ingber, D.E. 2003. Mechanosensation through integrins: Cells act locally but think globally. Proceedings of the National Academy of Sciences 100(4): 1472–1474.

Karpiński, S. and Szechyńska-Hebda, M. 2010. Secret life of plants. Plant Signaling & Behavior 5(11): 1391–1394.

Koonin, E.V. 2009. Darwinian evolution in the light of genomics. Nucleic Acids Research 37(4): 1011–1034.

Kramar, M. and Alim, K. 2021. Encoding memory in tube diameter hierarchy of living flow network. Proceedings of the National Academy of Sciences 118(10).

Lyon, P. 2015. The cognitive cell: bacterial behavior reconsidered. Frontiers in Microbiology 6(264).

Malkov, V.B., Chemezov, O.V. and Malkov, A.V. 2020. Aperiodic Schrodinger Crystals. *In*: IOP Conference Series: Materials Science and Engineering (Vol. 969, p. 012026). IOP Publishing.

Marchi, S., Patergnani, S. and Pinton, P. 2014. The endoplasmic reticulum–mitochondria connection: One touch, multiple functions. Biochimica et Biophysica Acta (BBA)—Bioenergetics 1837(4): 461–469.

Miller, W.B. 2013. The Microcosm Within: Evolution and Extinction in the Hologenome. Boca Raton, Florida, United States: Universal Publishers.

Miller, W.B. 2016. Cognition, Information Fields and Hologenomic Entanglement: Evolution in Light and Shadow. Biology 5(2): 21.

Miller, W.B. 2018. Biological information systems: Evolution as cognition-based information management. Progress in Biophysics and Molecular Biology 134: 1–26.

Miller, W.B., Baluška, F. and Torday, J.S. 2020a. Cellular Senomic Measurements in Cognition-Based Evolution. Progress in Biophysics and Molecular Biology 156: 20–33.

Miller, W.B. and Torday, J.S. 2018. Four domains: The fundamental unicell and Post-Darwinian Cognition-Based Evolution. Progress in Biophysics and Molecular Biology 140: 49–73.

Miller, W.B., Torday, J.S. and Baluška, F. 2019. Biological evolution as defense of "self". Progress in Biophysics and Molecular Biology 142: 54–74.

Miller, W.B., Torday, J.S. and Baluška, F. 2020b. The N-space Episenome unifies cellular information space-time within cognition-based evolution. Progress in Biophysics and Molecular Biology 150: 112–139.

Niemann, H.-J. 2021. Popper, Darwin, and Biology. pp. 231–256. *In*: Z. Parusniková and D. Merritt (eds.). Karl Popper's Science and Philosophy. Cham, Switzerland: Springer.

Nurse, P. 2020. What is Life?: Understand Biology in Five Steps. Oxford, UK: David Fickling Books.

Ogbunugafor, C.B. 2020. A Reflection on 50 Years of John Maynard Smith's "Protein Space". Genetics 214(4): 749–754.

Pinto, D. and Mascher, T. 2016. (Actino) Bacterial "intelligence": using comparative genomics to unravel the information processing capacities of microbes. Current Genetics 62(3): 487–498.

Pookottil, R. 2013. BEEM: Biological Emergence-based Evolutionary Mechanism: How Species Direct Their Own Evolution. Cornwall, UK: Fossil Fish Publishing.

Ramanathan, S. and Broach, J.R. 2007. Do cells think? Cellular and Molecular Life Sciences 64(14): 1801–1804.

Ramstead, M.J.D., Badcock, P.B. and Friston, K.J. 2018. Answering Schrödinger's question: A free-energy formulation. Physics of Life Reviews 24: 1–16.

Reber, A.S. 2019. The First Minds: Caterpillars, Karyotes, and Consciousness. Oxford, UK: Oxford University Press.

Reber, A.S. and Baluška, F. 2021. Cognition in some surprising places. Biochemical and Biophysical Research Communications 564: 150–157.

Reznikova, Z. and Ryabko, B. 2011. Numerical competence in animals, with an insight from ants. Behaviour 148(4): 405–434.

Rodríguez-Rosario, C.A., Frauenheim, T. and Aspuru-Guzik, A. 2013. Thermodynamics of quantum coherence.

Roskelley, C.D. and Bissell, M.J. 1995. Dynamic reciprocity revisited: a continuous, bidirectional flow of information between cells and the extracellular matrix regulates mammary epithelial cell function. Biochemistry and Cell Biology 73(7–8): 391–397.

Schrödinger, E. 1944. What is Life? The Physical Aspect of the Living Cell. Cambridge, UK: Cambridge University Press.

Schrödinger, E. 1984. General Scientific and Popular Papers. p. 334. *In*: Collected papers V. 4. General scientific and popular papers. Friedrich Vieweg Und Sohn, Braunschweig/Wiesbaden, Gremany.

Shapiro, J.A. 2007. Bacteria are small but not stupid: cognition, natural genetic engineering and socio-bacteriology. Studies in History and Philosophy of Science Part C: Studies in History and Philosophy of Biological and Biomedical Sciences 38(4): 807–819.

Shapiro, J.A. 2011. Evolution: A View from the 21st Century. Upper Saddle River, NJ: FT Press.

Shapiro, J.A. 2021. All living cells are cognitive. Biochemical and Biophysical Research Communications 564: 134–149.

Szathmáry, E. and Smith, J.M. 1995. The major evolutionary transitions. Nature 374(6519): 227–232.

Tasoff, J., Mee, M.T. and Wang, H.H. 2015. An Economic Framework of Microbial Trade. PLOS ONE, 10(7): e0132907.

Timsit, Y. and Grégoire, S.-P. 2021. Towards the Idea of Molecular Brains. International Journal of Molecular Sciences 22(21): 11868.

Torday, J.S. 2018. Quantum Mechanics predicts evolutionary biology. Progress in Biophysics and Molecular Biology 135: 11–15.

Torday, J.S. 2018. From cholesterol to consciousness. Progress in Biophysics and Molecular Biology 132: 52–56.

Torday, J. and Miller, W. 2020. Cellular-Molecular Mechanisms in Epigenetic Evolutionary Biology Cham, Switzerland: Springer.

Trewavas, A.J. and Baluška, F. 2011. The ubiquity of consciousness. EMBO Reports 12(12): 1221–1225.

Vallverdú, J., Castro, O., Mayne, R., Talanov, M., Levin, M., Baluška, F., ... Adamatzky, A. 2018. Slime mould: The fundamental mechanisms of biological cognition. Biosystems 165: 57–70.

Visick, K.L. and Fuqua, C. 2005. Decoding Microbial Chatter: Cell-Cell Communication in Bacteria. Journal of Bacteriology 187(16): 5507–5519.

Walker, S.I. and Davies, P.C.W. 2013. The algorithmic origins of life. Journal of The Royal Society Interface 10(79): 20120869.

Williams, P., Winzer, K., Chan, W.C. and Cámara, M. 2007. Look who's talking: communication and quorum sensing in the bacterial world. Philosophical Transactions of the Royal Society B: Biological Sciences 362(1483): 1119–1134.

Zhu, L., Kim, S.-J., Hara, M. and Aono, M. 2018. Remarkable problem-solving ability of unicellular amoeboid organism and its mechanism. Royal Society Open Science 5(12): 180396.

CHAPTER 4

The Cellular Information Cycle and Biological Information Management

The fundamental principle of biological action is the self-referential appraisal of information among the three cellular domains (Prokaryota, Archaea, Eukaryota) (Trewavas and Baluška 2011, Miller 2013, Lyon 2015, Miller 2016, 2018, Miller and Torday 2018, Torday and Miller 2020, Miller et al. 2020a,b). Biological inquiry thereby rests on understanding how organisms receive, assess, communicate, and deploy information under differing environmental conditions (Trewavas and Baluška 2011, Miller 2016, 2018, Miller et al. 2020a,b). Clearly, then, the essence of cellular life is centered within information management (Miller 2018). Therefore, a sounder understanding of the nature of biological information is essential to placing cellular life in its proper context. Accordingly, an exploration of biological information and how it drives multicellularity and concomitantly sustains individual cellular self-integrity would be of particular relevance.

The basic cell has remained the continuous physical unit of life for approximately four billion years (Papineau et al. 2022). Undoubtedly, that perpetual success derives from the embodied cognitive awareness that enables continuous flexible confrontation with environmental stresses (Miller and Torday 2018). Although that level of cognitive awareness and sentience is unlike our human-specific conscious palette, it is of sufficient magnitude to permit cells and cellular microbes to efficiently assess, communicate, and deploy information in communal combinations to protect individual homeorhetic preferences (Miller 2016, Baluška and Levin 2016, Marshall 2019). Cellular cognitive awareness is centered on the cellular ability to sufficiently measure informational inputs to support both free-standing life and more integrated communal states (Miller et al. 2020a). This capacity for information processing reveals biology's cardinal principle. Cellular life is information management (Miller 2016, 2018, Miller et al. 2020a,b).

Life requires information management because all the information that any cell can access is ambiguous (Miller 2016, Torday and Miller 2017). This assertion can be readily justified. For uncertain information to be usefully assessed, it must be measured according to some internal standard. After all, if the information that cells have available were precise, there would be no need to expend energy to measure it. Thus, cognitive cells can now be correctly appreciated as measuring instruments that energize the flow of ambiguous information from environmental stresses into biological outcomes (Miller 2016, 2018, Torday and Miller 2017, Baluška and Miller 2018, Miller et al. 2020a). Further, the purpose of those measurements can be readily discerned. All cellular measurements are ultimately directed toward cellular equipoise by maintaining cellular homeorhesis versus environmental cues and stresses (Baluška and Miller 2018, Miller et al. 2019, 2020a,b). Furthermore, and of signal importance, this information flow does not merely enable the cellular collaboration, cooperation, co-dependency, and competition that are the hallmarks of multicellular life but actually drives those living activities (Ben-Jacob and Levine 2006, Ben-Jacob 2009, Miller 2016, Miller and Torday 2018). Consequently, cognitive cells are measuring instruments to assess and manage information flow (Miller 2016, Miller and Torday 2018, Baluška and Miller 2018, Miller et al. 2020a).

All biological information is ambiguous. It is categorically unlike Shannon information typically used for modeling computer programming and telecommunication to assess the quantity and efficacy of binary information in a digital framework (Miller 2016, Torday and Miller 2017). Biological information differs from computer data through many interlinked causes. First and foremost, all of the information that any cell has from its external environment must travel across a physical environment and transit an energetic plasma membrane to gain entry into any cell. Further, in biological systems, communicative cell-cell signals are generally more of a broadcast than a targeted event. Typically, neither the sender of those signals nor the receiver of that incoming information is necessarily known to one another (Czárán and Hoekstra 2009, Witzany 2016).

Moreover, all communication signals, whether bioactive molecules or bioelectric discharges or their resulting fields, must travel from cell to cell through an intervening physical media (interstitial spaces) where they are subject to time and distance-related degradation or disruption. Further yet, once the external plasma membrane is transited, bioactive substances or energetic signals must travel the crowded, active environment of the cell, encountering intracellular structures like the cytoskelton, organelles and their endomembranes, and many other sources of potential loss of information density or quality. Accordingly, many intrinsic factors modulate cellular information, rendering it inherently 'noisy' and contributing to equivocal meaning (Miller 2016, 2018, Torday and Miller 2017, 2020).

The second major source of informational uncertainty is grounded in quantum mechanics and observer status. It has been experimentally proved that two observers will each experience a differential gap regarding the measurement of superimposed states (Proietti et al. 2019). This discontinuity relates to differing self-referential perspectives, yielding variances in the amplitudes, frequencies, or time signatures of informational cues as signals. Consequently, in a self-referential frame, cellular signals reference through various degrees of imprecision governed by an inherent 'uncertainty relation' (Gabor 1946).

This integral uncertainty relationship is an embedded reality of the living state and has been described as comprised of four categories of contextual observer-dependent 'antipodal' information (Matsui 2001, Miller et al. 2019). The first type of antipode conceptualizes information as a volume, such as a sphere rather than a point. When envisioned as a sphere, one side might represent the readily apparent information that we typically consider available to an observer. However, there is also an opposing aspect as the opposite 'hidden' side that is just as real in information space-time but unapparent to the same observer (Marijuán et al. 2015, Tozzi and Peters 2017). Consequently, volumetric information has a readily assessed side but also contains meaningful additional 'antipodal' information. However, and crucially, that antipodal information that is hidden to an original observer might represent readily observable information to some other cognitive observer in a different position, just as any two individuals might observe opposing aspects of a ball thrown in the air if standing on opposite sides of it. As another type of example, a child and an adult can both look at the moon. On initial observation, the child might believe it is disc in the sky. However, the adult knows it is a sphere with an opposite but unseen side. Consequently, the same object can have different information content to separate cognitive agents dependent on context.

The second category of antipodal information is the concept of 'distinction on the adjacents' described by Marijuán et al. (2015). This type of antipode is subjective supplementary informational inputs that represent the inescapable gaps between information reception and the process of its full assessment. The cellular appraisal of information requires its complete active internalization that can distort its measuring value. The closer these informational 'adjacents' can be brought together, the higher the measuring value, thereby permitting less uncertainty. Conceptually, this can be visualized akin to the dimple on a golf ball. The diameter across is shorter than a trip around its circumference. The shorter distance as a compaction of the informational gap narrows the uncertainty relationships. In living systems, this can be alternatively expressed as a compression of the range of anticipations and predictions based on any information set. How any cognitive agent assesses information depends to some extent on what that agent expects to derive from it. Unlike computers, biological systems never have a one-to-one match between reception and output. This is one of the crucial reasons why information must be measured so that this obligatory informational gap can be narrowed to permit contingent action.

The third antipodal form is frequently overlooked. It is the value of the spaces between points of information within any informational matrix. This is not empty space in an informational context and instead represents a meaningful 'Third State' that serves an organizational role that contributes order and meaning within an informational matrix (Forshaw 2016). These separations can be conceptualized by considering the spaces between words in this chapter. Understanding the meaning of the chapter would be seriously impaired without the spaces between the words. This effect is not theoretical in biology. The issue of spacing in information has been directly addressed experimentally by examining the effects of frequency, amplitude, and timing of stimuli on habituation and sensitization (Rankin et al. 2009).

The fourth type is implicit only among cognitive agents and exclusive to the 'knowing' self-referential state. This category is information that is anticipated but has not been received by a self-referential organism. That absence has meaning (Miller et al. 2020a). Self-referential organisms, including single-celled microbes, can predict and anticipate as an essential component of their cognitive toolkit and survival (Vallverdú et al. 2018). Therefore, at all scales, and similar to ourselves, anticipated information that remains absent is a vital aspect of information, just as in our own experience. Information that is anticipated but remains absent is significant and forms one contingent path towards biological action. This unsensed but highly relevant antipodal information is not just theoretical. Light-sensitive proteins are critical to retinal function but are not limited to it. Other non-visual light-receptor opsin proteins are found in tissue cells unrelated to vision (Sugihara et al. 2016). These melanopsins do not function in vision but affect alertness, mood, diurnal sleep patterns, and body temperature (Pilorz et al. 2016). Similar non-visual light-sensitive proteins in fat adipocytes participate in fuel metabolism, fuel storage, obesity, and mitochondrial respiration (Sato et al. 2020).

Several crucial derivatives emerge from this inherency of informational ambiguity in cellular life. Of prime importance, and perhaps the most surprising, is that all the information that any cell or any living creature has is self-produced and always represents a superposition of implicates. Although seemingly complex, the concept is readily understandable. Matter and energy are the only fundamental quantities in the universe. While it is true that matter and energy are interconvertible to information, that conversion is entirely dependent on its observation which equates with the self-referential state in biological terms. Living entities can detect the spatial and temporal dynamics of their environment. However, for those stimuli to have meaning, there must be self-referential cognitive discrimination to distinguish spatial and temporal differences that stimuli represent. This action can only be performed inside cells.

Many scientists declare that energy and information are interconvertible and information is actually primary in the universe. Accordingly, the interactions between matter and energy as information trigger 'causal' differences in our world and are relevant to the quantum realm that depend on observer context (Davies and Gregersen 2014). This differential gap equates with Bateson's definitional concept of information as 'a difference that makes a difference' as a generally accepted as a direct and practical definition of information.

Bateson's definition significantly impacts biology as it under pins the concept of infoautopoesis (Cárdenas-García 2020). For any organism at any scale, including cells, information from the outside must be internalized for assessment to have any meaning. This passage means that information must be measured to discern its difference. This requisite process of internal measurement and information analysis constitutes a cell's only real information. Consequently, all information is self-produced. Naturally, this living essential ramifies and explains our obliged circumstance and why each of us evaluates the same stimulus differently. Each independent observer/participant (every living entity) experiences information idiosyncratically due to all of the factors of informational ambiguity that have been enumerated. Accordingly, any set of informational uncertainties that any individual can settle into explicate action is obligatorily its own self-analysis. Each individual, be it a cell, bee, or human, will necessarily differ in that appraisal even compared to individuals of the same species. As a result, every living entity experiences its own specific individual reality. Directly put, cells can never experience any objective external reality as that is denied them and us. All external inputs must be interpreted and measured within cellular terms, which have distinctive limits (Miller 2018).

From this background, the origin of multicellularity clarifies. Cells measure information to survive and maintain their homeorhetic balance. Since the environment is constantly in flux, this action necessarily entails prediction and anticipation. Since all cellular information is ambiguous, the quality of information is the crucial aspect of its information management system. Consequently, biology is always directed toward a practical means of winnowing outward information space-time to permit the most accurate individual cellular self-referential predictions (Miller et al. 2019). To do this requires a shared evaluation of the available information, placing it into a manageable form that can permit a self-referential organism to make useful predictions. To achieve this outcome, cells communicate abundantly to share their individual interpretations of information (as their interpretation of external reality) to arrive at conjoint assessment as the 'wisdom of crowds'. This activity maximizes informational quality as Effective Information (*EI). The constant striving towards *EI is the defining feature of cellular life as the pivotal element of its information management system that is the basis of the organizational structure of the living state (Hall 2005, Miller 2018). Multicellularity is predicated on this living requirement, reiterating at successive scales to enable flexible holobionts (Miller et al. 2020a). It is only in this manner that the vast range of superpositions of possibilities imposed by ambiguous environmental cues can be settled into coordinate explicates as productive biological expressions. The direction of that information flow is the continuous defense of cellular homeorhesis and self-identity. Naturally, too, the optimization of information directly correlates with energy efficiencies to meet environmental flux.

Although the origin of self-reference is unknown, it is conditioned on observer/participant status, which implies its attachment to biological information space-time (Miller 2016, 2018, Miller et al. 2019). This critical attachment to information space-time highlights the differential crux between information in the inanimate sphere and information among the living. The transfer of information based on a relevant information system linked to retrievable and deployable memory anchors the cellular mechanisms that can sustain multicellular organisms. That conceptual bridge connecting the self-production of information by individual organisms to cell-wide information management is the unifying concept of the Pervasive Information Field (PIF) that permits a universal scale-free system of informational organization (Lloyd 2002, Miller 2016, 2018, Miller et al. 2020a,b). A PIF is an informational matrix representing a summation of all the potential sources of information available to a cell as an independent self-referential observer/participant. Similar in concept to an energy field, a PIF is an organizational architecture for any organism within a four-dimensional information field. In effect, it is the 'cloud' of all the superimpositions of possibilities that arise from the physical energy and materiality that present to any self-referential organism as sources of information. Thus, a PIF represents an organism's attachment to information space-time with its attendant ambiguities to effective information management, and in this volume is specifically applied to the cellular form.

Multicellular organisms can then be modeled as overlapping PIFs, thereby permitting a common informational architecture for communication and conjoint appraisal of environmental cues (Miller 2016, 2018). This model specifically emphasizes the distinctions between ambiguous biological information and other theoretical measures of information, such as the digital information that underlies Shannon communication systems (Walker and Davies 2013). For multicellular organisms to work together efficiently, there must be a shared information architecture for the conjoining assessment of the highly variable sources of information that are self-produced by each individual cell to be interpretable among them all. As a summation of a complex information space-time field, a PIF facilitates the handling of cellular information, serving as its interactive information platform, and forming a reference point for information assessment with other cells. In this manner, if fits the requirements for a "sender-receiver communicating compartment" as suggested by De Loof (2017) as necessary for information management. In this case, each cell is an individual sender-receiver with its own PIF, forming the basis of its individual interpretation of self-produced reality, overlapping with additional sender-receivers with their PIFs. These then unite into a shared information platform, an N-space Episenome, which will be described in detail in a following chapter. Together, these united PIFs have a shared attachment to an outward universal, scale-free informational space-time matrix.

With that background, a natural bridge exists between energy, information, and communication based on their thermodynamic interconvertibility in the self-referential frame, forming the basis of the inherent self-organizing properties of the living state (Deacon 2011, Miller 2018). Crucially, the reception of information by any self-referential cell initiates an obligatory information cycle based within thermodynamics (Miller 2018, Miller et al. 2020a,b). Briefly, any reception and measured assessment of information require an energy expenditure as work (Deacon 2011, Witzany 2011). In the self-referential frame, that work becomes an obligatory form of communicated information as an energetic signature to some other self-referential observer/participant (Miller 2018). In turn, that communication necessitates its further measured assessment, which is also work, resulting in a reiterating and self-reinforcing process.

The reasoning behind this interlinkage is direct. In the living frame where all participants are observer/participants as sender-receivers of communication, any reception of information is work done across an energy gradient. Thus, the reception of information by any observer/participant obliges work to assess it, which in turn consumes energy and thereby issues a secondary obligatory energetic signature. That stimulus constrains some other observer/participants to a reciprocal act of work to try to understand its significance, yielding a reiterative work channel (Deacon 2011) By default then, any of these actions become sources of information in a variety of other mediums, each issuing its own constraints within this obliged recursive work channel. Therefore, although thermodynamic processes impel biological activity through both classical and quantum means, the thermodynamics of biology fundamentally differs from the linear thermodynamics of Boltzmann (Ho 2008).

In sum, there is no energy use that does not impart information within a self-referential biological system as some reciprocal set of constraints. And therefore, every constraint, which necessarily requires a cellular energy expenditure as work, is information in a referential frame. The result is an obligatory work channel that specifically energizes collective cellular action. It is this reiterating cycle that forms the basis of biological self-organization and multicellularity (Miller 2018, Miller et al. 2020a,b). Therefore, a self-reinforcing cellular information cycle drives intercellular cooperative communication to better assess ambiguous environmental cues through 'a wisdom of crowds'. Crucially though, as this collaborative activity is dependent on the measurement of information and its shared communication, it necessarily represents a reciprocating, cooperative process that proceeds through the thermodynamic interconvertibility of energy-information-communication. Notably, and of the utmost importance, this obliged linkage is the essence of engineering, which is the 'act of working artfully to bring something about', which enables the multicellular niche constructions that become the tissue ecologies that form holobionts.

This fundamental, obliged interrelationship is the essential linchpin of multicellular biology. Triadic energy-information-communication sustains a self-reinforcing information cycle that permits the purposeful range of biological expressions that characterize life on the planet. From within this basal cellular process linking information to biological output, all evolutionary biology can now be appraised as representing the collective action of intelligent cells as problem-solving agents to meet ambiguous circumstances (Miller 2016, 2018, Miller and Torday 2018, Miller et al. 2020a,b). Consequently, triadic energy-information-communication is an embedded entanglement within self-referential information-space that defines life in every practical sense on the planet as the energizing agency that propels multicellularity and all holobionts.

Planetary biology is an essential living entanglement among the four domains (Prokaryota, Archaea, Eukaryota, Virome). These reciprocating connections result from the concordant use of shared information among cells, based on each cell's personalized information field (PIF) that shapes its self-produced information. Cells collaborate and collectively assess information to enhance the validity of their available information as Effective Information (*EI). The proper definition of * EI in cellular terms is straightforward. *EI is information that permits an organized system to perform non-random actions (Szostak 2003, Farnsworth et al. 2013). Therefore, multicellularity is based on the cellular imperative to maximize *EI. Improved information quality permits the cell to efficiently uphold its self-referential equipoise versus the external environment beyond random adjustments to successfully cope with the dual constraints of informational ambiguity and the intrinsic noise within all biological systems (Miller 2016, Torday and Miller 2017, Torday and Miller 2020) (Fig. 1).

There is an important codicil imposed by triadic energy-information-communication. As any self-referential cell processes information, it is inevitably doing so across an energy gradient. Accordingly, any measurement of information shifts is energy value and the energy status of that cell, which concomitantly shifts the meaning of information content. Thus, all information available to any organism is necessarily constrained through autopoietic self-production (Cárdenas-García 2020). There is never any opportunity for any communication of an absolute, objective measuring value of that information. Any cellular measurement is always self-interpreted. All the information that can be made available to any cell is the result of self-referential subjective assessment (self-production) and, therefore, has to be at least slightly adulterated through each individual's self-filtered energy-information experience. Simply put, any information that is received by a cell and assessed is

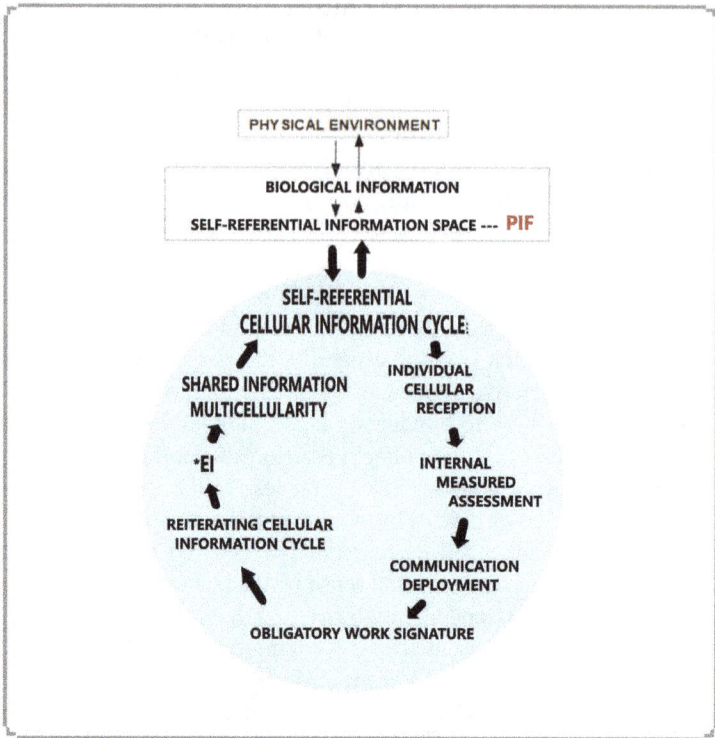

Fig. 1. The self-referential cellular information cycle leads to multicellularity

The external physical environment contributes ambiguous biological information-based cues to all cells. Each cell experiences an individualized attachment to information space-time as a Pervasive Information Field (PIF). Cellular information is received, measured for assessment, and potentially communicated to other cells. Irrespective of any direct communication, any information assessment and molecular deployments in a cell produce an obligatory work signal that initiates an iterative work channel energizing the collaborative appraisal of information. That obligatory work signature stimulates reiterating information cycles in other cells through sequential information reception and assessment. Cells share their assessment of environmental informational cues to improve their validity as Effective Information (*EI). This shared evaluation, communication, and deployment of information is the cornerstone of multicellularity.

inescapably altering it, which changes its communicated measuring value. Hence, ambiguity is the inescapable context of our living circumstance. In effect, it is the children's game of 'telephone' wired into planetary biology.

Therefore, the requirement to maximize information quality is the primary driver of biological activity. Indeed, this foundational requisite is so sweeping that even viruses collaborate to act more effectively (Santiana et al. 2018). It was believed until recently that viruses, such as enteric noroviruses or rotaviruses, act individually to infect cells. Instead, research reveals that viruses communicate and act collaboratively to become collective infectious agents (Miller et al. 2022). In this manner, they form a type of functional viral-cellular organism based on shared information.

One of the crucial elements of informational ambiguity is noise. Each of us is aware of how constant background noise is in our lives. Moreover, cells live in crowded, active environments, which affect any self-aware assessment of information space-time. However, background noise is not an unalloyed negative. We often put it to use. For instance, for a traveler in the jungle, any sudden and unanticipated silencing of noisy chatter in the canopy could be a meaningful warning of a lurking predator. So interruptions of background noise, or any other random noise patterns, can represent highly relevant information. Random informational inputs are constant but differ in amplitude. If these are effectively sorted between high amplitude informational inputs (with potential measuring value) from lower threshold more continuous random inputs, the results can be channeled into meaningful information as a 'harnessing of stochasticity' (Noble and Noble 2017). Again, this type of informational experience is not theoretical and has been demonstrated to be a crucial basis for habituation and sensitization among plants, animals, and single-celled organisms (Eisenstein and Eisenstein 2006, Gagliano et al. 2014).

Several cardinal elements of evolutionary development are clarified within a cognitive frame. First, cellular cognition confers the 'knowing' ability to measure ambiguous environmental stimuli. Secondly, and as a default of the first, multicellularity is driven by the cellular need to maximize the validity of environmental information through its collective measurement. Thirdly, the collaborative assessment of information, its shared communication, and the self-organizing collective deployment of resources categorically represent a form of collaborative action that is sufficient to drive processes of natural cellular engineering and mutualizing niche constructions that will be further discussed. Fourth, as an information system, biology must be underpinned by the cohering architecture of an informational space-time matrix to permit the conjoining assessment of information by individual self-referential observer/participants. Lastly, and ineluctably, all biological expression at all biological scales, including holobionts, must first express at the level of individual self-referential cells. Their self-generated interpretation of reality based on an attachment to informational space-time is the foremost biological reality.

Necessarily then, the self-referential cognition that enables adaptive evolution is defined by its exclusive attachment to biological information space through its 'knowing' of ambiguity. However, the activities of all cells are always narrowly directed towards defending their preferential homeorhetic state, which defines their self-identity. Consequently, adaptive evolution can now be understood as the continuous dynamic process of consistently reinforcing cellular self-referential states and protecting their essential self-identity. The defense of the 'self' depends on information quality. Succinctly then, evolution can be defined as the continuous defense of self-reference despite its originating constraint within ambiguity.

In 1994, the extraordinary physicist Mae-Wan Ho asserted that life can only be understood as the continuous active process of being an 'organizing whole'. Contemporary research now confirms that insight. Recognizing the inadequacy of prior concepts of evolutionary development, Marijuán and Navarro (2022) called for a new evolutionary synthesis that diverges from its classical emphasis on thermodynamic energy flows toward a parallel account centered on the flow of cellular information as organized cellular information management. Cognition-Based Evolution fulfills this preferred narrative, asserting that life represents a self-organizing whole based on the privilege of cellular cognition.

For reasons yet unknown, this 'knowing' state understands its conditional ambiguities. Self-referential awareness is the previously hidden glue that supports the multicellular drive towards the collective quantification of imprecise informational inputs that can successfully support that 'whole' versus a buffeting external environment. Thus, the essential feature of life clarifies. Life is information management (Farnsworth et al. 2013, Miller 2018). Building on that essential, Cognition-Based Evolution (CBE) argues that evolutionary biology represents an informational interactome centered within self-referential cellular dynamics. All biological action depends on the fundamental thermodynamic linkages between energy flows that transpose into measured information within cells and its further communication as a recursive, obliged work channel. This entailing linkage permits the self-reorganizing information cycle upon which life depends (Miller 2018, Miller and Torday 2018). Therefore, the study of evolutionary development should necessarily concentrate on the fundamental triadic linkage of energy/information/ communication. Through endless cycles of reiteration, contingent variations, and self-reinforcing constraints, macroorganic organisms are produced as seamlessly connected cell-based information architectural wholes as holobionts (Torday and Miller 2020).

Cognition Based Evolution (CBE) centers on the reciprocal assessment and communication of information by self-referential cells in reaction to environmental stresses. Therefore, in our requisite cell-based observer/participant frame, "information is the fabric of reality" (Dodig-Crnkovic 2014). Biology's crux is that self-referential information can only exist in any biological sense if information is deployed as a cognition-based solution to an environmental stimulus. Accordingly, biological information and its further communication exist only within the context of its yield as cellular problem-solving (De Loof 2015a,b, 2017, Miller 2016, 2018, Torday and Miller 2016). This summation further distills. All cellular information is purposed to the support of cellular self-identity as its preferential homeorhetic state (Miller et al. 2020a). Therefore, the flow of biological information clarifies. Cellular information flow is continually directed toward maintaining its originating self-awareness. All planetary life is the product of the three cellular domains and their partnering virome collaborating to perpetually sustain their originating self-identity. All multicellularity is viral-cellular information management to achieve that permanent outcome.

References

Baluška, F. and Levin, M. 2016. On Having No Head: Cognition throughout Biological Systems. Frontiers in Psychology 7.

Baluška, F. and Miller, Jr, W.B. 2018. Senomic view of the cell: Senome versus Genome. Communicative & Integrative Biology 11(3): 1–9.

Ben-Jacob, E. 2009. Learning from Bacteria about Natural Information Processing. Annals of the New York Academy of Sciences 1178: 78–90.

Ben-Jacob, E. and Levine, H. 2006. Self-engineering capabilities of bacteria. Journal of The Royal Society Interface 3(6): 197–214.

Cárdenas-García, J.F. 2020. The Process of Info-Autopoiesis—the Source of all Information. Biosemiotics 13(2): 199–221.

Czárán, T. and Hoekstra, R.F. 2009. Microbial Communication, Cooperation and Cheating: Quorum Sensing Drives the Evolution of Cooperation in Bacteria. PLoS ONE 4(8): e6655.

Davies, P. and Gregersen, N.H. (eds.). 2014. Information and the nature of reality: from physics to metaphysics. Cambridge University Press.

De Loof, A. 2015a. From Darwin's On the Origin of Species by Means of Natural Selection... to The evolution of Life with Communication Activity as its Very Essence and Driving Force (= Mega-Evolution). Life: The Excitement of Biology 3(3): 153–187.

De Loof, A. 2015b. Organic and Cultural Evolution can be Seamlessly Integrated Using the Principles of Communication and Problem-Solving: The Foundations for an Extended Evolutionary Synthesis (EES) as Outlined in the Mega-Evolution Concept. Life: The Excitement of Biology 2(4): 247–269.

De Loof, A. 2017. The evolution of "Life": A Metadarwinian integrative approach. Communicative & Integrative Biology 10(3): e1301335.

Deacon, T.W. 2011. Incomplete Nature. New York: W. W. Norton & Company.

Dodig-Crnkovic, G. 2014. Modeling Life as Cognitive Info-computation. pp. 153–162. *In*: Conference on Computability in Europe, Springer, Cham.

Eisenstein, E.M. and Eisenstein, D. 2006. A Behavioral Homeostasis Theory of Habituation and Sensitization: II. Further Developments and Predictions. Reviews in the Neurosciences 17(5).

Farnsworth, K.D., Nelson, J. and Gershenson, C. 2013. Living is Information Processing: From Molecules to Global Systems. Acta Biotheoretica 61(2): 203–222.

Forshaw, S.D. 2016. The Third State: Toward a Quantum Information Theory of Consciousness. NeuroQuantology 14(1): 49–61.

agliano, M., Renton, M., Depczynski, M. and Mancuso, S. 2014. Experience teaches plants to learn faster and forget slower in environments where it matters. Oecologia 175: 63–72.

Gabor, D. 1946. Theory of communication. Part 1: The analysis of information. Journal of the Institution of Electrical Engineers-part III: radio and communication engineering 93(26): 429–441.

Hall, W.P. 2005. Biological nature of knowledge in the learning organisation. The Learning Organization, 12(2): 169–188.

Ho, M.-W. 2008. The Rainbow and the Worm: The Physics of Organisms (2nd ed.). World Scientific Publishing Company.

Lloyd, S. 2002. Computational Capacity of the Universe. Physical Review Letters 88(23): 237901.

Lyon, P. 2015. The cognitive cell: bacterial behavior reconsidered. Frontiers in Microbiology 6(264).

Marijuán, P.C. and Navarro, J. 2022. The biological information flow: From cell theory to a new evolutionary synthesis. Biosystems 213: 104631.

Marijuán, P.C., Navarro, J. and del Moral, R. 2015. How the living is in the world: An inquiry into the informational choreographies of life. Progress in Biophysics and Molecular Biology 119(3): 469–480.

Marshall, W.F. 2019. Cellular Cognition: Sequential Logic in a Giant Protist. Current Biology 29(24): R1303–R1305.

Matsui, T. 2001. Gauge symmetry and neural networks. pp. 271–280. *In*: Fluctuating Paths and Fields. World Scientific.

Miller, W.B. 2013. The Microcosm Within: Evolution and Extinction in the Hologenome. Boca Raton, Florida, United States: Universal Publishers.

Miller, W.B. 2016. Cognition, Information Fields and Hologenomic Entanglement: Evolution in Light and Shadow. Biology 5(2): 21.

Miller, W.B. 2018. Biological information systems: Evolution as cognition-based information management. Progress in Biophysics and Molecular Biology 134: 1–26.

Miller, W.B., Baluška, F. and Torday, J.S. 2020a. Cellular senomic measurements in Cognition-Based Evolution. Progress in Biophysics and Molecular Biology 156: 20–33.

Miller, W.B., Reber, A.S., Marshall, P. and Baluška, F. 2022. Viral-Cellular Natural Engineering in Cognition-Based Evolution. Communicative and Integrative Biology. Accepted

Miller, W.B. and Torday, J.S. 2018. Four domains: The fundamental unicell and Post-Darwinian Cognition-Based Evolution. Progress in Biophysics and Molecular Biology 140: 49–73.

Miller, W.B., Torday, J.S. and Baluška, F. 2019. Biological Evolution as Defense of "Self". Progress in Biophysics and Molecular Biology 142: 54–74.

Miller, W.B., Torday, J.S. and Baluška, F. 2020b. The N-space Episenome unifies cellular information space-time within cognition-based evolution. Progress in Biophysics and Molecular Biology 150: 112–139.

Noble, R. and Noble, D. 2017. Was the Watchmaker Blind? Or Was She One-Eyed? Biology 6(4): 47.

Papineau, D., She, Z., Dodd, M.S., Iacoviello, F., Slack, J.F., Hauri, E., ... Little, C.T.S. 2022. Metabolically diverse primordial microbial communities in Earth's oldest seafloor-hydrothermal jasper. Science Advances 8(15).

Pilorz, V., Tam, S.K.E., Hughes, S., Pothecary, C.A., Jagannath, A., Hankins, M.W., ... Peirson, S.N. 2016. Melanopsin Regulates Both Sleep-Promoting and Arousal-Promoting Responses to Light. PLOS Biology 14(6): e1002482.

Proietti, M., Pickston, A., Graffitti, F., Barrow, P., Kundys, D., Branciard, C., ... Fedrizzi, A. 2019. Experimental test of local observer independence. Science Advances 5(9).

Rankin, C.H., Abrams, T., Barry, R.J., Bhatnagar, S., Clayton, D.F., Colombo, J., ... Thompson, R.F. 2009. Habituation revisited: An updated and revised description of the behavioral characteristics of habituation. Neurobiology of Learning and Memory 92(2): 135–138.

Santiana, M., Ghosh, S., Ho, B.A., Rajasekaran, V., Du, W.-L., Mutsafi, Y., ... Altan-Bonnet, N. 2018. Vesicle-Cloaked Virus Clusters Are Optimal Units for Inter-organismal Viral Transmission. Cell Host & Microbe 24(2): 208–220.e8.

Sato, M., Tsuji, T., Yang, K., Ren, X., Dreyfuss, J.M., Huang, T.L., ... Tseng, Y.-H. 2020. Cell-autonomous light sensitivity via Opsin3 regulates fuel utilization in brown adipocytes. PLOS Biology 18(2): e3000630.

Sugihara, T., Nagata, T., Mason, B., Koyanagi, M. and Terakita, A. 2016. Absorption Characteristics of Vertebrate Non-Visual Opsin, Opn3. PLOS ONE 11(8): e0161215.

Szostak, J.W. 2003. Functional information: Molecular messages. Nature 423(6941): 689.

Torday, J.S. and Miller, W.B. 2016. Life is determined by its environment. International Journal of Astrobiology 15(4): 345–350.

Torday, J.S. and Miller, W.B. 2017. The resolution of ambiguity as the basis for life: A cellular bridge between Western reductionism and Eastern holism. Progress in Biophysics and Molecular Biology 131: 288–297.

Torday, J. and Miller, W. 2020. Cellular-Molecular Mechanisms in Epigenetic Evolutionary Biology. Cham, Switzerland: Springer.

Tozzi, A. and Peters, J.F. 2017. What is it like to be "the same"? Progress in Biophysics and Molecular Biology 133: 30–35.

Trewavas, A.J. and Baluška, F. 2011. The ubiquity of consciousness. EMBO Reports 12(12): 1221–1225.

Vallverdú, J., Castro, O., Mayne, R., Talanov, M., Levin, M., Baluška, F., ... Adamatzky, A. 2018. Slime mould: The fundamental mechanisms of biological cognition. Biosystems 165: 57–70.

Walker, S.I. and Davies, P.C.W. 2013. The algorithmic origins of life. Journal of The Royal Society Interface 10(79): 20120869.

Witzany, G. 2011. The agents of natural genome editing. Journal of Molecular Cell Biology 3(3): 181–189.

Witzany, G. 2016. The biocommunication method: On the road to an integrative biology. Communicative & Integrative Biology 9(2): e1164374.

Witzany, G. and Baluška, F. (eds.). 2012. Biocommunication of Plants (Vol. 14). Berlin, Germany: Springer.

CHAPTER 5

The Senome
The Cellular Connection with Environmental Information

"An intelligent cell contains a compartment which is capable of collecting and integrating a variety of physically different and unforeseeable signals as the basis of problem-solving decisions".

Guenter Albrecht-Buehler, Professor Emeritus of Cell Biology
Northwestern University

A. The senome concept

The survival of all organisms depends on their self-referential adaptation to the stresses of environment flux (Agnati et al. 2009, Miller 2016). That environmental variability is a planetary constant and has an extraordinarily wide bandwidth compared to an organism's normative state of environmental preference. Consequently, all organisms must extract, process, and store effective information (*EI) as memory about their environment (Miller 2016, 2018, Miller and Torday 2017). As the previous chapters have enumerated, the process of the reception of information, its self-produced assessment, relevant communication, and further deployment is centered within a defined reiterative self-referential information cycle (Miller 2016, 2018, Miller et al. 2020a,b). That cycle undergirds all individual cellular life and governs all multicellular life. In that requisite frame, all genetic information and its further adjustment by epigenetic inputs must be viewed as essential elements of the cellular cognitive toolkit. All of these adjustments contribute to the cell-wide information cycle, which necessarily depends on the cellular reception of information as environmental cues, enabling cells to sense and adapt to environmental stresses (Miller 2018, Miller et al. 2020a).

Although the role of genetic material within a competent cell is undoubtedly crucial, it is obliged that any information that any organism has must be first perceived as sensory information deriving from relevant environmental cues. Therefore, sensory reception of information and the initiation of its assessment antecedes any action by the cellular genetic complement. Accordingly, cellular competence is crucially dependent on its sensory architecture in the continuous evaluation of environmental cues. Moreover, there is no indication that DNA directly participates in the acquisition of that sensory data (Baluška and Miller 2018). On the contrary, the cell's genetic material and its nuclear genome are dependent on an entire cellular sensory architecture. Therefore, it can now be understood that the deployment of all cellular genetic material is part of a larger process of information-dependent cell-based actions, all of which must derive from a cellular connection to the external environment (Agnati et al. 2009, Miller 2018, Marijuán et al. 2015, Marijuán and Navarro 2022).

Since all cells are cognitive, each must rely on its self-produced sensorial assessment of environmental cues to maintain homeorhetic balance and survive. These same cellular faculties are evident across the unicellular and multicellular realms. For example, bacteria have complex sensorimotor circuits that link plasma membrane-based sensors and their motor responses to flagellar activation (Stock and Levit 2000, Lyon 2015). Similarly, animal life depends on these same cellular processes. In the past, most scientists had not regarded plants as cognitive agents. Yet, plants have complex sensory systems that directly correspond to adaptive behaviors (Brenner et al. 2006, Baluška et al. 2009). Plant tropisms such as gravitropism and phototropism are not merely reflexive automatic reactions as once believed but are, instead, the product of a sophisticated, complex cellular evaluation of sensory cues whose coordinate responses link to motoric plant functions (Baluška et al. 2009, Stock and Levit 2000, Baluška and Mancuso 2013). Plant synapses integrate these sensorimotoric circuits by deploying bioelectric and biochemically-based cell-cell signaling through a system that resembles animal neuronal synapses and is organized along analogous principles (Baluška 2009, Baluška et al. 2009, Baluška and Mancuso 2013, 2014, 2018, 2021).

Given this requisite ordering of cellular activities, a codification of the complicated entirety of the sensory apparatus of the cell is conceptually necessary, recognizing that the cognitive awareness of a cell is a whole cell phenomenon. Therefore, just as the genome indicates the sum of the heritable endowment of cells, and the epigenome is used to designate the totality of all the chemical modifications to DNA and histone proteins that regulate the expression of genes deriving from environmental impacts, the senome has been conceived as the summation of the complete sensory apparatus of the cell and its totality of sensory experiences (Baluška and Miller 2018, Miller et al. 2020a,b, 2021). Cells use their senomes to maintain their preferential homeorhetic equipoise vis-a-vis all environmental impacts. Necessarily then, the senome is a characteristic of all cells.

For example, prokaryotes utilize highly varied and richly coordinated sensorimotor circuits that receive, measure, and contribute to the deployment of cellular information (Lyon 2015). These sensory inputs trigger a full range of homeorhetic responses, contributing to adaptive behaviors (Torday and Miller 2020). Thus, the senome is the cognitive gateway for the reception of all the varied sources of cellular information from the external environment, channeling it into the interior of the cell for its further synthesis to maintain homeorhetic flux (Baluska and Miller 2018). As the senome represents a critical component of the information management system of the cell, it is the primary biological nexus that connects the experiential aspect of cognitive awareness with cellular memory and bioactive molecular action (Miller 2016, 2016, Miller and Torday 2018, Miller et al. 2020a).

B. Senomic functioning

Multiple components of the senome have been described in relation to complex cellular activities. These include those aspects of memory that are encoded within the non-genomic structured arrangement of molecules or get transferred through membranous cell divisions across generations (Kováč 2000, 2008, Jose 2018). Importantly, the senome encompasses the panoply of the sensory fields of cells, including those bioelectric fields that participate in morphogenesis or vibratory fields elicited via mechano-transduction (Levin and Martyniuk 2018). All of these informational inputs store as senomic information, ultimately transmitting to the genome, intracellular genetic material, and the epigenome (Baluška and Miller 2018). This summation of sensory information gleaned from the environment interconnects with the plasma cell membrane to create a set of resonating senomic fields (Baluška and Miller 2018).

The plasma membrane is equipped with diverse sensors optimized to function in the retrieval of environmental information (Lintilhac 1999, Loewenstein 2013). Selective activation of sensors linked to plasma membrane pores is vital for transmembrane directional transport of various proteins and bioactive molecules such as Fibroblast Growth Factor 2, which participates in the formation of extracellular vesicles (Steringer and Nickel 2018). Therefore, the senome functions as a perceptual gateway that channels acquired sensory information into the cellular interior and concomitantly, functions in transmembrane transit of cellular molecules into extracellular spaces. In this way, environmental information is conveyed internally for its further measured assessment, and cells can contribute that information appraisal to partnering cells. All cellular structures participate.

Even the translational apparatus of the ribosome is now considered a link in the conversion of information as part of the cellular measuring process (Vitas and Dobovišek 2018). Given the boundary between the external environment and the internal cellular structures, it is likely that the outer boundary cellular plasma membrane assumes a substantial role in the active functioning of the senome. The plasma membrane is the sensory boundary of the cell and is known to play a role in both the acquisition and processing of cellular environmental information (Lintilhac 1999). A holographic model for the senome has been previously proposed likening the plasma membrane to a three-dimensional holographic projection screen that displays the outward environment through the selective activation of its varied sensory apparatus (Baluška and Miller 2018). In this way, environmental information is conveyed internally for measured assessment.

Clearly, then, the senome functions as an integral component of the information management system of the cell. Necessarily, other sensory processes beyond the exchange of bioactive molecules require senomic interaction as there are many additional sources of information exchange among cells, such as primary cilia and tunneling nanotubes. Primary cilia detect and transmit extracellular cues essential to cellular homeorhesis (Anvarian et al. 2019). The sensory capacity of the cilia are dependent on the coordination of specific sensors that can flexibly react to distinct developmental, environmental, and homeorhetic cues. Tunneling nanotubes are thin epithelial bridges that are vital to cell-cell communication. Known as tunneling nanotubes in animals and plasmodesmata in plants, their complex interactions depend on diverse sensory-based signaling pathways, including receptor-mediated and mechanotransductive signaling via intercellular gap junctions. Studies indicate that embryonic development is conditioned on TNT-dependent electrical coupling (Gerdes et al. 2013).

Connexins are assemblies of protein subunits, integral to membranes, that form channels between adjacent cells for the bidirectional exchange of ions, metabolites, and various signaling molecules essential to cellular homeorhesis (Trosko et al. 2000). Apoptopodia is a part of this communication process, enabling the formation of exosomes as membrane-bound extracellular vesicles extending from the plasma membrane (Atkin-Smith and Poon 2017). These vesicles can travel quite far from the plasma membrane, either as individual units or in combinations (Minton 2015). Their production is highly dependent on the cellular sensorium, and they actively participate in intercellular and cross-species communication and the protection of the cell from pathogens (Cai et al. 2015).

Bacteria travel by mechanotaxis which is their sense of touch (Kühn et al. 2021). Bacteria such as *Pseudomonas aeruginosa* migrate across surfaces by twitching, crawling-like motions coordinated via nanometers-wide filaments (type IV pili). Bacterial mobility represents a complex sensory process that connects mechanical surface characteristics to a complicated mixture of anchoring and movement that enables efficient motility. Thus, the bacterial sensory system enables spatially-resolved mechanosensing that harnesses two response regulators that sequentially activate antagonistic functions. A newly identified protein, paxillin, forms on the exterior of cells and participates in cellular surface focal adhesion and anchoring by enabling a cell to sense its surroundings and attach to the extracellular matrix (Ripamonti et al. 2021). As an integrin protein, paxillin relates to LIM domains (zinc-binding protein sequences) forming on the cell exterior, assisting in the acquisition of sensory information and triggering mechanical coupling and coordinate signaling to the cellular actin cytoskelton.

Slime molds (Physarum polycephalum) have a finely tuned sensory system that is extremely flexible in response to external environmental cues. Experiments indicate that slime molds will use environmental noise to find an optimal migrational path through a maze. A complex sensory system connects its nutritional status, communications, and avoidance of any noise-related noxious stimuli, such as intermittent light pulses (Meyer et al. 2017). Importantly, research confirms that the self-organized decision-making exhibited by slime molds is intimately dependent on its total sensorium, serving as a fundamental driver for the ability of all cells to flexibly adapt behavior to meet environmental flux (Meyer et al. 2017). Sensing random noise is not wasteful since it can be put to use. Random noise can create sensed stochastic resonances that can enhance the detection of weak information signals and improve biological information processing (Hänggi 2002, McDonnell and Ward 2011).

Each of these complex mechanisms is directly linked to the cellular senomic apparatus and its connection with the external environment to guide intracellular downstream effects. Further, all cellular activity links to bioelectric effects. Consequently, the genome, epigenome, and proteome must seamlessly relate to environmental cues, which is accomplished via the cellular senomic architecture and can be further conducted by bioelectrical and molecular signals through and among cells (Baluška and Reber 2021, Levin 2021). Cellular experiments have demonstrated that DNA does not directly specify the cellular spatial orientations required to enact biological forms (Levin and Martyniuk 2018, Miller et al. 2020b). Instead, multicellularity requires bioelectrical connections among cells for pattern regulation and plasticity. All cells participate, not just nerve cells, each sensing electrical patterning signaling circuits that link individual cellular behaviors to form. Necessarily, such connections are dependent on the cellular senomic architecture.

The senomes of individual cells must overlap and interlink to enable consonant multicellularity and collaborative reactions to environmental stresses. Multicellularity depends on cell-cell communication (Baluška and Witzany 2014, De Loof 2015, 2017, Witzany 2016) and cognition (Miller and Torday 2018). However, that crucial pathway travels across the senome, initiated by its proximate contact with the external environment compared to the internal cellular milieu. Thus, the senome is the functional interpretative, measuring nexus for cellular responsiveness to stimuli for collective information assessment and communication. Necessarily, it must be considered scale-free, operating at the level of cells, but still capable of acting as an aggregate in macroorganisms, including ourselves, exerting a substantial influence on behaviors and social networks (Parkinson et al. 2018, Dunbar 2018). Therefore, the senome directly relates to individual cellular perception and in multicellular aggregates such as ourselves contributes to mental signals and symbols from external environmental variables (memes). From these senomic stimuli, our cognitive behaviors produce emotions or feelings as experiences, ultimately participating in coordinated social cognition (Blackmore 2016, Schaden and Patin 2018, Parkinson et al. 2018, Dunbar 2018, Fortier et al. 2018, Wittmann et al. 2018).

Since cells are intimately interrelated with viruses, might viruses exert their own type of senomic structure? Although typically regarded as inert outside of target cells, when inside, viruses can actively manipulate cells and their senomic endowments, including their cellular membranes and cellular cytoskeleton. It may be through this senomic connection that viruses successfully gain access to the nucleus or form unique virus-derived vesicles to permit their replication or intercellular transmission (Claverie 2006, Hegde et al. 2009, Forterre 2010, Ruiz-Saenz and Rodas 2010).

C. Implications of the senome

The senome represents a total summation of all the sensory inputs of any cell derived from its entire sensory apparatus and sensory tools (Baluška and Miller 2018, Miller et al. 2020a). Therefore, it is through the senome that cells maintain critical homeorhesis. However, it is important to recognize that all the information that any cell possesses is actually through self-generated assessment. Although extracellular energy and matter exist, the cell can only know anything about these information sources through infoautopoiesis, that is, through self-generated internal analysis. Axiomatically then, all the information that any cell possesses is self-produced (Cárdenas-García 2020). Necessarily, the senome is the vital intercessory architecture that initiates that self-production of information, thereby defining its self-awareness. This critical linkage explains why the cell is a measuring instrument. To permit productive responses to environmental stresses, it must utilize its sensomic apparatus to begin its self-appraisal of information that constitutes its reality. Consequently, every cell will self-generate its slightly different version of that reality. This precise point explains why biological systems should not be broadly likened to machines. In biological organisms, there can never be pure data input that would produce identical outcomes if fed into like-kind machines the way computers and factory equipment operate. Every constituent cell is an individual observer/participant in information space. Variances in the self-appraisal of environmental inputs are biology's condition. Thus, cellular activities represent solutions to this embedded problem, defining the living condition as one other aspect of informational ambiguity.

No two cells have identical senomes or measuring tools. Accordingly, measurements will vary among cells based on intracellular dynamics and external variables relating to observer status concerning external environmental cues. Therein lies biology's crux. There is conditional independence between internal and external states, and this gap has been described as a 'Markov blanket' (Ramstead et al. 2018). A Markov blanket defines the statistical characteristics of the boundaries of a system, such as a cell, partitioning that system into internal and external states. The 'blanket' represents the state that separates the two, mediating that separation (Kirchhoff et al. 2018). Importantly, external and internal conditions can only influence one another through intermediary sensory and active states. Hence, "the internal states only 'see' the external states through the 'veil' of the Markov blanket" (Ramstead et al. 2018, p. 4). The concept of Markov blankets has proved helpful in understanding the statistical nature of inference and free energy flows in biological system as a useful adjunct in our understanding of the full role of the senome and as potential tool for its further scrutiny.

Further implications of Markov blankets are pertinent to the senome concept. Markov blankets directly relate to the Free Energy Principle. Both non-living and living systems can only remain within a non-equilibrium state by restricting themselves to a statistically limited number of possible states associated with the minimization of their free energy function. This free energy function is not equivalent to thermodynamic free energy but is meant to capture how perceptions link to inferences in living systems. That direction is towards a minimization of perceptual error or 'surprisal' (Demekas et al. 2020). Notably, surprisal within the cellular context can be equated with prediction. The biological impulse towards multicellularity thus clarifies. The external environment connects to the cellular milieu through an obligatory Markov blanket that 'veils' its information content. The senome is that intercessory and is necessary for cellular inferences about environmental cues. Since all intracellular information is self-produced, all information is ambiguous with respect to any external 'reality'. Consequently, cells devised multicellularity as their 'wisdom of crowds' to permit collective measurements to overcome these apparent difficulties by conjoining their senomic experiences and environmental self-evaluations.

To maintain homeorhetic equipoise, cells must construct an inferential model upon which a range of predictions might be made. As this is self-generated, the cell becomes a physical translation of the cell's self-interpretation of the environment over time. Thus, the direction of evolution as the consistent cellular assimilation of the environment can now be understood as the product of the continuous engagement of the cellular senome with the environment to support cellular prediction. Crucially, all these actions represent cognitive activity as the sine qua non of biological self-organization. Self-reference is self-organization (Miller et al. 2020a, 2021).

Biology depends on the information management system of cells (Miller 2018). That system is crucially dependent on the acuity of its senome that integrates the reception, assessment, communication, and deployment of all of its available information (Baluška and Miller 2018, Miller et al. 2020a). Any coordinated, highly integrated system of this type could not be a random one. Such coordination could never be contingent on random actions. Within this highly integrated and reciprocating narrative, the possibility that cognitive cells with their demonstrable faculties would center their adaptive future on purely random genetic mutations is untenable. Highly reciprocating processes such as these cannot be construed as originating from an endless series of random occurrences. Nor can biological processes be modeled according to digital systems that depend on precise inputs. There is no veiled information in computers. Computers have no doubts.

The senome translates physical signals from the outside world into the physicochemical language of cells. This highly linked senomic integration process is integral to intracellular and multicellular natural engineering. A senomic informational matrix serves as the ground state of individual unicellular self-identity (Miller et al. 2020a). Therefore, the senome is a crucial aspect of the common informational platform that energizes multicellularity, either as biofilms or holobionts. It follows that any adaptive reactions exhibited by cells in response to environmental shifts must first travel through the senomic apparatus of cells prior to becoming cellular phenotype. Clearly then, the senome directly links to adaptive variation.

As the senome is essential for self-reference and multicellularity, it is possible to consider the senome as constructed by individually definable constitutents. For example, just as specific DNA segments of the genome are identified as genes that specifically encode for particular biological products, the senome might be conceptualized as comprised of individual 'senes'. Each sene corresponds to a particular subjective sensory experience as a definable unit of self-produced information relating to specific environmental cues (Baluška and Miller 2018). Complex sensations and their corresponding bioactive responses would relate to the retrieval of combinations of senes which constitute bracketed constellations of heritable sensory memory as information to be reinterpreted. In this way, the senome links to the genome and epigenome in reciprocating responsiveness to the environment. In this framework, the strength of the coordinated senomic reaction to a stimulus can be conceptually compared to informational density. Just as chemotaxis provides a guidepost of concentration gradients for microorganisms signifying the availability of nutrients, information density plays a role similar to concentration in chemotaxis (Vergassola et al. 2007). Given this, cells faced with uncertainties might utilize problem-solving search processes based on information densities, thereby maximizing the rate of increase in information gain. Indeed, multicellularity would then be considered an infotaxis search process by which the rate of gain of information relates to the overlapping senomic capacities of constituent cells.

D. The senome as the crucial cellular cognitive portal

Cells employ a constellation of inter-relating determinative tools for the acquisition of information, its measured assessment, its storage as memory, and its contingent deployment. In addition to the familiar genome and epigenome, the novel concept of a co-existing senome is now added as the sovereign means by which the competent, measuring cell senses its external environment. Therefore, the cellular senome represents the cognitive portal that introduces sensory experiences into the cellular milieu for contingent action and for being encoded into memory. Through its reciprocating interactions with the genome and epigenome, the senome serves as a biological nexus through which sensory inputs can synthesize with sensory memory, deploy as bioactive molecules, and be further communicated across the cellular domains (Prokaryota, Archaea, Eukaryota) Together, this essential amalgamation forms a cell-wide informational architecture, enabling consistent cellular internalization of the external environment as perpetual compatibility with the planet's unyielding strictures.

Through this cognitive interface between the cell and the environment, the 20th century's dominant fixation on the genome can now be re-channeled and balanced by a complementary focus on the senome and its correspondent epigenome. All are critical, functioning parts of an integrated whole. However, as cognition drives biological actions based on accurate cellular measurement of environmental variables, the senome, serving as the summation of the cognitive cell's entire sensing apparatus, must be considered primary; all other cellular effects are downstream. The reception of senomic experiences begins the cognitive process of assessing biological information and its measurement. Any productive appraisal of contemporary biology must authorize that cognitive sensory linkages are antecedent to all further cellular activity. Thus, the senome represents a critical gateway to what, how, when, and with whom cells measure, representing an essential component of its sensorimotor connective tissue.

References

Agnati, L.F., Baluška, F., Barlow, P.W. and Guidolin, D. 2009. Mosaic, self-similarity logic and biological attraction principles. Communicative & Integrative Biology 2(6): 552–563.

Anvarian, Z., Mykytyn, K., Mukhopadhyay, S., Pedersen, L.B. and Christensen, S.T. 2019. Cellular signalling by primary cilia in development, organ function and disease. Nature Reviews Nephrology 15(4): 199–219.

Atkin-Smith, G.K. and Poon, I.K.H. 2017. Disassembly of the Dying: Mechanisms and Functions. Trends in Cell Biology 27(2): 151–162.

Baluška, F. 2009. Cell-Cell Channels, Viruses, and Evolution. Annals of the New York Academy of Sciences 1178: 106–119.

Baluška, F. 2014. Life is more than a computer running DNA software. World Journal of Biological Chemistry 5(3): 275.

Baluška, F. and Mancuso, S. 2013. Root Apex Transition Zone As Oscillatory Zone. Frontiers in Plant Science 4(354).

Baluška, F. and Mancuso, S. 2014. Synaptic view of eukaryotic cell. International Journal of General Systems 43(7): 740–756.

Baluška, F. and Mancuso, S. 2018. Plant Cognition and Behavior: From Environmental Awareness to Synaptic Circuits Navigating Root Apices. pp. 51–77. *In*: F. Baluska, M. Gagliano and G. Witzany (eds.). Memory and Learning in Plants. Cham, Switzerland: Springer.

Baluška, F. and Mancuso, S. 2021. Individuality, self and sociality of vascular plants. Philosophical Transactions of the Royal Society B: Biological Sciences 376(1821): 20190760.

Baluška, F., Mancuso, S., Volkmann, D. and Barlow, P. 2009. The 'root-brain' hypothesis of Charles and Francis Darwin. Plant Signaling & Behavior 4(12): 1121–1127.

Baluška, F. and Miller, Jr, W.B. 2018. Senomic view of the cell: Senome versus Genome. Communicative & Integrative Biology 11(3): 1–9.

Baluška, F. and Reber, A.S. 2021. CBC-Clock Theory of Life—Integration of cellular circadian clocks and cellular sentience is essential for cognitive basis of life. BioEssays 43(10): 2100121.

Baluška, F. and Witzany, G. 2014. Life is more than a computer running DNA software. World Journal of Biological Chemistry 5(3): 275.

Blackmore, S. 2016. Memes and the evolution of religion: We need memetics, too. Behavioral and Brain Sciences 39: e5.

Brenner, E.D., Stahlberg, R., Mancuso, S., Vivanco, J., Baluška, F. and Van Volkenburgh, E. 2006. Plant neurobiology: an integrated view of plant signaling. Trends in Plant Science 11(8): 413–419.

Cai, X., Wang, X., Patel, S. and Clapham, D.E. 2015. Insights into the early evolution of animal calcium signaling machinery: A unicellular point of view. Cell Calcium 57(3): 166–173.

Cárdenas-García, J.F. 2020. The Process of Info-Autopoiesis—the Source of all Information. Biosemiotics 13(2): 199–221.

Claverie, J.-M. 2006. Viruses take center stage in cellular evolution. Genome Biology 7(6): 110.

De Loof, A. 2015. From Darwin's On the Origin of Species by Means of Natural Selection... to The evolution of Life with Communication Activity as its Very Essence and Driving Force (= Mega-Evolution). Life: The Excitement of Biology 3(3): 153–187.

De Loof, A. 2017. The evolution of "Life": A Metadarwinian integrative approach. Communicative & Integrative Biology 10(3): e1301335.

Demekas, D., Parr, T. and Friston, K.J. 2020. An Investigation of the Free Energy Principle for Emotion Recognition. Frontiers in Computational Neuroscience 14.

Dunbar, R.I.M. 2018. The Anatomy of Friendship. Trends in Cognitive Sciences 22(1): 32–51.

Forterre, P. 2010. Defining Life: The Virus Viewpoint. Origins of Life and Evolution of Biospheres 40(2): 151–160.

Fortier, J., Besnard, J. and Allain, P. 2018. Theory of mind, empathy and emotion perception in cortical and subcortical neurodegenerative diseases. Revue Neurologique 174(4): 237–246.

Gerdes, H.-H., Rustom, A. and Wang, X. 2013. Tunneling nanotubes, an emerging intercellular communication route in development. Mechanisms of Development 130(6–8): 381–387.

Hänggi, P. 2002. Stochastic Resonance in Biology How Noise Can Enhance Detection of Weak Signals and Help Improve Biological Information Processing. ChemPhysChem 3(3): 285.

Hegde, N.R., Maddur, M.S., Kaveri, S.V. and Bayry, J. 2009. Reasons to include viruses in the tree of life. Nature Reviews Microbiology 7(8): 615–615.

Jose, A.M. 2018. Replicating and Cycling Stores of Information Perpetuate Life. BioEssays 40(4): 1700161.

Kirchhoff, M., Parr, T., Palacios, E., Friston, K. and Kiverstein, J. 2018. The Markov blankets of life: autonomy, active inference and the free energy principle. Journal of The Royal Society Interface 15(138): 20170792.

Kováč, L. 2000. Fundamental principles of cognitive biology. pp. 51–69. *In*: Evolution and Cognition. Oxford, UK: Oxford University Press.

Kovác, L. 2008. Bioenergetics—a key to brain and mind. Communicative & Integrative Biology 1(1): 114–122.

Kühn, M.J., Talà, L., Inclan, Y.F., Patino, R., Pierrat, X., Vos, I., … Persat, A. 2021. Mechanotaxis directs Pseudomonas aeruginosa twitching motility. Proceedings of the National Academy of Sciences 118(30).

Levin, M. 2021. Bioelectric signaling: Reprogrammable circuits underlying embryogenesis, regeneration, and cancer. Cell 184(8): 1971–1989.

Levin, M. and Martyniuk, C.J. 2018. The bioelectric code: An ancient computational medium for dynamic control of growth and form. Biosystems 164: 76–93.

Lintilhac, P.M. 1999. Toward a theory of cellularity——Speculations on the nature of the living cell. BioScience.

Loewenstein, W. 2013. Physics in Mind: A Quantum View of the Brain. New York: Basic Books.

Lyon, P. 2015. The cognitive cell: bacterial behavior reconsidered. Frontiers in Microbiology 6(264).

Marijuán, P.C. and Navarro, J. 2022. The biological information flow: From cell theory to a new evolutionary synthesis. Biosystems 213: 104631.

Marijuán, P.C., Navarro, J. and del Moral, R. 2015. How the living is in the world: An inquiry into the informational choreographies of life. Progress in Biophysics and Molecular Biology 119(3): 469–480.

McDonnell, M.D. and Ward, L.M. 2011. The benefits of noise in neural systems: bridging theory and experiment. Nature Reviews Neuroscience 12(7): 415–425.

Meyer, B., Ansorge, C. and Nakagaki, T. 2017. The role of noise in self-organized decision making by the true slime mold Physarum polycephalum. PLOS ONE 12(3): e0172933.

Miller, W.B. 2016. Cognition, Information Fields and Hologenomic Entanglement: Evolution in Light and Shadow. Biology 5(2): 21.

Miller, W.B. 2018. Biological information systems: Evolution as cognition-based information management. Progress in Biophysics and Molecular Biology 134: 1–26.

Miller, W.B., Baluška, F. and Torday, J.S. 2020a. Cellular Senomic Measurements in Cognition-Based Evolution. Progress in Biophysics and Molecular Biology 156: 20–33.

Miller, W.B., Enguita, F.J. and Leitão, A.L. 2021. Non-Random Genome Editing and Natural Cellular Engineering in Cognition-Based Evolution. Cells 10(5): 1125.

Miller, W.B. and Torday, J.S. 2017. A systematic approach to cancer: evolution beyond selection. Clinical and Translational Medicine 6(2).

Miller, W.B. and Torday, J.S. 2018. Four domains: The fundamental unicell and Post-Darwinian Cognition-Based Evolution. Progress in Biophysics and Molecular Biology 140: 49–73.

Miller, W.B., Torday, J.S. and Baluška, F. 2020b. The N-space Episenome unifies cellular information space-time within cognition-based evolution. Progress in Biophysics and Molecular Biology 150: 112–139.

Minton, K. 2015. Apoptotic beads on a string. Nature Reviews Molecular Cell Biology 16(8): 453–453.

Parkinson, C., Kleinbaum, A.M. and Wheatley, T. 2018. Similar neural responses predict friendship. Nature Communications 9: 332.

Ramstead, M.J.D., Badcock, P.B. and Friston, K.J. 2018. Answering Schrödinger's question: A free-energy formulation. Physics of Life Reviews 24: 1–16.

Ripamonti, M., Liaudet, N., Azizi, L., Bouvard, D., Hytönen, V.P. and Wehrle-Haller, B. 2021. Structural and functional analysis of LIM domain-dependent recruitment of paxillin to $\alpha v \beta 3$ integrin-positive focal adhesions. Communications Biology 4: 380.

Ruiz-Saenz, J. and Rodas, J. 2010. Viruses, virophages, and their living nature. Acta Virologica 54(2): 85–90.

Schaden, G. and Patin, C. 2018. Semiotic systems with duality of patterning and the issue of cultural replicators. History and Philosophy of the Life Sciences 40(1): 4.

Steringer, J.P. and Nickel, W. 2018. A direct gateway into the extracellular space: Unconventional secretion of FGF2 through self-sustained plasma membrane pores. Seminars in Cell & Developmental Biology 83: 3–7.

Stock, J. and Levit, M. 2000. Signal transduction: Hair brains in bacterial chemotaxis. Current Biology 10(1): R11–R14.

Torday, J. and Miller, W. 2020. Cellular-Molecular Mechanisms in Epigenetic Evolutionary Biology Cham, Switzerland: Springer.

Trosko, J.E., Chang, C.-C., Wilson, M.R., Upham, B., Hayashi, T. and Wade, M. 2000. Gap Junctions and the Regulation of Cellular Functions of Stem Cells during Development and Differentiation. Methods 20(2): 245–264.

Vergassola, M., Villermaux, E. and Shraiman, B.I. 2007. 'Infotaxis' as a strategy for searching without gradients. Nature 445(7126): 406–409.

Vitas, M. and Dobovišek, A. 2018. In the Beginning was a Mutualism—On the Origin of Translation. Origins of Life and Evolution of Biospheres 48(2): 223–243.

Wittmann, M.K., Lockwood, P.L. and Rushworth, M.F.S. 2018. Neural Mechanisms of Social Cognition in Primates. Annual Review of Neuroscience 41: 99–118.

Witzany, G. 2016. The biocommunication method: On the road to an integrative biology. Communicative & Integrative Biology 9(2): e1164374.

CHAPTER 6

The N-space Episenome
Concordant Cellular Information

There has been a long-standing assumption that the developmental bauplan for embryogenesis and morphogenesis is embedded in the nuclear genome (Rafiq et al. 2012). However, a growing body of research indicates that both morphogenesis and phenotype do not unfold as a direct function of the genomic complement (Miller et al. 2020a,b, Baverstock 2021, Levin 2021, Ildefonso 2021). Instead, multiple highly orchestrated and complex reciprocating feedback loops intertwine among cells, integrating their genome, epigenome, and correspondent senome for coordinated growth, development, and adaptation. The N-space Episenome concept has been introduced as a whole field informational projection that represents a shared information platform permitting that conjoining action. Every cell is connected to information space-time through its Pervasive Information Field as its circumscribed interrogation of information space as fundamental connections between matter and energy, limited by its sensory apparatus. This projection lattice serves as a concordant informational platform that self-referential cells require to coordinate complex morphogenetic development. Consequently, self-referential, self-organizing multicellular biological organization is based on dual reciprocating components consisting of its entire cell-wide biological materiality and an interlacing and overarching universal information-management space-time architecture (N-space).

Undoubtedly, there is effective and abundant cross-talk between the genome and cytoplasmic structures in morphogenesis. Cross-species cloning research confirms extensive cross-talk between maternal cytoplasm and the nuclear genome during development (Sun and Zhu 2014). Cross-species nuclear transfer from one species into a different species enucleated cell leads to an animal that does not fully resemble the genomic species, indicating that cytoplasmic participants and membranes have strong modulating influences on genetic regulation and the determination of phenotype (Sun and Zhu 2014, Zuo et al.2014). Embryogenesis requires seamless whole-cell coordination. If the correct cytoplasmic factors are unavailable, pre-implantation embryonic arrest will occur (Jiang et al. 2011, Lagutina et al. 2018).

Consequently, several alternative patterning systems have been proposed to account for both embryogenesis and morphogenesis. For example, the developmental bauplan might still remain centered in the genome but have distributed components.

Alternatively, it might be an emergent process based on networking cells, or it may rely on cell-generated and self-organizing fields that could be bioelectric or biophotonic, or a combination of these (Gilbert and Sarkar 2000, Tzambazakis 2015, Neuhof et al. 2016, Fels 2018, Levin et al. 2019). Cellular research indicates that a physiological system of cell-cell communication regulates patterning and shape during embryogenesis and governs regeneration (Sullivan et al. 2016). Consequently, it has been presumed that somatic cells coordinate networks by deploying shared neurotransmitters through common ion channels, electrical synapses, or extracellular 'morphogens'. These latter signaling proteins, such as Decapentaplegic or Wnt morphogen signaling proteins in Drosophila melanogaster, have been found to participate in regional patterning, growth, and development and thereby influence regional cellular transcription and phenotypes (Affolter and Basler 2007, Rogers and Schier 2011, Roy et al. 2014).

Although morphogens have consequential effects, the source of their regulation remains unclear as their effects must be exerted in concert across many orders of magnitude (Müller and Schier 2011). Accordingly, various energetic fields have been proposed to account for the coordination of morphogens, beginning with the introduction of the 'morphogenetic field' concept by Alexander Gurwitsch (Gurwitsch 1912). In theory, a radiating supracellular field mechanism exists that would control all molecular, cellular, and morphologic expression for developmental sequencing (Tzambazakis 2015). Although a variety of signaling gradients and energetic fields are certainly present, how these control developmental templating remains unresolved. Noting that the genome has no geometric specificity that could account for patterning. Levin and Martyniuk (2018) have offered bioelectric fields as a controlling agency in morphogenesis. Endogenous bioelectrical networks could store and release both genetic and non-genetic patterning information during development and regeneration (Levin 2014). For example, in *Xenopus laevis* embryos, when endogenous bioelectric patterns are interrupted, abnormal brain patterning is evident, resulting in malformations (Pai et al. 2015).

Further, evidence indicates that cell migration is guided by electrical fields dependent on distributed cellular-molecular mechanisms, some of which arise from the plasma membrane and convey directional information (Funk 2015). De Loof (2016) has asserted a cell-generated 'electrome' that co-exists with the genome and proteome and provides templating for growth and development. Others have proposed electromagnetic fields that create reinforcing waves that control cellular geometric ordering and patterning (Geesink and Meijer 2018). Conversely, Fleury and Gordon (2012) emphasize intercellular mechanical factors such as sheer stress, tensile stress, or cellular compression that induce mechanotransduction 'differentiation waves'.

Yet, despite the theories and research to date, the actual controlling agency for comprehensive embryogenesis and morphogenesis remains unknown. A productive answer can be provided by fully honoring that all biological actions of any kind are information-dependent and based within self-referential cellular measurement. Since morphogenesis requires exquisite shared cellular measurements to enable complex cellular ecologies, a heritable information-based supracellular platform must be accessible to permit the concordant measurements that patterning morphogenesis requires (Miller et al. 2020a). The reasoning is direct. Morphogenesis is cellular engineering. Holobionts are combinations of an enormous array of highly differentiated cells that must effectively communicate with one another and measure together. Further, the microbiome also participates in development and these are entirely different species and life forms. Since all engineering is measurement, there must be a common information platform serving as a mutualizing reference base.

To satisfy this implicit necessity in the self-referential frame, an N-space Episenome has been proposed as this requisite preexisting, heritable, and transferable information space-time matrix. The N-space Episenome serves as the consonant measuring platform for the concordant integration of the vast numbers of cells involved in the incredible complexities of morphogenesis, growth, and development (Miller et al. 2020a,b, 2021). Most significantly, this type of biological information field fulfills the requirements of all the prior research that attests to the richly coordinated cellular signaling networks that connect the cellular senome to intricate transcriptional pathways. Together, these enable the appropriate timing of genomic and epigenomic expressions that yield synchronized, productive biological activity (West and Greenberg 2011, Baluška and Miller 2018, Baluška and Reber 2019, Reber 2019, Yamada et al. 2019). Given this cell-wide complexity, a coordinating information matrix governing the entanglement between cellular information space-time, its genome, epigenome, and further cell-wide expressions is a precondition among self-referential cells. As each participating cell in any context is an independent self-referential agency whose information entirely consists of its own measured self-construction and self-generated interpretation of its meaning, a common informational platform for the cellular attachment to information space-time is required. Directly, if cellular information content is definitionally based within individual self-produced sensory measurements, a conjoining information space-time network that enables consonant, interpretable shared information measurement must be present to permit the seamless integration of collective and collaborative cellular actions (Miller et al. 2020a,b, 2021).

Absent this coordinating platform, the impeccable coordination among eukaryotic body cells in holobionts and its vast co-existent multispecies microbiome would not be conceivable. Crucially, this microbial fraction is not a placid appendage. Its contributions to holobionic life are highly nuanced, interlocking intimately with its eukaryotic partners. That relationship is so deeply interconnected that even changing a single microbial gene in only one bacterial strain is sufficient to alter the holobiont's metabolism. Deleting a single gene in the common mouse gut microbe *Bacteroides thetaiotaomicron* alters the circulating levels of the bile salt, tauro-β-muricholic acid. This shift results in reduced weight gain in mice, alters their circadian rhythms, and modulates respiratory exchange and immune parameters (Yao et al. 2018). If the change of a single gene in a holobionic partner can yield consequential physiological and metabolic shifts, there must be a coordinating platform for the exchange of cellular information for interspecies communication upon which holobionic life depends. None of the participants are automatons. Since embryogenesis and morphogenesis are dependent on intense, collaborative cell-cell communication, these processes must also rely on a shared informational platform. It is only by this means that widely differentiated eukaryotic cells or disparate cellular species could measure environmental stresses in concert to yield congruous living results.

The case for a common informational platform is strengthened by the recognition that all biological development is a function of symbiogenesis (Villarreal and Ryan 2019, Miller et al. 2020a). Chapman and Margulis (1998) have argued the importance of symbiogenesis in morphogenesis, emphasizing that experimental evidence indicates that morphogenesis and the specialized differentiation of certain eukaryotic cells are both dependent on symbionts. For example, the signals from heterotrophic bacteria are essential for the induction of legume root nodules and their proper development. Fungi can also affect morphological expressions in other organisms, influencing the relationships between holobionts and their interrelated microbial species (Bahram and Netherway 2022). Gilbert (2019) effectively summarizes this vital interdependence: "We do not develop as monogenomic organisms, instructed solely by DNA and the cytoplasm of the zygote. Rather, we are holobionts, symbiotic consortia containing numerous microbial genomes whose signals are critically important for our normal development. Microbes play crucial roles in forming and maturing animal guts, immune systems, nervous systems, and reproductive organs. In some species, they regulate such developmental phenomena as the proper orientation of the anterior-posterior axis and metamorphosis."

Given these justifications, exactly how does the common informational platform of the N-space Episenome operate? N-space represents a scale-free universal relational information matrix which will be further detailed in Chapter 18. N-space builds on the interconvertibility of matter, energy, and information. As noted in Chapter 4, information is considered by some physicists to be the primary fundamental of the universe. What matters to any living observer are the interactions between matter and energy rather than either per se. After all, any perception of matter is dependent on an energetic reaction with it. These reactions, either through classical Newtonian or quantum mechanisms, constitute a 'difference that makes a difference', which is the most practical definition of information and forms the conceptual basis of information in the living frame. It is the connections that create differences that make the universe since absent those, the universe would be inert, and there would be no difference, no perception, and no measurements. Consequently, N-space is the universe of all the possible connections between matter and energy, which represents an information fabric upon which life can build.

As introduced in Chapter 4, every cell is attached to universal information space through an individualized cellular Pervasive Information Field (PIF) (Miller 2016, 2018). That informational matrix represents all of the sources of information available to a cell by which its self-produced autopoietic information can be derived as correspondent internal measurements essential for cellular homeorhetic equipoise. Cell attaches to information space (the set of all external environmental cues) through their PIFs, receiving internal and external information cues through their senomes as a summation of all the sensory informational inputs of those cells (Baluška and Miller 2018). The senome of the cell serves as the interpretive gateway between the cell's relevant information space (cellular PIF) and all the cellular structures that enable the self-production of useful information for cells, including its genome and epigenome. In turn, the N-space Episenome represents an aggregation of all of the individualized information fields (PIFs) of the individual cells in multicellular organisms as a collective organism-wide summary informational architecture.

This coaligning summation field is the common information platform that permits concordant cellular measurements as the basis of its mutualized cellular information management, coordinating spatial orientation, development, and physiological concordance (Miller et al. 2020a). In this manner, the N-space Episenome is the major supporting structure that allows multicellular organisms with many disparate cell types to collectively assess environmental cues. Those cues emanate from within an external N-space as a global encompassing universal information matrix as a interrelating referencing informational field, representing the summation of all potential sources of information that might influence cells either in proximity or at a distance, either directly or indirectly (Fig. 1).

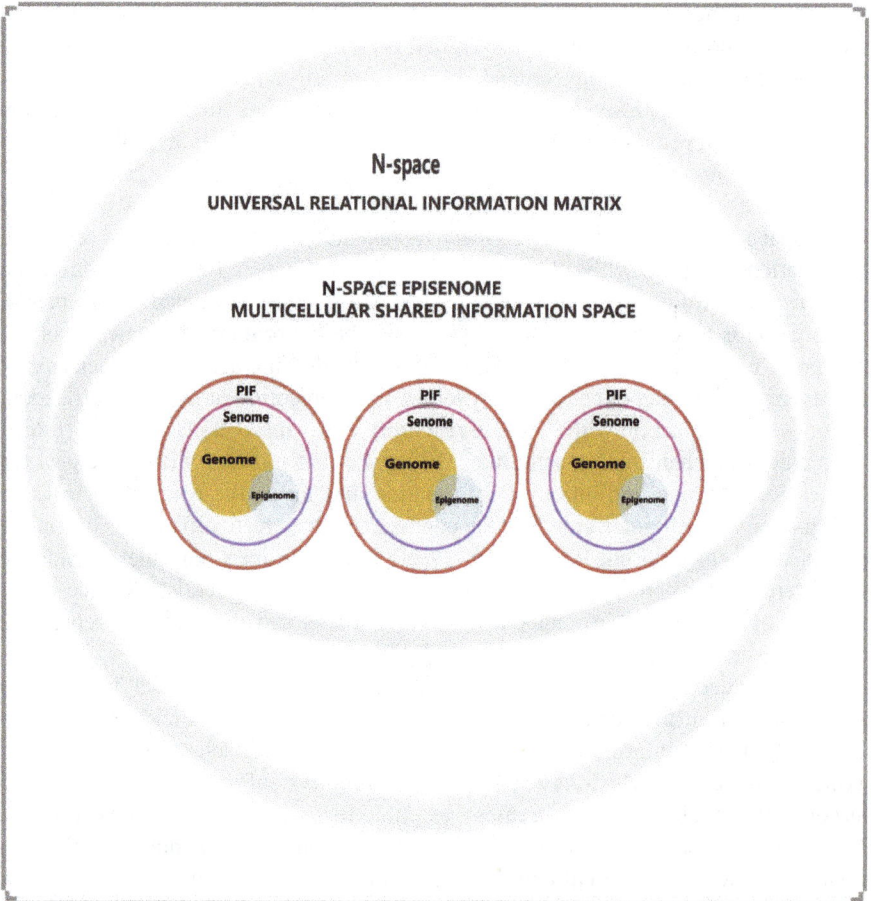

Fig. 1. N-space architecture.

Every self-referential cell is attached to information space-time through its individual Pervasive Information Field (PIF), integrating its senome, genome, and epigenome. Each cellular PIF connects to a system-wide N-space Episenome as a shared informational platform intrinsic to its multicellular architecture. The N-space Episenomes further attaches to the external environment as a universal relational information space-time matrix (N-space).

The effective sharing of information at every scale ranging over many orders of magnitude extending from independent unicells to biofilms, holobionts, and social groups of holobionts, mandates a common platform for information measurement. Intelligible communication, the effective trading of resources, and the complementary deployment of resources to effectively deal with environmental stresses all depend on a common referencing system. It is the same with our own engineering activities. Consequently, an informational matrix exists, and in multicellularity represents a preexisting transferable and heritable developmental information space architectural template for biological development and morphogenesis (Miller et al. 2020a). Successful aggregate multicellular life requires more than the memory encoded in a genetic code or plasma membrane. Although these biological structures are integral to cellular memory, the pertinent issue is that working information is self-produced. It is always individualized and distinct to every cell. Therefore, that self-generated information can only be invested with contextual meaning when juxtaposed with the environmental measures of other like-kind cells with their own self-generated information if there is a common referencing platform. In human engineering, standardized weights and measures are an absolute necessity. Cells that internally generate their own information require the same. It is only in that way that the meaning of mutually experienced environmental cues can coalesce into productive biological expression. Therefore, a common referencing architecture is present and sufficient to permit flexible concordant cellular responses across the orders of magnitude that constitute holobionic life with its multitude of species contributors.

The complicated relationship between self-referential biological information and how it corresponds with physical reality is pertinent to this contention of the self-referential necessity for a common measuring lattice (Walker et al. 2016). The general presumption about information is that it represents the type of structured transmissible data that computers utilize to receive, process, and communicate. However, biological information is fundamentally different. Biological information is physical since it directly ties to physical representations and, further, denotes a superposition of possibilities and embodied constraints (Landauer 2018). It has been vigorously maintained that information not only represents a physical quantity just as crucial as energy or matter but, further, that the real world is actually composed of information. Consequently, matter and energy are merely incidental epiphenomena (Bekenstein 2003). Biological information has a direct physical character since it is the crucial link between cellular genetic material, its proteome, and its functional energy characteristics (Baverstock 2013). Furthermore, any actual biological meaning can only be achieved within physical degrees of freedom; that is, matter and energy only exist for a cell insofar as they constitute interrelational contextual information that achieves physical form (Davies and Walker 2016). According to physicists Davies and Walker (2016, p. 4): "practical biology demands a rough-and-ready type of dualism—of matter and information" (Davies and Walker 2016, p. 4).

Nonetheless, there is no requirement that the physicality of information has to conform to the causal order that we assume represents external physical reality (Oreshkov et al. 2012). Indeed, the principles of quantum mechanics explicitly undermine our intuitive sense that physical properties are absolute, well-defined, and independent of measurement. In the self-referential frame, all measurements inevitably change the status of the system. Therefore, in biological systems where all information is a self-produced measurement, any available information only indirectly correlates with an external reality whose shape is debatable.

Because of differences in scales that seamlessly combine in biology, biological information exists through both quantum and classical correlations. However, it has been argued that quantum correlations are stronger than classical ones (Pawłowski et al. 2009). In that case, the principle of Information Causality can be considered as applying, permitting self-referential agents to access information from other self-referential entities that extends beyond local resources or the classical boundaries of information (Pawłowski et al. 2009). Although the specifics are outside of the scope of this chapter, in some interpretations of quantum theory, a broader range of informational cues can be enlisted between self-referential observer/participants than classical conditions would imply, extending beyond one-to-one correspondences. In biology, information is volumetric and has antipodes that are observer-dependent. The pertinence of Information Causality applying between separate self-referential observers is that there have to be global causal relations for the channeling of quantum information in space-time. Causal relations among independent self-generating information observers requires a common referential framework (Pawlowski et al. 2009). Importantly, however, this relationship must be preexisting to their interrelationship. It cannot simply emerge since, in the absence of the preexisting frame, there would never be any meaning to the very first information exchange, and nothing meaningful could follow. Biological information upends the traditional concept of causal order through its constraint of self-production that distances any information in the living frame from any purely objective external reality. The principle of Information Causality explains how causal order in information can be re-established, but for that to be the case, a preexisting global causal relational channel has to exist.

Accordingly, for information to be effectively available among independent self-referential agents across scales, the traditional view of information must be rethought. For there to be causal order in biology, a heritable N-space architectural information space-time matrix is a requirement. In sum, biological self-organization requires a preexisting templating mechanism. It is this preexisting heritable information channel that enables a common measuring nomenclature among self-referential cells across the living scale. At the level of holobionts, this preexisting architecture is the multicellular N-space Episenome as a transferable form of embedded memory that guides morphogenesis, growth, and development. The same matrix would equally influence heritable patterns of behavior. After all, vulnerable animals like wildebeest calves rise to their feet immediately after birth and suckle. There is no opportunity for instruction from the mother. It is an instantaneous reaction. It is no mere reflex since that coordinated movement is not just tendon or muscle but a preexisting behavioral pattern. Therefore, it defaults that there has to be an informational N-space that supplies a heritable memory that permits the correct complex coordinated behavior by the new organism, for whom all information is self-generated, to productively respond to its entirely novel post-natal surroundings. Thus, an N-space Episenome architectural information matrix is an inherited feature of the unicellular zygote. This preexisting scaffolding is a critical aspect of its informational repository as a heritable self-referential attachment to information space-time, thus serving as its morphogenetic template (Miller et al. 2020a).

It is essential to regard any form of informational architecture, whether a PIF or an N-space Episenome, as a read-write component of the cell-wide information management system (Miller et al. 2020a). The concept that informational systems in biology are read-write is not new, having been fully articulated as an embedded feature of all genomes (Witzany 2010, Shapiro 2017). In CBE, it has been previously proposed that the genome read-write architecture is in continuous reciprocation with N-space Episenome and, thereby, with the entire cell structure (Miller et al. 2020a). For any global causal information system to be of current use for multicellular life, its informational space-time matrix must have a read-write capacity for ongoing information management. Further, and by definition, to be a global source of information causality among individual self-referential infoautopoietic agents, it must be transferable through mitosis, heritable through meiosis, and adjust in tandem with cellular variations.

The N-space model depends on conceptual links to several other well-defined field concepts. N-space architecture strongly resembles the non-physical *k*-space, which is the information space matrix populated by acquired data during Magnetic Resonance Imaging (MRI) scans. The same information architecture is also used in photoelectron spectroscopy for devising nanostructure materials (Lev et al. 2018). MRI *k*-space is an extension of mathematical Fourier space, representing a specific type of compaction of both 2D and 3D spatial frequency information (Moratal et al. 2008). By using information acquired from orthogonal axes, a highly accurate anatomic spatial map can be electronically acquired. In MRI, *k*-space data is a temporary image information matrix that can be mathematically reconstructed using Fourier transforms, creating a spatial representation of acquired MRI signals which can be mapped to specific anatomic structures for patient diagnosis or surgical planning (Fig. 2).

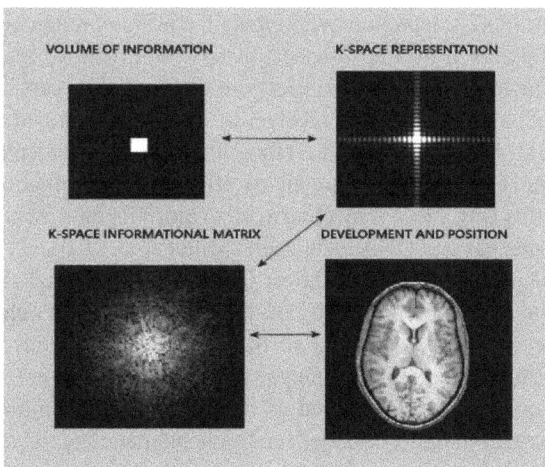

Fig. 2. Image formation in *k*-space.

In MRI scanning, a volume of information is acquired as a corresponding representation in a holographic-like *k*-space informational matrix architecture. Information acquisition is obtained along orthogonal axes filling in a *k*-space informational templating matrix, permitting the spatial dimensionality necessary for reconstructing images of precise anatomic locations.

An important property of *k*-space is that there is no precise one-to-one correspondence between any point in *k*-space and any resulting image. Every point in *k*-space both receives and delivers contributions to and from every other point in *k*-space, functioning like a hologram (Mezrich 1995). Therefore, every point in *k*-space is a summation of contributions from the entirety of *k*-space. Separate points in k-space have different relative information densities based on location but still have contributions from all other points. This information space model correlates precisely with the features of a biological N-space information matrix. The only conceptual difference is that in MRI scanning, *k*-space is a temporary record of an information matrix. In contrast, N-space is sufficiently permanent to enable a transferable and heritable N-space Episenome as the indispensable intercessory between cells and the external environment.

There is another applicable overlap between the techniques of the formation of MRI *k*-space and biological N-space. In MRI imaging, the sharpness of the image directly relates to the number of data inputs that create the relevant summary *k*-space data points. The validity of the data improves with the number of acquisitions. Greater numbers of *k*-space inputs translate into reduced noise and, ultimately diagnostic validity. This imaging circumstance yields an effective analogy with biological information systems. In cellular dynamics, a greater number of measuring participants in the assessment of any relevant environmental cue directly relates to the validity of the measured biological information (decreased noise and lowered ambiguity). Multicellularity is biology's answer to unsharp information and is also adroitly commensurate with basic thermodynamic requirements. Better information means less wasted energy. Correlatively, higher information densities correspond with the minimization of variational free energy as higher predictive value and suppression of surprisal, reflecting energy efficiency (Friston et al. 2006, Torday and Miller 2016).

Furthermore, to gain more precise spatial information in MRI, an orthogonal electrical gradient field is deployed overlapping the primary magnetic field to obtain positional information from the *k*-space matrix. Pertinently, research indicates that DNA is a fractal antenna within electromagnetic fields, capable of electronic conduction as both a transmitter and receiver of electromagnetic information (Blank and Goodman 2011, Sepehri 2017). Therefore, it may not be incidental that DNA has a helical structure. Helical antennae are highly efficient for information reception and transmission. In this way, DNA may contribute to morphogenesis and patterning in a manner that extends beyond DNA code itself (Miller et al. 2020a).

A *k*-space architecture has been shown to have an additional feature of innate symmetries that allow for the mirroring of data in *k*-space (Mezrich 1995). In MRI imaging, *k*-space can be interrogated so that *k*-space data points have mirroring symmetries that can be productively mapped to distant locations in *k*-space. In effect, k-space information can be considered as having both a presenting value and an antipodal (symmetric but opposite-sided) component, matching the characteristics of biological information and its conditional ambiguities (Miller et al. 2020a).

Specific features of N-space overlap other significant field concepts. Notably, the multi-dimensional topological features of Euclidean vector space closely analogizes with N-space biological information space architecture. Significantly, the architecture of Euclidean vector space has already been validated as a useful tool in understanding morphometrics and modeling phenotype (Huttegger and Mitteroecker 2011).

An important component of N-space is that its features permit both information content and three-dimensional spatial orientation, closely mirroring how *k*-space operates in MRI imaging. MRI data acquisition follows a compartment model. Subsections of *k*-space are sequentially filled as more data points are acquired. The availability of the subsections to be data-filled can be controlled but always follows a systematic logic. This feature fortuitously matches biology's requirements for both information quality and geometric patterning since it permits a productive analogy with biological development. Morphogenetic development, growth, development, and maturation can be likened to the sequential filling of a preexisting N-space information architecture. During development, cells must be ready to contribute information to the relevant N-space compartment. In turn, that compartment must be available to be filled, thereby controlling morphogenesis, growth, and development by the timed opening of N-space compartments. N-space compartments must be in concert with the cellular readiness to supply the measured information required to fill it. It is argued that this essential reciprocation has been the elusive essential missing element that can satisfactorily account for the unfolding of morphogenesis (Fig. 3).

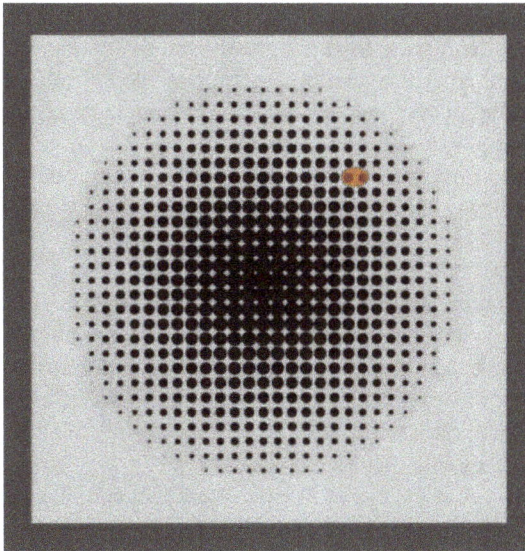

Fig. 3. Two-dimensional representation of multi-dimensional N-space compartments within an N-space Episenome information matrix.

N-space represents a holographic density/position information matrix, which is partitioned in multicellularity as an N-space Episenome. The developmental process can be conceptualized as sequentially filling in of a series of localized areas of *N*-space (orange circle) as templating informational compartments. The opening of any compartment is a function of both appropriate cellular information to fill it (physiological maturation) and the concurrent morphogenetic timing availability of that relevant *N*-space compartment. Morphogenetic development unfolds based on this reciprocation between cellular cues and available templating *N*-space compartmentaliztions.

Bayes' Theorem describes the probability of the occurrence of an event based on prior knowledge of conditions that might be related to that outcome. This systematic, mathematized method of reasoning has had a prodigious impact on our understanding of statistical inference (Efron 2013). Notably, Bayes' Theorem can be helpful in conceptualizing N-space. The constraints of Bayes' Theorem assert that any individual cellular assessment of informational cues, even though self-generated, does not exist in a void divorced from its history. Self-produced informational cues do not arise within a vacuum. Instead, new evidence is used to update prior beliefs. Pertinently, that new evidence does not completely undermine prior beliefs. Therefore, within Bayes' Theorem, prior beliefs are not expunged, just nudged into a different direction. Thus, the multicellular 'wisdom of crowds' is the cellular consummation of adjusted cellular predictions from within a generally normative range. It follows that within a self-referential matrix where all cellular information is self-generated, a consonant platform must exist where cellular 'beliefs' are encoded within memory that can be shared as useful predictions and meaningfully updated. As a further conspicuous derivative, Bayes' Theorem can be productively applied to cellular information since, in that framework, the statistical validity of information directly correlates with sample size. Notably, this statistical relationship justifies multicellularity and explains why planetary life is not based on only a few large, highly capable, and efficient cells. Instead, myriads of small conjoined, individual, communicating self-referential cells link into multilevel biological outcomes through a common platform where individual inferences can productively meld. In this way, self-produced information emerges from the restricted matrix of theoretically limitless N-space possibilities, permitting congruent cellular informational appraisals that can become functioning predictions of anticipated future environmental conditions. From this narrowed set of superimposed possibilities, implicate biological potentials can settle into functional explicates through a Bayesian process of the 'wisdom of crowds'.

The reasoning behind the insistence on the existence of an N-space architecture is quite direct. Every cell is a self-referential individual. Any information it has is the result of its own measurement. For those measurements to be shared, each as the result of idiosyncratic self-production, there has to be a mutual measuring platform to gauge the relative concordant validity of those individual measurements. A templating information-space architecture meets that absolute requirement for both individual and collaborative noetic survival. Coordinate life requires concordant measurements (Miller et al. 2020b).

N-space architecture is the common latticework that conjoins cellular 'beliefs' as shared, collective, and potentially predictive measuring values. It represents, then, a multicellular forum from which collective inference can be nudged into productive multicellular predictions. In N-space, cellular inferences become multicellular predictions that are used to settle the superposition of possibilities in informational space-time into discrete biological expressions. In the self-referential cognitive frame, this inferential assessment is imperative for functional biological expression (Friston 2018, Miller et al. 2020a). Thus, cellular life can achieve useful multicellular concurrences by being 'nudged' through Bayesian 'wisdom of crowds' towards advantaged biological outcomes. Through these means, the information content of environmental cues gets translated across crucial cellular linkages from the sensory apparatus of the senome, to the genome, and then to proteomic deployment. Thus, the N-space Episenome represents a replete informational landscape in consistent reciprocation with biological materialism as a guiding field of potentials and constraints in information space-time. In this way, the complete panoply of cellular tools (senome, genome, epigenome, cytoplasmic components, endomembranes) can reciprocate with the external environment, permitting necessary real-time responsiveness to environmental stresses to ensure continuous assimilation. Accordingly, the N-space Episenome is the essential multicellular conjoining read-write, preexisting and heritable informational matrix that connects and coordinates the structural complexity of a self-referential cell to other cells, enabling multicellular life as an informational template for biological development and morphogenesis.

References

Affolter, M. and Basler, K. 2007. The Decapentaplegic morphogen gradient: from pattern formation to growth regulation. Nature Reviews Genetics 8(9): 663–674.

Bahram, M. and Netherway, T. 2022. Fungi as mediators linking organisms and ecosystems. FEMS Microbiology Reviews 46(2).

Baluška, F. and Miller, Jr, W.B. 2018. Senomic view of the cell: Senome versus Genome. Communicative & Integrative Biology 11(3): 1–9.

Baluška, F. and Reber, A. 2019. Sentience and Consciousness in Single Cells: How the First Minds Emerged in Unicellular Species. BioEssays 41(3): 1800229.

Baverstock, K. 2013. The role of information in cell regulation. Progress in Biophysics and Molecular Biology 111(2–3): 141–143.

Baverstock, K. 2021. The gene: An appraisal. Progress in Biophysics and Molecular Biology 164: 46–62.

Bekenstein, J.D. 2003. Information in the Holographic Universe. Scientific American 289(2): 58–65.

Blank, M. and Goodman, R. 2011. DNA is a fractal antenna in electromagnetic fields. International Journal of Radiation Biology 87(4): 409–415.

Chapman, M.J. and Margulis, L. 1998. Morphogenesis by symbiogenesis. International Microbiology: The Official Journal of the Spanish Society for Microbiology 1(4): 319–326.

Davies, P.C.W. and Walker, S.I. 2016. The hidden simplicity of biology. Reports on Progress in Physics 79(10): 102601.

De la Fuente, I.M. 2021. Correspondence insights into the role of genes in cell functionality. Comments on "The gene: An appraisal" by K. Baverstock. Progress in Biophysics and Molecular Biology 167 152–160.

De Loof, A. 2016. The cell's self-generated "electrome": The biophysical essence of the immaterial dimension of Life? Communicative & Integrative Biology 9(5): e1197446.

Efron, B. 2013. Bayes' Theorem in the 21st Century. Science 340(6137): 1177–1178.

Fels, D. 2018. The Double-Aspect of Life. Biology 7(2): 28.

Fleury, V. and Gordon, R. 2012. Coupling of Growth, Differentiation and Morphogenesis: An Integrated Approach to Design in Embryogenesis. pp. 385–428. *In*: Cellular Origin, Life in Extreme Habitats and Astrobiology Origin(s) of Design in Nature. Netherlands: Springer.

Friston, K. 2018. Am I Self-Conscious? (Or Does Self-Organization Entail Self-Consciousness?). Frontiers in Psychology 9.

Friston, K., Kilner, J. and Harrison, L. 2006. A free energy principle for the brain. Journal of Physiology-Paris 100(1–3): 70–87.

Funk, R.H.W. 2015. Endogenous electric fields as guiding cue for cell migration. Frontiers in Physiology 6.

Geesink, H.J.H. and Meijer, D.K.F. 2018. Mathematical Structure for Electromagnetic Frequencies that May Reflect Pilot Waves of Bohm's Implicate Order. Journal of Modern Physics 9(5): 851–897.

Gilbert, S.F. 2019. Towards A Developmental Biology Of Holobionts. pp. 13–22. *In*: G. Fusco (ed.). Perspectives On Evolutionary And Developmental Biology. Padova: Padova University Press.

Gilbert, S.F. and Sarkar, S. 2000. Embracing complexity: organicism for the 21st century. Developmental Dynamics: An Official Publication of the American Association of Anatomists 219(1): 1–9.

Gurwitsch, A. 1912. Biologisches Zentralblatt. Die Vererbung As Verwirklichungsvorgang 32: 458–486.

Huttegger, S.M. and Mitteroecker, P. 2011. Invariance and Meaningfulness in Phenotype spaces. Evolutionary Biology 38(3): 335–351.

Jiang, Y., Kelly, R., Peters, A., Fulka, H., Dickinson, A., Mitchell, D.A. and St. John, J.C. 2011. Interspecies Somatic Cell Nuclear Transfer Is Dependent on Compatible Mitochondrial DNA and Reprogramming Factors. PLoS ONE 6(4): e14805.

Lagutina, I., Lazzari, G. and Galli, C. 2018. 35 Interspecies Somatic Cell Nuclear Transfer Embryos that Form Nucleoli do not Always Activate Mitochondrial Functional Differentiation at the Time of Embryonic Genome Activation. Reproduction, Fertility and Development 30(1): 157.

Landauer, R. 2018. Information is inevitably physical. pp. 77–92. *In*: Feynman and computation. CRC Press.

Lev, L.L., Maiboroda, I.O., Husanu, M.-A., Grichuk, E.S., Chumakov, N.K., Ezubchenko, I.S., ... Strocov, V.N. 2018. k-space imaging of anisotropic 2D electron gas in GaN/GaAlN high-electron-mobility transistor heterostructures. Nature Communications 9: 2653.

Levin, M. 2014. Endogenous bioelectrical networks store non-genetic patterning information during development and regeneration. The Journal of Physiology 592(11): 2295–2305.

Levin, M. 2021. Bioelectric signaling: Reprogrammable circuits underlying embryogenesis, regeneration, and cancer. Cell 184(8): 1971–1989.

Levin, M. and Martyniuk, C.J. 2018. The bioelectric code: An ancient computational medium for dynamic control of growth and form. Biosystems 164: 76–93.

Levin, M., Pietak, A.M. and Bischof, J. 2019. Planarian regeneration as a model of anatomical homeostasis: Recent progress in biophysical and computational approaches. Seminars in Cell & Developmental Biology 87: 125–144.

Ildefonso, M. 2021. Correspondence insights into the role of genes in cell functionality. Comments on "The gene: An appraisal" by K. Baverstock. Progress in Biophysics and Molecular Biology 167: 152–160.

Mezrich, R. 1995. A perspective on K-space. Radiology 195(2): 297–315.

Miller, W.B. 2016. Cognition, Information Fields and Hologenomic Entanglement: Evolution in Light and Shadow. Biology 5(2): 21.

Miller, W.B. 2018. Biological Information Systems: Evolution as cognition-based Information Management. Progress in Biophysics and Molecular Biology 134: 1–26.

Miller, W.B., Baluška, F. and Torday, J.S. 2020b. Cellular senomic measurements in Cognition-Based Evolution. Progress in Biophysics and Molecular Biology 156: 20–33.

Miller, W.B., Enguita, F.J. and Leitão, A.L. 2021. Non-Random Genome Editing and Natural Cellular Engineering in Cognition-Based Evolution. Cells 10(5): 1125.

Miller, W.B., Torday, J.S. and Baluška, F. 2020a. The N-space Episenome unifies cellular information space-time within cognition-based evolution. Progress in Biophysics and Molecular Biology 150: 112–139.

Moratal, D., Valles-Luch, A., Marti-Bonmati, L. and Brummer, M. 2008. k-Space tutorial: an MRI educational tool for a better understanding of k-space. Biomedical Imaging and Intervention Journal 4(1): e15.

Müller, P. and Schier, A.F. 2011. Extracellular Movement of Signaling Molecules. Developmental Cell 21(1): 145–158.

Neuhof, M., Levin, M. and Rechavi, O. 2016. Vertically- and horizontally-transmitted memories—the fading boundaries between regeneration and inheritance in planaria. Biology Open 5(9): 1177–1188.

Oreshkov, O., Costa, F. and Brukner, Č. 2012. Quantum correlations with no causal order. Nature Communications 3: 1092.

Pai, V.P., Lemire, J.M., Pare, J.-F., Lin, G., Chen, Y. and Levin, M. 2015. Endogenous Gradients of Resting Potential Instructively Pattern Embryonic Neural Tissue via Notch Signaling and Regulation of Proliferation. Journal of Neuroscience 35(10): 4366–4385.

Pawłowski, M., Paterek, T., Kaszlikowski, D., Scarani, V., Winter, A. and Żukowski, M. 2009. Information causality as a physical principle. Nature 461(7267): 1101–1104.

Rafiq, K., Cheers, M.S. and Ettensohn, C.A. 2012. The genomic regulatory control of skeletal morphogenesis in the sea urchin. Development 139(3): 579–590.

Reber, A.S. 2019. The First Minds: Caterpillars, Karyotes, and Consciousness (Illustrate). New York, NY, USA: Oxford University Press.

Rogers, K.W. and Schier, A.F. 2011. Morphogen Gradients: From Generation to Interpretation. Annual Review of Cell and Developmental Biology 27: 377–407.

Roy, S., Huang, H., Liu, S. and Kornberg, T.B. 2014. Cytoneme-Mediated Contact-Dependent Transport of the Drosophila Decapentaplegic Signaling Protein. Science 343(6173).

Sepehri, A. 2017. A mathematical model for DNA. International Journal of Geometric Methods in Modern Physics 14(11): 1750152.

Shapiro, J. 2017. Living Organisms Author Their Read-Write Genomes in Evolution. Biology 6(4): 42.

Sullivan, K.G., Emmons-Bell, M. and Levin, M. 2016. Physiological inputs regulate species-specific anatomy during embryogenesis and regeneration. Communicative & Integrative Biology 9(4): e1192733.

Sun, Y.-H. and Zhu, Z.-Y. 2014. Cross-species cloning: influence of cytoplasmic factors on development. The Journal of Physiology 592(11): 2375–2379.

Torday, J.S. and Miller, W.B. 2016. On the Evolution of the Mammalian Brain. Frontiers in Systems Neuroscience 10(31).

Tzambazakis, A. 2015. The evolution of the biological field concept. pp. 1–27. *In*: D. Fels, M. Cifra and F. Scholkmann (eds.). Fields of the Cell. Kerala, India: Research Signpost.

Villarreal, L.P. and Ryan, F. 2019. Viruses in the Origin of Life and Its Subsequent Diversification. In Handbook of Astrobiology. Boca Raton, Florida, United States: CRC Press.

Walker, S.I., Kim, H. and Davies, P.C.W. 2016. The informational architecture of the cell. Philosophical Transactions of the Royal Society A: Mathematical, Physical and Engineering Sciences 374(2063): 20150057.

West, A.E. and Greenberg, M.E. 2011. Neuronal Activity-Regulated Gene Transcription in Synapse Development and Cognitive Function. Cold Spring Harbor Perspectives in Biology 3(6): a005744–a005744.

Witzany, G. 2010. Biocommunication and natural genome editing. World Journal of Biological Chemistry 1(11): 348.

Yamada, T., Yang, Y., Valnegri, P., Juric, I., Abnousi, A., Markwalter, K.H., … Bonni, A. 2019. Sensory experience remodels genome architecture in neural circuit to drive motor learning. Nature 569(7758): 708–713.

Yao, L., Seaton, S.C., Ndousse-Fetter, S., Adhikari, A.A., DiBenedetto, N., Mina, A.I., ... and Devlin, A.S. 2018. A selective gut bacterial bile salt hydrolase alters host metabolism. Elife 7.

Zuo, Y., Gao, Y., Su, G., Bai, C., Wei, Z., Liu, K., … Li, G. 2014. Irregular transcriptome reprogramming probably causes thec developmental failure of embryos produced by interspecies somatic cell nuclear transfer between the Przewalski's gazelle and the bovine. BMC Genomics 15(1): 1113.

CHAPTER 7

Natural Cellular Engineering as Multicellular Problem-solving

"How selfish soever man may be supposed, there are evidently some principles in his nature which interest him in the fortune of others".

Adam Smith

Recent research suggests that the transition to multicellularity was not as daunting as it has been portrayed. Undoubtedly, its origin is ancient, beginning at or nearly coincident with the instantiation of cellular life at least 3.7 billion years ago and perhaps over 4 billion years ago (Nutman et al. 2019, Papineau et al. 2022). These primordial prokaryotes were metabolically diverse and capable of colonial life since their discovery hinges on finding microfossils which are evidence of collective life. Individual free-standing organisms leave no trace. Furthermore, recent experimental research with unicellular life forms indicate that multicellularity can emerge with surprising facility in only a few hundred generations (Bernardes et al. 2021).

To account for this ubiquitous impulse and its evolutionary origins, widely varied theories have been proposed. The assumption that planetary oxygen levels might have governed multicellular interplay is prominent among these speculations. Some scientists have suggested that it took time for planetary oxygen levels to rise sufficiently to support true multicellular complexity. Until then, low oxygen availability suppressed multicellular networks by forcing them to remain small (Bozdag et al. 2021). Others believe the opposite. Butterfield (2018) argues that low oxygen levels propelled complex multicellular life since these life forms were better equipped to harvest nutrients through improved motility.

Significantly, once multicellular life was triggered, there was little impetus to disaggregate back to unicellular life (Libby and Ratcliff 2014). As cells adapted within multicellular clusters, they began to sub-specialize and developed mutual dependencies creating a 'ratcheting' process that led to the adoption of traits that entrenched the multicellular lifestyle and stabilized the resultant networks, increasing multicellular complexity. Effectively, the increased division of labor that multicellularity permits was not a consequence of multicellular life but its actual driver, thereby suggesting that natural selection favors co-dependence (Cooper and West 2018). However, it is vital to get the order correct in biology. For cellular resources to be traded and sub-specializations to develop, intelligible information must be first shared. Any such activity requires an initiating impulse, and further, there has to be an applicable platform to support the mutual exchange of that information. Necessarily then, unicellular organisms must be initially equipped to enable any originating, first tentative steps to mutual association. Indeed, genetic experiments have confirmed that the genetic architecture necessary for multicellularity already existed in unicellular organisms. As an example, single-celled choanoflagellates have the genes for tyrosine kinases and certain essential proteins that multicellular animals depend upon for parts of their metabolism (Levin et al. 2014). Further, unicellular organisms encode for cell growth regulators that can only be directly useful among multicellular organisms

Consequently, many of the subspecialized functions in multicellular organisms were not novel to them and were exapted from unicellular genes to support multicellular life. For instance, genomic analysis of the protist *Capsaspora owczarzaki*, a unicellular relative of animals, reveals much of the transcription-factor networks and extensive network interconnections necessary for animal development (Sebé-Pedrós et al. 2016). To a significant extent, multicellular life can be considered a form of fine-tuning.

Four factors must be addressed to suitably explain the origin of multicellularity. First, why is multicellularity so prevalent that it is an apparently inevitable outcome in cellular life? Second, what factors might account for the long period of stasis between collaborative multicellular life as biofilms and the further entwining of multicellular life represented by plants and animals? For example, sponges are considered one of the most 'primitive' living animals, and these appear to date back only 750 million years, appearing billions of years after unicellular life began. Third, why would microbial communities in biofilms that contain several different species coexist? They share similar metabolic abilities and nutrient sources which should increase competitive drive. And fourth, what factors might justify the permanent fluidity of association and even the exchange of genetic resources that continues to be exhibited between unicellular and multicellular forms? For instance, microbial biofilms freely disperse and re-coalesce cells. Even some animals, such as corals, routinely displace large numbers of endosymbiotic dinoflagellate cells from their multicellular form into a free-standing status that will re-coalesce back with its multicellular form in a continual dynamic interplay (Bellantuono et al. 2019).

Fortunately, a ready explanation of these biological phenomena can be offered by understanding the processes that underlie natural cellular dynamics. Although previously unrecognized, the cellular glue impelling either the colonial living form or discrete multicellularity is the conditional ambiguity of information that requires self-referential participants to cooperate to improve the validity of their available information. There is no perfect information in biology. Consequently, cells must measure an entire range of available information to determine its predictive utility to deploy limited resources properly. This pattern of collective measurement, communication, and collaborative problem-solving is represented across all living scales and expresses as natural cellular engineering.

Basic research conducted on prokaryotes illustrates many faculties that reiterate among multicellular eukaryotes. Experiments with prokaryotes under stress indicate extensive cell-cell signaling that supports a wide range of social and collective actions (Ben-Jacob and Levine 2006). Within biofilms, participants transfer information along multiple paths. Quorum sensing is a process of ubiquitous communication of the measured assessment of information (De Loof 2015a). By this means, information self-assessment conducted by each participant contributes to a collective cellular appraisal of environmental stress (De Loof 2015b).

A pertinent example is the transfer of antibiotic resistance among microbial participants in biofilms (Odling-Smee et al. 2003). In any biofilm, some microbes have the correct prophage genetic sequences for antibiotic resistance, while others may not. Under antibiotic-induced stress, antibiotic-resistant bacterial participants can conduct a plasmid-vectored genetic transfer that confers antibiotic resistance to genetically-deficient bacteria. The plasmid exchange between the donor and the recipient is conducted via a thin needle-like extension tube, a conjugative pilus, that enables the two-way transfer. *Entercoccus faecalis* biofilms operate in this manner under antibiotic stress (Baluška et al. 2006). Antibiotic resistance is delivered to susceptible recipient cells from plasmid-carrying donor cells through conjugative plasmid pCF10 transfer in exchange for nutrients.

This complex action is far more than a simple exchange among automatons. It can only occur when both participants are self-aware of their status vis-a-vis others within the multicellular collaborative. Further, such a complicated exchange depends on an entire suite of faculties that enable this trading of resources, encompassing complex self-assessment involving measurement, negotiation, memory, anticipation, and prediction. All are linked to a sophisticated apparatus for the contingent deployment of cellular tools. Clearly, in such a process, information is being assessed and communicated among cooperative participants to engage in mutualized problem-solving. All these actions constitute a form of collaborative biofilm engineering that is simultaneously directed toward the protection of individual self-integrity and the common defense of the collaborative structure (Miller 2016, 2018, Miller et al. 2019).

Biofilms consistently engage in other forms of joint activities that are forms of natural cellular engineering. Microorganisms such as *Escherichia coli* will construct discrete channels within biofilms to aid in the acquisition of nutrients and their intra-colony transport (Trewavas and Baluška 2011). Analogous to human engineering with physical materials, these cellular architectural structures within biofilms can be orders of magnitude larger than the individuals constructing them. In this engineering process, *E. coli* construct a polymeric extracellular matrix architecture that dwarfs the size of individual cells, thereby facilitating large-scale cellular communities (Baluška and Levin 2016).

Indeed, this same type of coordinated response to environmental stress can be identified down to the level of the cell itself as a form of internal engineering. Immotile strains of *Pseudomonas fluorescens* have been bioengineered to lack a flagellum by the deletion of *fleQ*, a critical regulatory gene (Vallverdú et al. 2018). The bacteria regained their flagellum and motility within 96 hours by re-purposing alternative genetic pathways. The expression of *NtrC,* a nitrogen regulatory protein, was amplified, and *fleQ* target genes were stimulated, boosting their expression. The net result was the reclamation of bacterial motility. This complex internal re-wiring is a form of non-random internal engineering and problem-solving. The cell is self-aware and has a sense of its available tools to the limits of its scale. And just as with human engineering, the solution comes with a cost. Improved motility directly correlates with a deterioration in nitrogen regulation (Vallverdú et al. 2018).

Such coordinate action can yield further forms of phenotype. For example, in biofilms of *Bacillus subtilis,* the participating cells can sub-specialize to elaborate an optimal extracellular matrix for either cell surface adhesion or colonial mobility (Lyon 2015). The composition of the respective matrices, each representing a differing biofilm phenotype, enables either adhesion or motility dependent on surface conditions. To accomplish this, the constituent partner cells must accommodate sub-specializations, unlike those characteristics exhibited in their free-living form (Dexter et al. 2019). Sub-specialization of this type requires some cellular participants to voluntarily cede some capacities necessary for survival to support the collective whole. In exchange for these subspecialized metabolic resources, the essential products from the sacrificed pathways are contributed back by recipients to the donor cells. In effect, each sub-specialized cell is betting its existence on a high level of traded resources and mutualistic reciprocations (Miller 2016, 2018, Dexter et al. 2019).

Within these background faculties, the unicellular realm establishes the essential patterns from which multicellular life can emerge. The narrative pathway is quite direct. Each cell has self-referential awareness and maintains its homeorhetic equipoise through self-referential measurement. Since eukaryotic cells and cellular microbes can measure and communicate, they can engineer collaboratively. That engineering impulse is the direct forward expression of the conditional cellular requisite to assess the validity of imprecise biological information through the collective form. This coordinated action expresses among prokaryotes in the colonial form as biofilms and then becomes the further collective expression among eukaryotic and microbial partners as holobionts. Thus, co-engineering is no mere epiphenomenon. It is a singular principle of cellular life, permitting the advantaged assessment of ambiguous information and the downstream energy efficiencies that ensue (Miller et al. 2021).

Bacillus subtilis biofilms have been demonstrated to remarkably organize into ring patterns through developmental pathways similar to those that drive vertebral embryological development and segmentation (Chou et al. 2022). To aid this process, bacterial communities propagate electrodynamic communication signals that can extend over long distances (Larkin et al. 2018). Using capacities such as these, microbes effectively engage in community-level problem-solving that closely mirrors our human engineering efforts at our scale. Necessarily, such community-level problem solving is memory dependent. Bacteria use light-induced changes in potassium channels as a membrane potential to form robust and complex memory patterns (Yang et al. 2020). In a self-similar manner, cellular membrane potentials are triggered in the metazoan brain to form and retrieve memories. From within these unicellular patterns, all life on the planet reiterates in its myriad variations that lead from the unicellular domains to the vast complexity of multicellular eukaryotic life as holobionts.

The problem-solving creativity within the unicellular realm that leads to remarkably complex biofilms and other microbial associations offers clarity about the emergence and predominance of multicellularity. Exaptation is a term used in evolutionary biology to describe a trait that has been co-opted for a use other than its original purpose. Developmental exaptation extends that concept further to include preexisting characteristics of a developmental program. Most commonly, exaptation has applied to genes or genetic regulatory elements, but that concept has more recently been broadened to include the timing of developmental events, aspects of morphogenesis, tissue differentiation, and cellular behavior (Chipman 2021). In theory, developmental exaptation exerts its influences by co-opting genes subject to natural selection. However, the major exaptation in biology has been the seamless transference of cellular rules of engagement present in the unicellular realm that extended from preexisting patterns of unicellular collective behavior to equally apply and amplify in the multicellular realm. Thus, developmental exaptations apply smoothly from the unicellular scale forward as fundamental coordinating factors stimulating and enabling natural cellular engineering. Accordingly, the same channeling impulses have equivalency across the unicellular and multicellular realms. Only their biological manifestations differ.

Natural cellular engineering is 'natural' since it follows a well-delineated, self-organizing, and reiterating pattern. Among both unicells and multicellular entities, cells sustain individualized states of homeorhetic balance through the cooperative assessment and communication of ambiguous information to improve its predictive value. Therefore, cells assess information at the individual level through self-production and collaboratively as a collective 'wisdom of crowds' (Miller et al. 2020a). This requisite activity energizes a self-organizing and self-reinforcing obligatory information cycle among all participant/observers (Miller 2016, 2018, Miller et al. 2019). Any reception of information by a cell stipulates its assessment as a measuring function. However, that measuring action is cellular work, expending energy as a thermodynamic requirement. In immediate consequence, that cellular work becomes an available energetic signature as a source of information and communication to any other cellular observer/participants within the same information field (Miller 2018). Thus, an obligatory information cycle propels collective measurement and shared communication that becomes the gateway to multicellular energy optimization and the trading of cellular resources. Accordingly, both information requirements and obligatory thermodynamic reactions establish an obligatory work channel that specifically energizes collective cellular action. Expressly, this informational interactome drives natural cellular engineering, establishing an obliged, reiterating cycle that forms the basis of biological self-organization and multicellularity (Miller 2016, 2018, Miller et al. 2019). From this base, natural cellular engineering drives the formation of the multicellular tissue ecologies and mutualized niche constructions that enable holobionts (Fig. 1).

Niche construction is the biological process by which organisms actively reshape their environment as they adapt to create a preferential habitat (Odling-Smee et al. 2003, Laland et al. 2011, Miller and Torday 2017). These modifications proceed

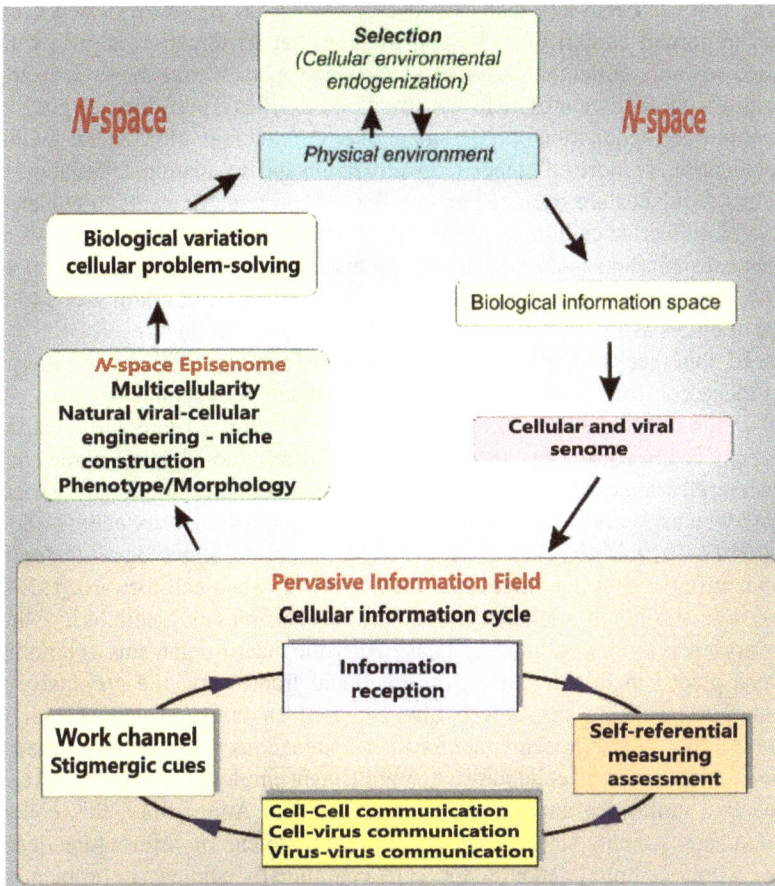

Fig. 1. An informational interactome governs natural cellular engineering.

Environmental cues from the information space of the physical environment (N-space) impact the senomes of individual cells and initiate a reiterating biological information cycle. The cellular information cycle is dependent on each cell's Pervasive Information Field. Since all cellular information is ambiguous, the self-referential assessment of information requires a corresponding measuring assessment, expending energy as work. That work becomes an obligatory signature as a communication signal to any other self-referential organism within the relevant information field. The process reiterates since this obligatory communication necessitates further measuring assessment by other recipients in the information field, which also constitutes work. The resulting work channel stimulates collaborative natural cellular engineering and niche constructions, which depends on collective measurement, necessitating a coordinating multicellular N-space Episenome. Various environmental stresses impact multicellular organisms, including genomic shifts, leading to biological variations as alternative phenotypes. These phenotypes represent cellular problem-solving expressed through coordinated natural cellular engineering. The results are filtered by natural selection, assuring continuous organismal-environmental complementarity.

through purposeful self-directed actions effecting a reciprocal physical change in the organism's immediate environment. Cells adapt collectively to the evolving niche, shifting their homeorhetic flux to meet environmental constraints, thereby enabling cellular behaviors that can reverberate across an entire nested ecology. It is an important codicil of niche construction that its direction is not explicitly purposed towards collective survival as Neodarwinism would insist. Instead, the most significant motivator of niche construction activities is self-directed goals, such as the optimized acquisition or assessment of environmental information, facilitated communication, or more efficient energy utilization and dissipation. Ultimately, these individual requisites are balanced among the collective participants to enhance the chances of collective cellular survival. Importantly though, the primary focus of all cells is to sustain their individually self-appraised states of homeorhetic preference. Notably, cells seem to innately sense that their best chance of equipoise is to be in service to others as full participants within the networked cellular ecology.

Indeed, this specific cellular stance is the crux of biology. This biological impulse makes niche construction the keystone activity of self-referential organisms. When placed within the proper framework, this is not surprising. When niche construction is correctly centered within the context of an informational interactome, natural cellular engineering can be accurately appraised as natural informational engineering as the integral expression of the information management systems of cells (Miller and Torday 2017, 2019, Miller et al. 2019, 2020a,b). Therefore, all biology is undergirded by an identifiable aligning process from the first competent cell forward. This same process reiterates in self-similarity across scales and enables multicellularity, whether as prokaryotic biofilm or as multicellular eukaryotic macro-organisms as holobionts.

Once placed in this cognitive, informational framework, the propulsive flow of evolution clarifies. This flow is always based on the requisites of individual living cells that choose to aggregate for the advantageous maintenance of their own equipoise. This perpetual collaborative entanglement employs the abundant exchange of bioactive molecules and cell-cell communication. And further, this collective reception, assessment, communication, and deployment of information leads to biology's final common denominator; cells communally engineer to problem-solve to meet environmental exigencies, just as we humans do (Miller 2016, 2018, Torday and Miller 2016, Miller et al. 2019). Most importantly, their niche construction activities never have to process beyond their individual needs. The driving impulse towards concordant natural cellular engineering and its further expression as mutualized niche constructions is the cellular information cycle. This crucial cycle balances and reinforces the living requirement for the collective measurement of cellular environmental cues amid thermodynamic strictures. As an integral derivative, all natural cellular engineering and its attendant niche constructions are always centered and sustained at the level of each individual cellular participant. Each cellular ecological constituent measures environmental cues through its own senomic reception apparatus and construes its particular self-produced assessment of all its available environmental information (Miller et al. 2020a,b).

Stigmergy is a vital sub-process within the energizing construct of natural cellular engineering and niche constructions. This networking process connects participants, coordinating them by both direct and indirect signaling cues based on the traces of information in the environment, incurring through the action of any participant within a stigmergic network (Heylighen 2015). Consequently, stigmergy represents a feedback loop among participants; each leaves a trace of its action that some other participant might follow within the common information space. Heylighen (2015, p. 5) defined stigmergy as ".... an indirect, mediated mechanism of coordination between actions, in which the trace of an action left on a medium stimulates the performance of a subsequent action".

The robustness of this process can be recognized through the self-organization seen in the construction of termite mounds. None of the participants is in charge, and each has its individual goals. Yet, stigmergic cues lead to a common result in which global order can arise from local actions without the need for central planning. Complexity builds based on the stream of information in which each of the self-referential participants is directed towards sustaining itself. Because stigmergic cues are a broadcast and there is no central authority, no resistance is triggered, and furthermore, no errors are being made in some putative central plan (Heylighen 2001). The process is designed to be a general drift towards a consensual accommodative outcome, yielding concrete biological results. This same process operates at all scales, even among ourselves. Shared communications, as direct and indirect stigmergic cues among self-referential competent cellular agents within a shared information space, process through natural cellular engineering to yield cellular niche constructions that ultimately produce our human phenotypes (Miller et al. 2020a,b, Torday and Miller 2020).

Cells utilize tools to accomplish these cooperative engineering patterns. Extracellular vesicles are one means of extracellular communication and are deployed by cells to synchronize cellular phenotypes among differentiated cells beginning in embryogenesis (Minakawa et al. 2021). Even mitochondria exhibit engineering patterns. These life-sustaining organelles populate our cytoplasm and demonstrate remarkable sociability. They communicate actively with one another and the cellular nucleus, form interdependent group formations, synchronize behaviors, collaborate, and specialize in enabling specific cell functions as a type of coordinate intracellular engineering (Picard and Sandi 2021).

Nor is it different in the unicellular realm when cells coalesce to form protective biofilms. The nutrient channels within mature biofilm macro-colonies of *Escherichia coli* emerge through collective action as intercellular engineering resembling human-engineered canals. Cells in biofilms use these channels to transport particles and liquids, and assist in acquiring and distributing colony-wide nutrients, just as humans use canals for transport (Rooney et al. 2020). In such biofilms, distributed information is shared among an estimated 10^9–10^{12} bacteria and must be communally deployed. Accordingly, bacteria engineer biofilms through collective measurement and share their individual assessment through quorum-sensing, chemotactic signaling, and plasmid exchange (Ben-Jacob and Levine 2006).

Creative engineering is seen in all planetary environments. Ocean microbes will form co-engineering clumps to confront nutritional stresses. In low-nutrient environments, marine microbes will aggregate, teaming up with smaller microbes with vibrating, hairlike appendages (cilia) on their surface. The coordinated beating of these cilia creates microcurrents that draw in ten times more nutrients than would otherwise be available to the partnering microbes (Kanso et al. 2021). By this means, complex colonial forms emerge, and the same processes, utilizing different tools, reiterate across scales to the level of entire holobionts, expressed further as their own forms of social-based engineering (Miller et al. 2020a). At all levels and in all planetary environments, natural cellular engineering enables mutualizing niche constructions to enact cellular solutions to environmental stresses.

A crucial component of cellular engineering is a responsive and reciprocating genome as natural genetic engineering (Witzany 2010, 2011, Shapiro 2011, 2017). In this process, genomes monitor and edit themselves, actively participating in solutions to environmental problems by initiating flexible responses. This well-documented feature of flexible and reciprocating genomes was first recognized by Nobel prize-winner Barbara McClintock over three decades ago, and the concept has been meticulously validated over the years (McClintock 1984). In this way, the genome accommodates diverse impacts in which individual contributors to the genome, such as retroviral transcriptional regulators, can monitor and flexibly alter their own insertions and genomic amplifications (Shapiro 2017). For instance, in embryonic stem cells, the transcription of proviruses that have accumulated through infectious events, or even the expression of native genes, can be adjusted by the regulatory apparatus of the genome (Schlesinger and Goff 2015). Consequently, the genome is a lively co-participant in an entire cell-wide narrative that enables competent cells to flexibly adjust to environmental strictures through cooperative engineering with partnering cells of many types. This capacity is critical since the nuclear genome contains a substantial fraction of the total information content of the cell. Consequently, it is imperative that its information architecture complements that of the cell as a whole, thereby ensuring a unified whole. Through these constant reciprocations, cells and their genomes can shape their own evolution through non-random mechanisms in continuous response to environmental stresses (Oliver and Greene 2009, Schrader et al. 2014, Miller et al. 2021).

Considering that all processes in biology reiterate across scales, it should not be surprising that the cell itself is regarded as the first niche construction (Torday 2016). Indeed, the extreme complexity of the crowded, active intracellular environment is undoubtedly an engineering marvel. Since life began with self-referential cognition, the cells in which it is embodied should be regarded as the epitome of natural informational engineering, which underlies further biological expression as natural cellular engineering and resultant niche constructions. Therefore, just as the genome is considered a focus of natural genetic engineering, then the same engineering principle should apply to other vital intracellular structures such as mitochondria, ribosomes, and the endoplasmic reticulum. The plasma membrane should be similarly considered since it is the crucial intermediary between the external environment and the cellular cytoplasm. In this way, cells self-modify and can flexibly uphold themselves over evolutionary space-time as continuous, direct participants in their evolution based on their extensive developmental intracellular repertoire. With this background, the long period in the fossil record between the emergence of life approximately 4 billion years ago and highly integrated multicellular life nearly 3 billion years later can be rationalized. That was the period that was needed for the level of intracellular natural engineering and niche construction to become a competent eukaryotic cell with its embedded organelles capable of a higher level of collaborative engineering. Since the eukaryotic cell is the product of the endosymbiotic merger of one prokaryote with another, each with its own 'self', that merger might have been protracted

(Baluška et al. 2021). As in all engineering endeavors, the first steps are typically the most likely to lead to delays, complications, and dead ends.

There is no doubt that cells are equipped with sufficient faculties to support engineering capabilities. Certainly, all cells are cognitive agents. (Shapiro 2011, Lyon 2015, Miller 2016, Reber 2019, Shapiro 2021, Reber and Baluška 2021, Baluška et al. 2021). Undoubtedly too, engineering is predicated on measurements and communication. Notably, though, any assessment of information in the self-referential frame is necessarily a measurement, and although not obvious, communication is also a form of measurement. Any communication is received and measured as a potential prediction by another self-referential agent that has expended energy to evaluate it (Miller et al. 2020a,b). It is only through the measurement of the available information that cells determine which tools to deploy (Miller et al. 2021). All of these actions objectively mirror our own human engineering interactions. To engage in natural cellular engineering and niche constructions, cellular actions include cooperation, collaboration, co-dependence, the trading of resources, subspecializations, and competition. We exhibit the same in our human engineering actions. All these behaviors are required for successful engineering and niche construction in our human frame, which is the same for our cells.

The narrative of multicellular life follows a readily understood progression. The self-referential living condition is sustained by a definable information cycle that triggers thermodynamic constraints. This reciprocating action stimulates an obligatory work channel among cells expressed as collaborative natural cellular engineering and niche construction. In a cognitive framework, the information cycle yields its own self-reinforcing engineering cycle (Miller 2018). Since cells can measure information and communicate their interpretation towards a common goal, they can engineer. All biology serves as proof.

References

Baluška, F. and Levin, M. 2016. On Having No Head: Cognition throughout Biological Systems. Frontiers in Psychology 7.

Baluška, F., Miller, W.B. and Reber, A.S. 2021. Biomolecular Basis of Cellular Consciousness via Subcellular Nanobrains. International Journal of Molecular Sciences 22(5): 2545.

Baluška, F., Volkmann, D., Hlavacka, A., Mancuso, S. and Barlow, P.W. 2006. Neurobiological View of Plants and Their Body Plan. pp. 19–35. In: Communication in Plants. Berlin, Heidelberg: Springer Berlin Heidelberg.

Bellantuono, A.J., Dougan, K.E., Granados-Cifuentes, C. and Rodriguez-Lanetty, M. 2019. Free-living and symbiotic lifestyles of a thermotolerant coral endosymbiont display profoundly distinct transcriptomes under both stable and heat stress conditions. Molecular Ecology 28(24): 5265–5281.

Ben-Jacob, E. and Levine, H. 2006. Self-engineering capabilities of bacteria. Journal of The Royal Society Interface 3(6): 197–214.

Bernardes, J.P., John, U., Woltermann, N., Valiadi, M., Hermann, R.J. and Becks, L. 2021. The evolution of convex trade-offs enables the transition towards multicellularity. Nature Communications 12: 4222.

Bozdag, G.O., Libby, E., Pineau, R., Reinhard, C.T. and Ratcliff, W.C. 2021. Oxygen suppression of macroscopic multicellularity. Nature Communications 12: 2838.

Butterfield, N.J. 2018. Oxygen, animals and aquatic bioturbation: An updated account. Geobiology 16(1): 3–16.

Chipman, A.D. 2021. Developmental Exaptation. pp. 29–38. In: L.N. de la Rosa and G.B. Müller (eds.). Evolutionary Developmental Biology. Springer Cham.

Chou, K.-T., Lee, D.D., Chiou, J., Galera-Laporta, L., Ly, S., Garcia-Ojalvo, J. and Süel, G.M. 2022. A segmentation clock patterns cellular differentiation in a bacterial biofilm. Cell 185(1): 145–157.e13.

Cooper, G.A. and West, S.A. 2018. Division of labour and the evolution of extreme specialization. Nature Ecology & Evolution 2(7): 1161–1167.

De Loof, A. 2015a. From Darwin's On the Origin of Species by Means of Natural Selection… to The evolution of Life with Communication Activity as its Very Essence and Driving Force (= Mega-Evolution). Life: The Excitement of Biology 3(3): 153–187.

De Loof, A. 2015b. How to deduce and teach the logical and unambiguous answer, namely L = ∑C, to "What is Life?" using the principles of communication? Communicative & Integrative Biology 8(5): e1059977.

Dexter, J.P., Prabakaran, S. and Gunawardena, J. 2019. A Complex Hierarchy of Avoidance Behaviors in a Single-Cell Eukaryote. Current Biology 29(24): 4323–4329.e2.

Heylighen, F. 2001. The science of self-organization and adaptivity. The Encyclopedia of Life Support Systems 5(3): 253–280.

Heylighen, F. 2015. Stigmergy as a Universal Coordination Mechanism: Components, varieties and Applications. Lewis, T. and Marsh, L. (eds.). *In*: Human Stigmergy: Theoretical Developments and New Applications. New York, Springer.

Kanso, E.A., Lopes, R.M., Strickler, J.R., Dabiri, J.O. and Costello, J.H. 2021. Teamwork in the viscous oceanic microscale. Proceedings of the National Academy of Sciences 118(29).

Laland, K.N., Sterelny, K., Odling-Smee, J., Hoppitt, W. and Uller, T. 2011. Cause and Effect in Biology Revisited: Is Mayr's Proximate-Ultimate Dichotomy Still Useful? Science 334(6062): 1512–1516.

Larkin, J.W., Zhai, X., Kikuchi, K., Redford, S.E., Prindle, A., Liu, J., … Süel, G.M. 2018. Signal Percolation within a Bacterial Community. Cell Systems 7(2): 137–145.e3.

Levin, T.C., Greaney, A.J., Wetzel, L. and King, N. 2014. The rosetteless gene controls development in the choanoflagellate S. rosetta. ELife 3.

Libby, E. and Ratcliff, W.C. 2014. Ratcheting the evolution of multicellularity. Science 346(6208): 426–427.

Lyon, P. 2015. The cognitive cell: bacterial behavior reconsidered. Frontiers in Microbiology 6.

McClintock, B. 1984. The Significance of Responses of the Genome to Challenge. Science 226(4676): 792–801.

Miller, W.B. 2016. Cognition, Information Fields and Hologenomic Entanglement: Evolution in Light and Shadow. Biology 5(2): 21.

Miller, W.B. 2018. Biological Information Systems: Evolution as cognition-based Information Management. Progress in Biophysics and Molecular Biology 134: 1–26.

Miller, W.B., Baluška, F. and Torday, J.S. 2020a. Cellular senomic measurements in Cognition-Based Evolution. Progress in Biophysics and Molecular Biology 156: 20–33.

Miller, W.B., Enguita, F.J. and Leitão, A.L. 2021. Non-Random Genome Editing and Natural Cellular Engineering in Cognition-Based Evolution. Cells 10(5): 1125.

Miller, W.B. and Torday, J.S. 2017. A systematic approach to cancer: evolution beyond selection. Clinical and Translational Medicine 6(2).

Miller, Jr, W.B. and Torday, J.S. 2019. Reappraising the exteriorization of the mammalian testes through evolutionary physiology. Communicative & Integrative Biology 12(1): 38–54.

Miller, W.B., Torday, J.S. and Baluška, F. 2019. Biological evolution as defense of "self". Progress in Biophysics and Molecular Biology 142: 54–74.

Miller, W.B., Torday, J.S. and Baluška, F. 2020b. The N-space Episenome unifies cellular information space-time within cognition-based evolution. Progress in Biophysics and Molecular Biology 150: 112–139.

Minakawa, T., Matoba, T., Ishidate, F., Fujiwara, T.K., Takehana, S., Tabata, Y. and Yamashita, J.K. 2021. Extracellular vesicles synchronize cellular phenotypes of differentiating cells. Journal of Extracellular Vesicles 10(11).

Nutman, A.P., Bennett, V.C., Friend, C.R.L., Van Kranendonk, M.J., Rothacker, L. and Chivas, A.R. 2019. Cross-examining Earth's oldest stromatolites: Seeing through the effects of heterogeneous deformation, metamorphism and metasomatism affecting Isua (Greenland) ≈ 3700 Ma sedimentary rocks. Precambrian Research 331: 105347.

Odling-Smee, F.J., Laland, K.N. and Feldman, M.W. 2003. Niche Construction: The Neglected Process in Evolution. Princeton, New Jersey: Princeton University Press.

Oliver, K.R. and Greene, W.K. 2009. Transposable elements: powerful facilitators of evolution. BioEssays 31(7): 703–714.

Papineau, D., She, Z., Dodd, M.S., Iacoviello, F., Slack, J.F., Hauri, E., … Little, C.T.S. 2022. Metabolically diverse primordial microbial communities in Earth's oldest seafloor-hydrothermal jasper. Science Advances 8(15).

Picard, M. and Sandi, C. 2021. The social nature of mitochondria: Implications for human health. Neuroscience & Biobehavioral Reviews 120: 595–610.

Reber, A.S. 2019. The First Minds: Caterpillars, Karyotes, and Consciousness (Illustrate). New York, NY, USA: Oxford University Press.

Reber, A.S. and Baluška, F. 2021. Cognition in some surprising places. Biochemical and Biophysical Research Communications 564: 150–157.

Rooney, L.M., Amos, W.B., Hoskisson, P.A. and McConnell, G. 2020. Intra-colony channels in E. coli function as a nutrient uptake system. The ISME Journal 14(10): 2461–2473.

Schlesinger, S. and Goff, S.P. 2015. Retroviral Transcriptional Regulation and Embryonic Stem Cells: War and Peace. Molecular and Cellular Biology 35(5): 770–777.

Schrader, L., Kim, J.W., Ence, D., Zimin, A., Klein, A., Wyschetzki, K., … Oettler, J. 2014. Transposable element islands facilitate adaptation to novel environments in an invasive species. Nature Communications 5: 5495.

Sebé-Pedrós, A., Ballaré, C., Parra-Acero, H., Chiva, C., Tena, J.J., Sabidó, E., … Ruiz-Trillo, I. 2016. The Dynamic Regulatory Genome of Capsaspora and the Origin of Animal Multicellularity. Cell 165(5): 1224–1237.

Shapiro, J.A. 2011. Evolution: A View from the 21st Century. Upper Saddle River, NJ: FT Press.\

Shapiro, J.A. 2017. Exploring the read-write genome: mobile DNA and mammalian adaptation. Critical Reviews in Biochemistry and Molecular Biology 52(1): 1–17.

Shapiro, J.A. 2021. All living cells are cognitive. Biochemical and Biophysical Research Communications 564: 134–149.

Torday, J. 2016. The Cell as the First Niche Construction. Biology 5(2): 19.

Torday, J. and Miller, W. 2016. Phenotype as Agent for Epigenetic Inheritance. Biology 5(3): 30.

Torday, J. and Miller, W. 2020. Cellular-Molecular Mechanisms in Epigenetic Evolutionary Biology. Cham, Switzerland: Springer.

Trewavas, A.J. and Baluška, F. 2011. The ubiquity of consciousness. EMBO Reports 12(12): 1221–1225.

Vallverdú, J., Castro, O., Mayne, R., Talanov, M., Levin, M., Baluška, F., … Adamatzky, A. 2018. Slime mould: The fundamental mechanisms of biological cognition. Biosystems 165: 57–70.

Witzany, G. 2010. Biocommunication and natural genome editing. World Journal of Biological Chemistry 1(11): 348.

Witzany, G. 2011. The agents of natural genome editing. Journal of Molecular Cell Biology 3(3): 181–189.

Yang, C.-Y., Bialecka-Fornal, M., Weatherwax, C., Larkin, J.W., Prindle, A., Liu, J., … Süel, G.M. 2020. Encoding Membrane-Potential-Based Memory within a Microbial Community. Cell Systems 10(5): 417–423.e3.

Chapter 8

Non-random Variations in Natural Cellular Engineering

"How remarkable is life? The answer is: very. Those of us who deal in networks of chemical reactions know of nothing like it. How could a chemical sludge become a rose, even with billions of years to try?"

George Whitesides (2004)

Previous chapters enumerated the forces that yield coordinate multicellularity as a product of natural cellular engineering and niche constructions as cellular problem-solving. Crucially, all such actions ensue from the activities and behaviors of self-directed cognitive cells to sustain individual homeorhetic equipoise, best achieved in the multicellular form. This recasting of the centrality of biological activity is the defining separation between Cognition-Based Evolution and the currently conceived Modern Synthesis. Moreover, the exact crux between these disparate stances is a readily identifiable focal question. Are the genetic variations that lead to biological outcomes essentially random or not? Neodarwinism remains firmly affixed to its originating narrative that evolutionary variations are sourced from random genetic mutations. However, once evolutionary development is correctly placed into a cognitive framework in which collective cellular engineering parameters yield consonant biological expressions, the concept of evolution by random genetic errors collapses.

Five evolutionary mandates spring from this requisite cognitive frame in which the collective measurement of information, cell-cell communication, and the fruitful collaborative deployment of that information dominate. First, since phenotype is the product of natural cellular engineering and niche constructions, it is the result of largely non-random cellular problem-solving in response to environmental and epigenetic impacts (Miller et al. 2021). Secondly, as all multicellular eukaryotes are holobionts with constituencies from each of the four domains (Prokaryota, Archaea, Eukaryota, Virome), all phenotypic variation is a product of the co-engineering participation among all of these domains. Consequently, the virome has to be viewed as a significant co-participant in cellular engineering processes.

Third, cellular engineering cannot be adequately understood as an exclusively genetic enterprise. It naturally must encompass the entire cellular apparatus to achieve its information-based goals. Therefore, gene-centric Neodarwinism must yield to a re-envisioning of evolutionary development as a comprehensive and reciprocating organism-wide cognition-based informational interactome directed to sustaining cells against environmental impacts (Miller 2018). Fourth, since multicellular biological expression and phenotypes are the product of natural cellular engineering and engineering activities are defined by cooperation, collaboration, co-dependence, and mutualizing competition, symbiosis is the primary active planetary process rather than natural selection. Selection can now be properly placed as post-facto to variations, as Darwin intended. Moreover, given the fundamental role of cooperation in biology, contrary to our prior prejudices, the symbiotic instinct must be extended to include the great majority of cellular interactions with the virome (Villarreal and Ryan 2019). And fifth, despite the trillions of cells that collaborate to create seamless holobionts, all co-engineering activities remain rooted in the continuous protection of the self-identities of each cellular participant (Miller et al. 2019). The root of all multicellular life is the protection of each cell's homeorhetic equipoise. Multicellularity is the most advantaged means for attaining that discrete cellular goal.

It is now well-recognized that all the essential aspects of multicellular life emerged from unicellular roots. All the basic fundamental principles that govern cellular interactions have extended forward across billions of years. All biology achieves reiterating congruence through the agency of self-referential awareness that obliges the measuring assessment of information from relevant information fields, its shared communication, and collaborative deployment. All of this activity reduces to a single straightforward dynamic. All of biology and evolution represent contingent cellular problem-solving in the face of environmental pressures. In sum, all biology is problem-solving. From this actuality, one of biology's central debates resolves. When biological and evolutionary development depends on collective processes and the consensual deployment of resources, it cannot be random. Necessarily too, these non-random particulars must have their origin within unicellular roots that further reiterate to express across multicellularity to yield the great varieties of creatures that populate the planet.

Experiments on stressed prokaryotes confirm this perspective (Asfahl and Schuster 2017). In biofilms, information among unicellular organisms is individually assessed and shared for the collaborative appraisal of environmental stress along numerous pathways to enable quorum sensing (Popat et al. 2015). The willing transfer of antibiotic-resistance genes among biofilm participants is an example. Antibiotic-resistant microbial constituents will offer to donate a plasmid with the appropriate genetic material to other cells that are lacking in exchange for nutrients (Goldenfeld and Woese 2011). The deployment of a conjugative pilus between the parties enables the two-way transfer. Each participant must know its own status and communicate and cooperate with others to effect this consensual trading of resources. Goldenfeld and Woese (2011) note that this conjugative plasmid transfer of antibiotic resistance is controlled by highly coordinated cell–cell communications and linked actions. Undoubtedly, this type of action cannot be random activity insofar as the entire process is communication-dependent and soundly based on the collective measurement of information and cellular prediction. This type of unicellular adaptation is an excellent example of natural genetic editing as a sub-process within non-random natural cellular engineering dedicated to collaborative problem-solving (Miller et al. 2021).

These same types of genetic transfers are not confined to trading among unicellular participants. *Agrobacterium tumefaciens* is an aerobic, gram-negative bacterium that causes crown gall disease in many types of plants. It is also known to be a natural genetic engineer and frequently transfers genetic material such as Ti plasmids to plant cells (LaCroix and Citovsky 2019). Little is understood about the specifics of the process, but it is known to require the synthesis of complex membrane fibrils to effect plant cellular attachment (LaCroix and Citovsky 2019). This complex process is based on the non-random integration of information derived from a replete sensory apparatus, triggering contingent transcriptional activities in elaborate synchrony as natural cellular engineering (Matveeva and Otten 2019). Consequently, it is untenable that these sophisticated information-based biological activities could have arisen merely through accidental genetic variations. Nonetheless, there remains a surprisingly stubborn insistence that single-nucleotide polymorphisms comprise the mainstay of selectable variation (Wellenreuther et al. 2019).

On the contrary, contemporary research has disclosed multiple diverse mechanisms for genetic variation. Villarreal and Ryan (2019) catalog four major mechanisms for triggering heritable shifts: mutational copying errors, epigenetics, symbiosis, and hybridization. Furthermore, increasing emphasis is being placed on significant viral interactions with genomes through horizontal genetic transfer (HGT) among eukaryotes since viral transfer and transposon activation are now known to be critical elements within gene regulatory networks, acting as a source of new genetic code (Broecker and Moelling 2019). Notably, critical eukaryotic evolutionary transitions have been directly linked to the non-random actions of transposable elements (Sultana et al. 2017).

Our emerging grasp of the broader scope of this genetic interplay has led to a much better understanding of our genome. It is not a sacrosanct compartment as long believed, nor is it even fundamentally stable. Our contemporary view accredits the genome as an inherently flexible and reciprocating informational palette, permitting organisms to "co-construct and co-evolve with their environments" (Laland et al. 2014). Indeed, there has been a thorough reanalysis of the genome, which is no longer believed to be a settled read-only programmatic code. Rather, recent research confirms that the genome is a flexible 'read-write' genetic toolbox that can individually react to environmental stresses. Genomes are fully capable of epigenetic formatting or shifts in DNA sequencing that impact biological expression and function (Witzany 2011, Shapiro 2013, 2017). Significantly, these genetic interactions and their resulting biological variations are no longer considered random occurrences. Instead, they constitute a complex, coordinated cellular signaling process that includes horizontal genetic transfers, active dispersal of mobile genetic elements, and integrated symbiogenesis. Together, these processes comprise complex and responsive natural genetic engineering.

Witzany (2011) has placed this natural genetic engineering activity into the context of biocommunication. As such, natural genetic editing can now be seen as a non-random mechanism for the direct internalization of cellular stresses, yielding directed cell-based phenotypic adaptive expression (Witzany 2011, Shapiro 2017). Consequently, as tissue ecologies in holobionts are composed of constituents from among all domains, integrated natural genetic engineering expresses through cell-based epigenetic adjustments in context-dependent patterns of natural cellular engineering in which the virome also plays a substantial symbiotic role (Witzany 2011, Villarreal and Ryan 2019). Consequently, as opposed to the passive coding library that the genome was once thought to represent, where changes were nearly exclusively due to random genetic coding errors or the insertion of transposable elements at random genomic sites, the contemporary context of the genome is as an active and pliant participant in non-random evolutionary development (Miller et al. 2021).

Among the most active natural genomic editing activities, horizontal gene transfer (HGT) has played a major role in the evolution of genomes. Estimates suggest that up to three-quarters of bacterial genes give evidence of an HGT event at some point in their evolution (Arkhipova and Yushenova 2019). Until recently, HGT was thought to be rare among eukaryotes. Recent research indicates otherwise, primarily due to transposable elements and an extensive suite of other potential HGT events. Cells frequently transfer genetic code to viruses; conversely, viral transfer and ultimate endogenization has been frequent in the evolution of eukaryotes (Gilbert and Cordaux 2017). These transfers can have adaptive value as part of an extensive genetic mobilome that vigorously propels multicellular eukaryotic adaptation in non-random directions (Zamai 2020). Interestingly, many of these genetic transfers originate from within the eukaryotic microbiome (Soucy et al. 2015). Indeed, HBT is 25-fold more likely to be triggered from the eukaryotic mammalian microbiome than any other source (Soucy et al. 2015).

Contemporary research indicates that the role of transfers from prokaryotes to eukaryotes has been substantially underestimated. Recent analysis indicates that HGT episodes hundreds of millions of years ago from bacteria, fungi, and viruses to green algae drove the evolution of land plants (Ma et al. 2022). Nearly 600 gene families in modern plants were acquired through HGT and became fully integrated, thereby facilitating the water-to-land transition. This genetic migration involved at least two major episodes, contributing to the rapid development of new traits that serve essential biological functions in plants. For example, the ammonium transporter gene that assists plants in absorbing soil nitrogen for growth and development was acquired from fungi.

Retroviral incursions and their endogenizations are critical components in cellular evolutionary development, although many other viral types can do similarly (Feschotte and Gilbert 2012). Consequently, viral insertions are now acknowledged as dominant drivers of evolutionary variation and adaptation by targeting regulatory components of the genome. These intermittent viral incursions account for a substantial fraction of the eukaryotic proteome, contributing to genome plasticity and providing genetic resources that can be recruited for developmental and physiological novelty, such as the formation of syncytiotrophoblastic tissues and the placenta in mammals (Koonin 2009, Ryan 2009, Black et al. 2010, Enard et al. 2016, Roberts et al. 2021). These types of genomic events trigger retrotranspositions that stimulate new regulatory patterns (Popov et al. 2018). Evidence indicates that these episodes are targeted, non-random interactions involving ribosomal RNA genes as infectious interchanges (Martinez et al. 2014). Furthermore, accumulating evidence indicates that the genetic effects of viral HGT do not conform to a simple stochastic distribution (Wilke and Sawyer 2016). Consequently, their cumulative evolutionary impact is substantial. For example, it is estimated that nearly 30% of all human protein adaptations since humans diverged from chimpanzees have been driven by viral genetic transfers (Enard et al. 2016).

Contemporary research confirms that non-vertical transmission of genetic information is a critical force in evolutionary development, representing one aspect of an integrated genetic mobilome with frequent interchanges both within the unicellular domains and among eukaryotes (Wallau et al. 2018). Consequently, an updated view of the genome is conceptually modeled as a responsive ecosystem and genetic habitat in which diverse communities of transposable elements (TEs) reside in non-random distributions along with many other diverse participants (Wei et al. 2016). Pertinently, the same features that characterize other ecosystems, such as cooperation and competition among constituents, have counterparts within genomes (Bourque et al. 2018). Surprisingly, TEs occupy a larger fraction of the genomic DNA content in eukaryotes than prokaryotes, contributing the majority of the total genome in vertebrates and plants due to repeated episodes of horizontal expansion (Arkhipova and Yushenova 2019). Notably, there is significant evidence that these expansions and their distribution are not random occurrences, instead relating to targeting site preferences (Sultana et al. 2017). For instance, more recent TEs in fungal genomes are likely to be active and cluster in non-random distributions at preferential target sites (Muszewska et al. 2019). This non-random site-preferential distribution of TEs in prokaryotic and eukaryotic genomes supports the view of the genome as an active mobilome, actively seeking phenotypic solutions to environmental stresses by being repurposed to perform target organism functions as parts of novel protein-coding genes, serving adaptive creativity (Jangam et al. 2017).

Nonetheless, there is no simple one-to-one relationship between genomes and biological expression, even though non-random influences dominate. Instead, there is a complex relationship between TEs and genes and their biological outputs. Variations arise from interactions with a wide variety of non-coding RNAs originating from TEs. Significantly, these non-coding RNAs are transcribed into bioactive molecular protein complexes that reciprocally influence regulatory effects on transposons through non-random processes (Mustafin 2018, Miller et al. 2021). Accordingly, TEs directly affect genome plasticity and genomic transcription into a bioactive proteome that ultimately translates into adaptive biological variation.

It is not only the central genome that participates in reciprocating and responsive intracellular and intercellular interchanges. A sizable portfolio of other genetic entities circulates in cells, including small RNAs, plasmids, RNA stem-loops, and other non-chromosomal cellular genetic constituents. In toto, these supply a substantial fraction of the suite of genetic capacities within cells. These are crucial participants in cellular life, critically dependent on microRNAs (miRNAs), including rRNA, tRNA, and mRNA for DNA replication (Villarreal and Witzany 2019). Messenger RNAs (mRNAs) serve a significant function as epigenetic sources. Other microRNAs and non-coding RNAs, such as circRNAs, endo-siRNAs, piRNAs, antisense and long ncRNAs, are essential for cellular adaptation and flexible evolutionary novelty (Smalheiser 2014). It is known that infectious incursions of non-coding RNAs non-randomly target non-coding DNA regions that govern regulatory features (Villarreal and Witzany 2018). Furthermore, non-coding virus-interfering RNAs (viRNAs) contribute to cellular immune responses, providing protection against pathogens (Smalheiser 2014, Palmer et al. 2018).

Extra-chromosomal circular DNAs, microRNAs, and circular RNAs are also vital genetic contributors. They are widely distributed across species, serving cells by producing short regulatory RNAs whose functional genetic expression is also non-random (Paulsen et al. 2019). All these genetic constituents are part of the information management system of cells, providing vital information and essential functions. None of these initially originated by direct vertical transmission but, once acquired, were endogenized, becoming heritably transmitted and forming parts of the encompassing suite of cellular information that can be non-randomly deployed to meet environmental stresses. Consequently, all of these cellular genetic participants constitute a continual evolutionary force (Koonin 2009, Villarreal and Witzany 2018, Miller et al. 2021).

Beyond circulating RNAs, cells adjust to environmental stresses through structural genetic changes. Chromatin regulation of DNA and RNA stem-loops link to cellular responses to stress and phenotypic variations (Waszak et al. 2015). Both DNA and RNA can form stem-loops, serving of particular importance in RNA communication as catalytic signaling drivers in protein sequencing (Villarreal and Witzany 2019). Stem-loop RNAs also function as consortia as competent agents in natural genome editing (Witzany 2014).

Although genetic content expresses proteins reciprocally, the proteome contributes to genetic regulation. The Krüppel-associated box domain zinc finger proteins (KRAB-ZFPs) family of proteins link to TEs and act as transcription regulators for their silencing and derepression in higher vertebrates, affecting cell proliferation, differentiation, and apoptosis (Ecco et al. 2017, Lupo et al. 2013). In addition, myriad virus-interacting proteins contribute to phenotypic adaptation in a significant proportion of humans and other mammals.

Prions are another source of protein-centered variation. These are self-replicating transmissible proteins known to cause neurodegenerative disorders, such as Creutzfeld-Jakob (Mad Cow) disease. However, they are also a heritable source of phenotypic variations in several species, such as yeasts, and participate in adaptive memory (Shorter and Lindquist 2005).

The virome is a voluminous source of adaptive variation among multicellular eukaryotes as essential partners in natural viral and cellular co-engineering and flexible adaptations to stress (Miller et al. 2022). Another potent source of non-random variation comes from the microbiome, exerting its heritable influence through intermittent episodes of HGT and consistent epigenetic contributions (Torday and Miller 2020a). Although once discounted, epigenetic environmental impacts of many kinds substantially impact heritable genetic expression. For example, there is documentation of non-random transgenerational effects in humans from starvation (Jobson et al. 2015), hypoxia (Brown and Rupert 2014), alternate sources of dietary intake (Stegemann and Buchner 2015), and critical life-affecting changes from generalized stress (Skinner 2014). In holobionts with their obligatory microbial fraction, all development is co-development (Rook et al. 2017). Importantly, this principle extends further. All evolution is co-evolution, proceeding along non-random pathways in organism-wide responses to environmental stresses (Miller et al. 2020a).

The cellular toolkit has an expansive range of non-genome-centered mechanisms for flexible, non-random problem-solving in reaction to environmental stresses. The origination of Eukaryota by endoymbiosis was just such an example (Martin et al. 2015, Archibald 2015). In endocytosis, a cellular membrane forms from the cell proper, engulfing bioactive molecules or genetic material, including viruses (Sobhy 2017). Entosis is the complete engulfment of one living cell by another, accompanied by the whole-scale transfer of the entire genetic informational architecture of the acquired cell (Overholtzer et al. 2007, Janssen and Medema 2011). Surprisingly, among non-neural organisms, not only genetic code but learned patterns of behavior can be acquired by this mechanism (Vogel and Dussutour 2016).

On these grounds, it is fully justifiable to dismiss the Neodarwinian narrative that heritable variation arises and evolution proceeds by endless rounds of random genetic errors. In its place, cells can be correctly accorded as information management systems that deliver outputs non-randomly. Cells deploy their flexible resources among a myriad of potential inputs on a contingent basis to meet environmental exigencies (Miller 2016, 2018, Miller and Torday 2019, Miller et al. 2020a,b).

Although not often emphasized but of absolute significance, any variations of any kind must first occur and express at the cellular level. Yet, its macroorganic expression must exert in an organism-wide coordinated manner if it is to eventually yield productive biological expression. At the genomic level, this cellular coordination is the product of natural genetic engineering as a facet of natural cellular engineering, yielding integrated, environmentally responsive, organism-wide outputs. Through these well-documented mechanisms and pathways of natural genetic engineering, non-random genome editing contributes to adaptive solutions to cellular problems as a reciprocating participant in cell-wide activities as a part of the continuum of cellular problem-solving that grants survival ((Witzany 2011, Shapiro 2016a,b). Consequently, genes and their associated, plentiful intracellular genetic partners can now be acknowledged as tools of larger cellular faculties (Miller et al. 2019).

If the genome and intracellular genetic constituents are correctly appraised as tools of competent, cognitive cells, then all aspects of cellular life participate in information acquisition, measuring assessment, storage as memory, and deployment. Accordingly, the extracellular matrix must play a reciprocating role in the cellular maintenance of information assessment, communicative transfer, and its ultimate collective deployment (Milstein and Meiners 2011). Since each cell self-produces its individual interpretation of all available information cues, every aspect of cellular life is a requisite participant in sustaining cells in their continuous confrontation with environmental stresses. Successful adaptation to environmental stresses requires cellular reciprocations at all levels. A non-random distribution of influences must bridge each cellular ecological level within an entire holobiont to yield collaborative results that energize the successful assimilation of the cellular outside in complementarity to its internal homeorhetic balance (Miller 2016, Miller et al. 2021). That continuous internalization of the external environment is organism-wide, permitting phenotypic expression through concordant natural cellular engineering. To effect such unifying results, all constituent populations within holobionts must cooperate to grant survival. Any belief that these consummately seamless results can result from random forces at any level, especially the zygote, is no longer reasonable. Accordingly, with natural cellular engineering and its mutualistic niche constructions at the core of phenotypic expression, any lasting, heritable biological variation must represent a deliberate, harmonious adjustment to environmental cues among trillions of cellular participants in holobionts.

Research demonstrates a significant discrepancy between the standard model of time-homogeneous genetic mutation rates and what can be observed through phylogenetics and discernable phenotypic changes (Martinez et al. 2021). Since this problem is recognized, it has been counter-proposed that random mutations themselves accelerate mutations. However, this solution is plainly problematic. If accidental mutations accelerated the general rate of genomic mutation, then there is nothing to prevent this process from becoming a runaway, uncontrollable one (Miller et al. 2021). Moreover, there are well-documented robust error-correction mechanisms to prevent accumulating errors.

Four overarching issues contradict the theory that random mutations propel evolutionary development (Miller et al. 2021). First, the probability of random changes yielding biologically consonant results is itself improbable. This obstacle has been long recognized. Over fifty years ago, Sir Ernst Boris Chain, having won the Nobel Prize for Physiology or Medicine in 1971, addressed this specific issue: "To postulate, as the positivists of the end of the 19th century and their followers here have done, that the development of survival of the fittest is entirely a consequence of chance mutations, or even that nature carries out experiments by trial and error through mutations in order to create living systems better fitted to survive, seems to me a hypothesis based on no evidence and irreconcilable with the facts" (Chain 1971). Now many decades later, this unflinching declaration is still conspicuously accurate, fully supported by voluminous evidence (Villarreal and Witzany 2018). Importantly, it is not that random errors during genetic replication do not occur, simply that interconnected, effective cellular policing policies rigorously expunge them. Indeed, some errors are not only tolerated but are desirable in context. This elegant principle has been elaborated as the 'harnessing of stochasticity' (Noble 2018). Some genetic errors may occur, and some of these can be productively deployed towards constructive engineering goals, participating in flexible adaptations to environmental adaptations (Noble 2018, Miller 2018, Miller et al. 2019).

A second consequential deficiency of random genetic mutation theory is readily apparent by considering the fossil record. If genetic mutations were indeed random and thus ever-ongoing, no species could resist the accumulated weight of random genomic errors. Species drift over time would be the inescapable norm. However, the morphology of horseshoe crab has remained essentially identical for nearly 445 million years (Rudkin and Young 2009), and crocodiles are stable living forms for over 200 million years (Stockdale and Benton 2021). Consequently, there are robust and effective mechanisms to eliminate random genetic errors that specifically account for long-term species stability (Witzany 2020).

Well-documented instances of reverse evolution constitute a third direct argument against predominately random mechanisms in evolution. In specific instances, forms with retrograde primitive features have derived from immediate ancestors, such as the regression of conodonts during the Permian-Triassic period due to sub-lethal stress (Kilic et al. 2016). Similar phenomena have been documented in the retrograde polymorphisms exhibited during the Cretaceous by the planktonic foraminiferal lineage, Ticinella-Thalmanniell, as well as those seen in ammonites in the Middle and Late Toarcian (Guex 2020). If random genetic events propelled evolution, no backwardization of forms would yield identical phenotypic outputs compared to ancestral forms as there would be no consonant pathway (Torday and Miller 2020b).

A further foundational argument stems from the accumulated evidence supporting the actions of non-random genomic editing throughout evolution (Shapiro 2017, Witzany 2020). Shapiro (2016a,b) asserts that most significant evolutionary variations generating phenotypic adaptations cannot be accounted for based on random genetic errors. There is no way that complex 'read-write' genomes could yield productive evolutionary results if based on random mechanisms (Shapiro 2019). Instead, variations originate from within reciprocations among intelligent cells and their flexible and responsive 'read-write' genomes, resulting from a complex cell-centered process that links to genomic expression as non-random outputs (Shapiro et al. 2017).

Beyond these reasons, there is one further compelling argument. Our human bioengineering, although still fledgling, uses retroviral tools for targeted gene therapy. If humans can co-opt viral vectors as bioengineering tools to exert targeted genomic changes, then nature had previously devised that means to target and integrate genetic information at preferential sites (Ambrosi et al. 2008). These considerations are by no means conjectural. HPV genomic integration triggers cervical cancer by targeting genomic hotspots linked to specific cytogenetic bands (Schmitz et al. 2012).

A comprehensive challenge to the belief that evolution arises from non-random mechanisms has been presented by Zamai (2020), who has deftly enumerated many non-random genome editing pathways stimulated by environmental stresses. Specifically, viral-cellular symbiotic interactions induce the activation of environmentally-responsive CRISPR-Cas systems, triggering cellular antiviral strategies through retrotransposon-guided mutagenic enzymes. The result is stress-induced non-random genome editing, preferentially constrained at 'hyper-transcribed' genetic foci. Similarly, other environmental stresses stimulate the non-random targeting of specific immunoglobulin genes within the genome of eukaryotes. Although some genetic variations may be random, the specific targets are not, nor are their biological expression, as they specifically drive productive cellular adaptive responses to stress.

Although a previous mainstay of evolutionary thought about the source of biological variations, even single nucleotide point mutations are not random in distribution. Recent research has uncovered non-random directionality (Habig et al. 2021). That evidence confirms that organisms can occasionally experience stress-induced alterations in their rates of mutations. In the fungus *Zymoseptoria tritici,* a pest of wheat, researchers have uncovered a causal relationship between stress-induced protein modifications and subsequent mutation rates (Habig et al. 2021). Further, a study of the de novo origination of the human HbS mutation that generates sickle-cell trait and confers some protective effects against malaria indicates that it not only arose much faster than expected based on a purely random basis but also did so much more quickly in its target sub-Saharan population than in other geographic locales (Melamed et al. 2022). This surprisingly non-random pattern affirms that complex information accumulates in a genome through successive generations in response to environmental stresses, impacting the likelihood of relevant mutations. Furthermore, surveys of de novo mutations in the plant *Arabidopsis thaliana* confirm non-random genetic variation with differential mutation biases within some genetic segments, yet constrained in others, primarily related to epigenomic influences that suppress deleterious mutations (Monroe et al. 2022). Nor are these epigenomic influences themselves random in the manner in which they impact DNA. Consequently, neither their distribution nor their epigenetic methylation patterns are random. For example, histones have distinct preferences for one strand of DNA compared to another. The weight of this preferential targeting is only partially counterbalanced by the actions of Minichromosome Maintenance Complex Component 2 (MCM2), a protein that forms part of the pre-replicative process (Petryk et al. 2018).

Based on this voluminous evidence, the ingrained concept that mutation is a directionless evolutionary force is no longer credible. Unquestionably, DNA is a crucial aspect of cellular information. However, it is one among many cellular tools in service to cellular cognition in its continual confrontation with environmental tensions. Consequently, the concept of the genome as a linear inviolate 'blueprint' of genetic information has been discarded (Ball 2016). Beyond DNA, essential aspects of evolutionary development center within RNA agents that purposefully function to regulate, coordinate, and communicate to abet multicellular tissue ecologies. In the same regard, the role of viral and sub-viral interactions as a form of communication among cells is being increasingly recognized (Villarreal and Witzany 2019). Furthermore, cytogenetic and genomic studies confirm that TEs are key drivers of evolutionary novelty, enabling major evolutionary transitions (Shapiro 2022). Extensive restructuring of genomes can occur in eukaryotes, reaching novel combinations of templated and non-templated DNA sequences, and permitting rapid evolutionary change. Many of these transcriptional templates consist of non-coding DNA, previously and incorrectly considered "junk" DNA (Shapiro 2022). Consequently, the prior dominating concept of random genetic errors as the prime evolutionary driver must be supplanted by a contemporary narrative that affirms the established role of non-random genetic content editors with complex reciprocal working relationships among cells.

There are myriad genetic exchanges among cells at all scales, including TEs, viruses, circular DNAs, RNA networks, and stem-loops. Each transfer confers genetic information and the further opportunity for regulatory control and immune competence. Together, their non-random conjoining effects are evolution in action, dominating evolutionary transitions as tools of the cell (Witzany 2020). When natural cellular engineering becomes the framework of active biological expression, the process of evolutionary development dramatically shifts. It is no longer a narrative of non-random genetic variations steeped within selective pressures. Instead, 21st century evolution is viewed as the continuous cellular measuring assessment of environmental information impacting cellular information architecture through non-random means (Miller 2018, Miller et al. 2019, 2020a,b, 2021).

References

Ambrosi, A., Cattoglio, C. and Di Serio, C. 2008. Retroviral Integration Process in the Human Genome: Is It Really Non-Random? A New Statistical Approach. PLoS Computational Biology 4(8): e1000144.

Archibald, J.M. 2015. Endosymbiosis and Eukaryotic Cell Evolution. Current Biology 25(19): R911–R921.

Arkhipova, I.R. and Yushenova, I.A. 2019. Giant Transposons in Eukaryotes: Is Bigger Better? Genome Biology and Evolution 11(3): 906–918.

Asfahl, K.L. and Schuster, M. 2017. Social interactions in bacterial cell–cell signaling. FEMS Microbiology Reviews 41(1): 92–107.

Ball, P. 2016. The problems of biological information. Philosophical Transactions of the Royal Society A: Mathematical, Physical and Engineering Sciences 374(2063): 20150072.

Black, S.G., Arnaud, F., Palmarini, M. and Spencer, T.E. 2010. Endogenous Retroviruses in Trophoblast Differentiation and Placental Development. American Journal of Reproductive Immunology 64(4): 255–264.

Bourque, G., Burns, K.H., Gehring, M., Gorbunova, V., Seluanov, A., Hammell, M., ... Feschotte, C. 2018. Ten things you should know about transposable elements. Genome Biology 19: 199.

Broecker, F. and Moelling, K. 2019. Evolution of Immune Systems From Viruses and Transposable Elements. Frontiers in Microbiology 10.

Brown, C.J. and Rupert, J.L. 2014. Hypoxia and Environmental Epigenetics. High Altitude Medicine & Biology 15(3): 323–330.

Chain, E.B. 1971. Social Responsibility and the Scientist in Modern Western Society. Perspectives in Biology and Medicine 14(3): 347–369.

Ecco, G., Imbeault, M. and Trono, D. 2017. KRAB zinc finger proteins. Development 144(15): 2719–2729.

Enard, D., Cai, L., Gwennap, C. and Petrov, D.A. 2016. Viruses are a dominant driver of protein adaptation in mammals. ELife 5.

Feschotte, C. and Gilbert, C. 2012. Endogenous viruses: insights into viral evolution and impact on host biology. Nature Reviews Genetics 13(4): 283–296.

Gilbert, C. and Cordaux, R. 2017. Viruses as vectors of horizontal transfer of genetic material in eukaryotes. Current Opinion in Virology 25: 16–22.

Goldenfeld, N. and Woese, C. 2011. Life is Physics: Evolution as a Collective Phenomenon Far From Equilibrium. Annual Review of Condensed Matter Physics 2(1): 375–399.

Guex, J. 2020. The Controversial Cope's, Haeckel's and Dollo's Evolutionary Rules: The Role of Evolutionary Retrogradation. pp. 13–22. *In*: Morphogenesis, Environmental Stress and Reverse Evolution. Cham: Springer International Publishing.

Habig, M., Lorrain, C., Feurtey, A., Komluski, J. and Stukenbrock, E.H. 2021. Epigenetic modifications affect the rate of spontaneous mutations in a pathogenic fungus. Nature Communications 12: 5869.

Jangam, D., Feschotte, C. and Betrán, E. 2017. Transposable Element Domestication As an Adaptation to Evolutionary Conflicts. Trends in Genetics 33(11): 817–831.

Janssen, A. and Medema, R.H. 2011. Entosis: aneuploidy by invasion. Nature Cell Biology 13(3): 199–201.

Jobson, M.A., Jordan, J.M., Sandrof, M.A., Hibshman, J.D., Lennox, A.L. and Baugh, L.R. 2015. Transgenerational Effects of Early Life Starvation on Growth, Reproduction, and Stress Resistance in Caenorhabditis elegans. Genetics 201(1): 201–212.

Kiliç, A.M., Plasencia, P., Ishida, K., Guex, J. and Hirsch, F. 2016. Proteromorphosis of Neospathodus (Conodonta) during the Permian–Triassic crisis and recovery. Revue de Micropaléontologie 59(1): 33–39.

Koonin, E.V. 2009. Darwinian evolution in the light of genomics. Nucleic Acids Research 37(4): 1011–1034.

Lacroix, B. and Citovsky, V. 2019. Pathways of DNA Transfer to Plants from Agrobacterium tumefaciens and Related Bacterial Species. Annual Review of Phytopathology 57: 231–251.

Laland, K., Uller, T., Feldman, M., Sterelny, K., Müller, G.B., Moczek, A., ... Strassmann, J.E. 2014. Does evolutionary theory need a rethink? Nature 514(7521): 161–164.

Lupo, A., Cesaro, E., Montano, G., Zurlo, D., Izzo, P. and Costanzo, P. 2013. KRAB-Zinc Finger Proteins: A Repressor Family Displaying Multiple Biological Functions. Current Genomics 14(4): 268–278.

Ma, J., Wang, S., Zhu, X., Sun, G., Chang, G., Li, L., ... Huang, J. 2022. Major episodes of horizontal gene transfer drove the evolution of land plants. Molecular Plant 15(5): 857–871.

Martin, W.F., Garg, S. and Zimorski, V. 2015. Endosymbiotic theories for eukaryote origin. Philosophical Transactions of the Royal Society B: Biological Sciences 370(1678): 20140330.

Martinez, G., Castellano, M., Tortosa, M., Pallas, V. and Gomez, G. 2014. A pathogenic non-coding RNA induces changes in dynamic DNA methylation of ribosomal RNA genes in host plants. Nucleic Acids Research 42(3): 1553–1562.

Martinez, J., Klasson, L., Welch, J.J. and Jiggins, F.M. 2021. Life and Death of Selfish Genes: Comparative Genomics Reveals the Dynamic Evolution of Cytoplasmic Incompatibility. Molecular Biology and Evolution 38(1): 2–15.

Matveeva, T.V. and Otten, L. 2019. Widespread occurrence of natural genetic transformation of plants by Agrobacterium. Plant Molecular Biology 101(4–5): 415–437.

Melamed, D., Nov, Y., Malik, A., Yakass, M.B., Bolotin, E., Shemer, R., ... Livnat, A. 2022. *De novo* mutation rates at the single-mutation resolution in a human HBB gene region associated with adaptation and genetic disease. Genome Research 32(3): 488–498.

Miller, W.B. 2016. Cognition, Information Fields and Hologenomic Entanglement: Evolution in Light and Shadow. Biology 5(2): 21.

Miller, W.B. 2018. Biological Information Systems: Evolution as cognition-based Information Management. Progress in Biophysics and Molecular Biology 134: 1–26.

Miller, W.B., Baluška, F. and Torday, J.S. 2020b. Cellular senomic measurements in Cognition-Based Evolution. Progress in Biophysics and Molecular Biology 156: 20–33.

Miller, W.B., Enguita, F.J. and Leitão, A.L. 2021. Non-Random Genome Editing and Natural Cellular Engineering in Cognition-Based Evolution. Cells 10(5): 1125.

Miller, W.B.J., Reber, A.S., Marshall, P. and Baluška, F. 2022. Viral-Cellular Natural Engineering in Cognition-Based Evolution. Communicative & Integrative Biology. Accepted in press.

Miller, Jr, W.B. and Torday, J.S. 2019. Reappraising the exteriorization of the mammalian testes through evolutionary physiology. Communicative & Integrative Biology 12(1): 38–54.

Miller, W.B., Torday, J.S. and Baluška, F. 2019. Biological evolution as defense of "self". Progress in Biophysics and Molecular Biology 142: 54–74.

Miller, W.B., Torday, J.S. and Baluška, F. 2020a. The N-space Episenome unifies cellular information space-time within cognition-based evolution. Progress in Biophysics and Molecular Biology 150: 112–139.

Milstein, J.N. and Meiners, J.-C. 2011. On the role of DNA biomechanics in the regulation of gene expression. Journal of The Royal Society Interface 8(65): 1673–1681.

Monroe, J.G., Srikant, T., Carbonell-Bejerano, P., Becker, C., Lensink, M., Exposito-Alonso, M., ... Weigel, D. 2022. Mutation bias reflects natural selection in Arabidopsis thaliana. Nature 602(7895): 101–105.

Mustafin, R.N. 2018. Hypothesis on the Origin of Viruses from Transposons. Molecular Genetics, Microbiology and Virology 33(4): 223–232.

Muszewska, A., Steczkiewicz, K., Stepniewska-Dziubinska, M. and Ginalski, K. 2019. Transposable elements contribute to fungal genes and impact fungal lifestyle. Scientific Reports 9: 4307.

Noble, D. 2018. Central Dogma or Central Debate? Physiology 33(4): 246–249.

Overholtzer, M., Mailleux, A.A., Mouneimne, G., Normand, G., Schnitt, S.J., King, R.W., ... Brugge, J. S. 2007. A Nonapoptotic Cell Death Process, Entosis, that Occurs by Cell-in-Cell Invasion. Cell 131(5): 966–979.

Palmer, W.H., Hadfield, J.D. and Obbard, D.J. 2018. RNA-Interference Pathways Display High Rates of Adaptive Protein Evolution in Multiple Invertebrates. Genetics 208(4): 1585–1599.

Paulsen, T., Shibata, Y., Kumar, P., Dillon, L. and Dutta, A. 2019. Small extrachromosomal circular DNAs, microDNA, produce short regulatory RNAs that suppress gene expression independent of canonical promoters. Nucleic Acids Research 47(9): 4586–4596.

Petryk, N., Dalby, M., Wenger, A., Stromme, C.B., Strandsby, A., Andersson, R. and Groth, A. 2018. MCM2 promotes symmetric inheritance of modified histones during DNA replication. Science 361(6409): 1389–1392.

Popat, R., Cornforth, D.M., McNally, L. and Brown, S.P. 2015. Collective sensing and collective responses in quorum-sensing bacteria. Journal of The Royal Society Interface 12(103): 20140882.

Popov, M., Kolotova, T. and Davidenko, M. 2018. Endogenous Retroviruses as Genetic Modules that Shape the Genome Regulatory Networks During Evolution. Journal of V.N. Karazin Kharkiv National University, Series "Medicine" 36: 80–95.

Roberts, R.M., Ezashi, T., Schulz, L.C., Sugimoto, J., Schust, D.J., Khan, T. and Zhou, J. 2021. Syncytins expressed in human placental trophoblast. Placenta 113: 8–14.

Rook, G., Bäckhed, F., Levin, B.R., McFall-Ngai, M.J. and McLean, A.R. 2017. Evolution, human-microbe interactions, and life history plasticity. The Lancet 390(10093): 521–530.

Rudkin, D.M. and Young, G.A. 2009. Horseshoe Crabs—An Ancient Ancestry Revealed. pp. 25–44. *In*: Biology and Conservation of Horseshoe Crabs. Boston, MA: Springer.

Ryan, F. 2009. Virolution. New York, NY, USA: HarperCollins Publishers.

Schmitz, M., Driesch, C., Jansen, L., Runnebaum, I.B. and Dürst, M. 2012. Non-Random Integration of the HPV Genome in Cervical Cancer. PLoS ONE 7(6): e39632.

Shapiro, J.A. 2013. How life changes itself: The Read–Write (RW) genome. Physics of Life Reviews 10(3): 287–323.

Shapiro, J. 2016a. Nothing in Evolution Makes Sense Except in the Light of Genomics: Read–Write Genome Evolution as an Active Biological Process. Biology 5(2): 27.

Shapiro, J.A. 2016b. The basic concept of the read–write genome: Mini-review on cell-mediated DNA modification. Biosystems 140: 35–37.

Shapiro, J.A. 2017. Living Organisms Author Their Read-Write Genomes in Evolution. Biology 6(4): 42.

Shapiro, J.A. 2019. No genome is an island: toward a 21st century agenda for evolution. Annals of the New York Academy of Sciences 1447: 21–52.

Shapiro, J.A. 2022. What we have learned about evolutionary genome change in the past 7 decades. Biosystems 215–216: 104669.

Shorter, J. and Lindquist, S. 2005. Prions as adaptive conduits of memory and inheritance. Nature Reviews Genetics 6(6): 435–450.

Skinner, M.K. 2014. Environmental stress and epigenetic transgenerational inheritance. BMC Medicine 12(1): 153.

Smalheiser, N.R. 2014. The RNA-centred view of the synapse: non-coding RNAs and synaptic plasticity. Philosophical Transactions of the Royal Society B: Biological Sciences 369(1652): 20130504.

Sobhy, H. 2017. A comparative review of viral entry and attachment during large and giant dsDNA virus infections. Archives of Virology 162(12): 3567–3585.

Soucy, S.M., Huang, J. and Gogarten, J.P. 2015. Horizontal gene transfer: building the web of life. Nature Reviews Genetics 16(8): 472–482.

Stegemann, R. and Buchner, D.A. 2015. Transgenerational inheritance of metabolic disease. Seminars in Cell & Developmental Biology 43: 131–140.

Stockdale, M.T. and Benton, M.J. 2021. Environmental drivers of body size evolution in crocodile-line archosaurs. Communications Biology 4: 38.

Sultana, T., Zamborlini, A., Cristofari, G. and Lesage, P. 2017. Integration site selection by retroviruses and transposable elements in eukaryotes. Nature Reviews Genetics 18(5): 292–308.

Torday, J. and Miller, W. 2020a. Cellular-Molecular Mechanisms in Epigenetic Evolutionary Biology. Cham, Switzerland: Springer.

Torday, J.S. and Miller, W.B.Jr. 2020b. Evolution as a timeless continuum. *In*: J. Guex, J.S. Torday and W.B.J. Miller (eds.). Morphogenesis, Environmental Stress and Reverse Evolution. Cham, Switzerland: Springer.

Villarreal, L.P. and Ryan, F. 2019. Viruses in the Origin of Life and Its Subsequent Diversification. *In*: V.M. Kolb (ed.). Handbook of Astrobiology. Boca Raton, Florida, United States: CRC Press.

Villarreal, L.P. and Witzany, G. 2018. Editorial: Genome Invading RNA Networks. Frontiers in Microbiology 9(581).

Villarreal, L.P. and Witzany, G. 2019. That is life: communicating RNA networks from viruses and cells in continuous interaction. Annals of the New York Academy of Sciences 1447: 5–20.

Vogel, D. and Dussutour, A. 2016. Direct transfer of learned behaviour via cell fusion in non-neural organisms. Proceedings of the Royal Society B: Biological Sciences 283(1845): 20162382.

Wallau, G.L., Vieira, C. and Loreto, É.L.S. 2018. Genetic exchange in eukaryotes through horizontal transfer: connected by the mobilome. Mobile DNA 9: 6.

Waszak, S.M., Delaneau, O., Gschwind, A.R., Kilpinen, H., Raghav, S.K., Witwicki, R.M., ... Dermitzakis, E.T. 2015. Population Variation and Genetic Control of Modular Chromatin Architecture in Humans. Cell 162(5): 1039–1050.

Wei, B., Liu, H., Liu, X., Xiao, Q., Wang, Y., Zhang, J., ... Huang, Y. 2016. Genome-wide characterization of non-reference transposons in crops suggests non-random insertion. BMC Genomics 17: 536.

Wellenreuther, M., Mérot, C., Berdan, E. and Bernatchez, L. 2019. Going beyond SNPs: The role of structural genomic variants in adaptive evolution and species diversification. Molecular Ecology 28(6): 1203–1209.

Whitesides, G. 2004. The improbability of life Fitness of the Cosmos for Life: Biochemistry and Fine-Tuning vol xiii, ed J D Barrow et al. Cambridge: Cambridge University Press.

Wilke, C.O. and Sawyer, S.L. 2016. At the mercy of viruses. ELife, 5.

Witzany, G. 2011. Natural Genome Editing from a Biocommunicative Perspective. Biosemiotics 4(3): 349–368.

Witzany, G. 2014. RNA Sociology: Group Behavioral Motifs of RNA Consortia. Life 4(4): 800–818.

Witzany, G. 2020. Evolution of Genetic Information without Error Replication. pp. 295–320. *In*: Theoretical Information Studies: Volume 11. World Scientific Series in Information Studies.

Zamai, L. 2020. Unveiling Human Non-Random Genome Editing Mechanisms Activated in Response to Chronic Environmental Changes: I. Where Might These Mechanisms Come from and What Might They Have Led To? Cells 9(11): 2362.

CHAPTER 9

The Virome and Natural Viral-Cellular Engineering

Modern metagenomic studies have established the ubiquity of the virome (Zhang et al. 2019). Viruses are planetary-wide, extending across all land surfaces. For example, the marine virome is highly diverse and globally distributed (Rohwer and Thurber 2009). One kilogram of marine sediment has 1 million separate viral genotypes, and 100 liters of seawater has over 5,000 viral genotypes (Rohwer and Thurber 2009). Even the deep-earth environment supports a rich virome, extending at least 400 meters below the surface and consistently interacts with extensive sub-terrestrial prokaryotic life (Holmfeldt et al. 2021). Indeed, phage viruses that infect bacterial cells and transfer genetic material to them are considered the dominant organisms in the biosphere (Clokie et al. 2011). Phages can reprogram bacterial cell metabolism and translation machinery, contributing different ecosystem-wide potentials with broad planetary consequences. Yet, for all that has been discovered about their flexibility and extent, only a small percentage of their total population has been identified, just as with cellular microbes (Mirzaei et al. 2021).

Given our rapidly progressing technological capacities, many new viruses and viral types have recently been found, substantially altering our understanding of their collective influence on the planet and its evolution. The relatively recent discovery of giant viruses is illustrative. The giant Pandoravirus has an astoundingly large viral genetic complement. Unlike other viruses, its genome is as large as some bacteria, encoding a broad range of proteins. Many of these genetic sequences function across a comprehensive biological spectrum in cells, including photosynthesis, translation, the support of the cellular cytoskeleton, intracellular protein transportation, and many cellular metabolic processes (Cheng et al. 2021). Their genomes are so different from either typical viruses or other microbes that it is wondered whether they represent a distinct evolutionary path.

The 2003 discovery of the first giant virus to be identified, *Acanthameba polyphaga* mimivirus, common to amoeba, was heralded as a "new era in microbiology" (Fischer 2016). These extraordinary double-stranded DNA giant viruses, termed nucleocytoplasmic large DNA viruses (NCLDVs), have had a consequential evolutionary impact. It is well-recognized that viral horizontal gene transfers have been a significant source of eukaryotic genetic variation and novelty (Shapiro 2016). For example, approximately 10% of the mammalian genome consists of elements of a prior endogenous retroviral invasion (Chuong 2013). However, an unexpected finding was the facility with which the relatively enormous genomes of NCLDVs, known to be abundant in the biosphere, have integrated into eukaryotic genomes via large-scale horizontal transfers (Moniruzzaman et al. 2020). For example, in some algae species, giant endogenous viral elements can total one million base pairs constituting up to 10% of their genome. Importantly, this same type of endogenization process can be found across Eukaryota.

On the other end of the size spectrum, an unexpected variety of subviral particles have also been identified that can aggressively transfer genetic information among the cellular domains. Transposable elements, including transposons, retrotransposons, and retroviruses that constitute nearly half of our human genome are examples. Additionally, virophages, Polintons, and transpovirons are a separate class of mobile genetic elements (Koonin and Krupovic 2017). Virophages are small satellite DNA viruses that replicate in giant viruses. Polintons are a related form of transposon, widespread in the genomes of diverse eukaryotes (Koonin and Krupovic 2017). Transpovirons are double-stranded DNA elements that are giant virus parasites. As separate, additional genetic sources, plasmids are small bits of DNA that exist in cells separate from the cellular nuclear genome and can transmit between cells by horizontal gene transfer. Prophages are a segment of phage virus DNA that can exist either as a plasmid or integrate with bacterial DNA. Even isolated RNA molecules can transmit between cells as ribozymes that can function as enzymatic catalysts. (Serganov and Patel 2007). All of these mobile elements form a dynamic network that can integrate into the genomes of target cells.

All planetary biology can be divided into two major categories: 'capsid-encoding organisms' (viruses) and 'ribosome-encoding organisms' (cells) (Forterre 2017). Each viral type encodes its own capsids, and these protein capsules protect the viral genome and serve a correlative role in viral dynamics similar to the role of the cell membrane. Capsid proteins are remarkably flexible and may have initially derived from cells (Koonin and Krupovic 2017). They can even self-assemble within cells into virus-like particles (VLPs) as supramolecular complexes that resemble complete viruses (Mohsen et al. 2018). However, as they contain no viral genetic material, VLPs are not infectious; nonetheless, they can stimulate innate and adaptive immune responses comparable to those of complete viruses. Further, and remarkably, even though a viral genome is absent, VLPs can take up and package nucleic acids (RNAs) within target cells that participate in biological expression and immune responses.

Forterre (2016) points out that viruses have been typically defined by the virus nucleic acid genome inside its outer protein capsid shell. However, this should be regarded as an artifact of the timeline of their discovery rather than being based on viral properties. Some scientists assert that viruses represent dual-state organisms with two distinctive phenotypic phases (Bândea 1983). A vegetative phase in which the virus is dispersed within cells begins a stage of viral protein synthesis as an essential aspect of viral metabolism, growth, and reproduction. This stage is followed by a reproductive-assembly stage for the reproductive reconstitution of nucleic acids and viral sub-components that comprise the protective layer, completing the entire virus. Thus, some contend that viruses are complex 'living' entities in which virions are actually viral spores as a 'life-cycle' stage, suited for existence outside of cells, and the virus proper, is the infected cell as a 'virocell' (Forterre 2013). Notably, the properties of cells and their viruses as 'virocells' distinctly overlap.

The virocell concept focuses on the intracellular phase of the viral reproductive life cycle and pertinently, implies that this combination of indwelling virus and its target cell becomes a new type of fused cellular organism (Forterre 2013). In this manner, the virus and the cell become a new co-respondent entity as a functioning combination of a capsid-encoded entity and a ribosomal cellular one as a riboviral-cell (Forterre 2013). Importantly, whether or not the virocell concept is a precisely accurate depiction of the status between the virus and the cell, what is not in doubt is that the result of this union is exploited by one or both participants with a range of outcomes that extend from cellular death to long-term symbiosis as a virocell. In all instances, these results represent biological engineering by one or the other of the two biological entities, or both as co-engineering.

Cognition-Based Evolution maintains that viruses are also capable of engineering, most particularly within cells but also outside of cellular boundaries. Viruses avidly form collective infectious units that travel as a group for joint transmission to their target cells (Thoulouze and Alcover 2011, Sanjuán and Thoulouze 2019). This activity necessitates virion aggregation and their collecting and deploying various specific extracellular components for their mutual encasement in protein matrices. These groups can inhabit lipid vesicles and collectively aggregate on cell surfaces. As this action is conducted in the extracellular spaces, it represents extracellular Natural Viral Engineering (NVE) that is not specifically part of Natural Cellular Engineering. Furthermore, viruses can independently leverage cellular resources such as cellular membranes and contingently compartmentalize some of their ongoing viral functions such as genome replication and particle assembly (Schoelz and Leisner 2017). All of these processes should also be considered a form of NVE.

When intracellular, viruses can exploit several types of viral compartments, including inclusion bodies, viroplasms, and viral factories that offer further support for the NVE concept. Viral factories are viral clusters in perinuclear or other cytoplasmic locations that can disassemble and assemble by recruiting cellular organelles, such as mitochondria and cytoskeletal structures. Through coordinated viral action, these clusters can produce various cellular proteins necessary for viral replication (Novoa et al. 2005, de Castro et al. 2013). In that process, numerous virus-induced vesicles are also produced that partially derive from the membranes of the endoplasmic reticulum in close proximity or at the site of viral assembly (den Boon et al. 2010). These 'viroplasms' or inclusion bodies represent highly organized viral factory sites that transport along cellular microtubules and whose products are deployed in viral genome replication, capsid packaging, and providing protection for viruses against cellular antiviral defenses (Patton et al. 2006, Netherton and Wileman 2011, Niehl et al. 2013).

Claverie (2006) has argued that the "virus factory should be considered the actual virus organism when referring to a virus" (Claverie 2006, p. 110.4). There is no doubt that this complex process represents engineering since the operations themselves define the pertinent viral form, requiring divisions of labor and communication among the participants, both viral and cellular to enable their products. All such activities are supported by bi-directional viral-cellular HGT. Thus, the virocell, or 'riboviro-cell', represents a unique metabolome that combines both encoding types. Even though the number of viral genes may be few, they are sufficient to support this symbiotic intermediate 'carrier' state (Forterre 2017).

The concept of viral engineering in which viruses are contingent participants and not mere non-discretionary entities has been greatly strengthened by recent research on a range of viral faculties that are quite remarkable and unexpected. Consequently, an entirely new field of 'sociovirology' has formed (Claverie and Abergel 2016). Individual virions do not establish a foothold in target cells, and it is now understood that successful viral infection requires collective intracellular units that can cooperate and compete just like cells (Sanjuán 2017, Trinh et al. 2017). Indeed, such virus-virus interactions are now understood to be pervasive (Díaz-Muñoz et al. 2017). Within these collective actions, there are distinct features of complex viral sociality characterized by fundamental social rules, including cooperation, communication, conflict, and cheating (Domingo-Calap et al. 2019). The trading of viral products, such as genetic material, capsids, or structural proteins, operates within these social dynamics (Dolgin 2019). Further, there is a surprisingly robust level of interviral communication among intracellular viruses, which has been categorized as 'chit-chat', mediated by a six amino acid long protein, arbitrium (Dolgin 2019). This interviral communication is highly complex and fluid, functioning analogously to bacterial quorum-sensing.

In this way, viruses can cooperate to co-infect cells, spread between them, determine lysogeny vs. lysis, and cooperate to evade cellular antiviral immune defenses (Wang et al. 2018). Most significantly, viral communication enables an 'understanding' of a cell's communication processes that can form the basis for viral-cellular co-engineering. All of these social capacities yield previously unexpected conclusions that viruses can engage in contingent decision-making and are, therefore, capable of natural engineering.

Indeed, it is clear that these viral capacities of communication and cooperation can even extend outside of cells (Díaz-Muñoz et al. 2017, Witzany 2020, Abedon 2020). For example, when a temperate phage infects a target cell, it can either initiate a lytic cycle or become a dormant prophage. This lysis-lysogeny 'decision' coordinates through an intricate network of transcriptional repressors or activators that involves RNA degradation and proteolysis (Ofir and Sorek 2018). This distinction between two remarkably different pathways must rely on the viral reception and interpretation of cellular cues that inform it about the metabolic state of the cell and the presence of co-infecting phages. It defaults that this complex coordinated activity requires measurements of status, which depends on various peptide concentrations forming an 'arbitrium' communication system involving three phage genes (Erez et al. 2017). Furthermore, this information can be communicated among successive populations of intracellular phages (Erez et al. 2017). Thus, once comfortably presumed to be automatons, viruses can receive information, cooperate or compete, communicate with each other, and engage in coordinated contingent decision-making (Witzany 2020). Thus, viruses can distinguish self- from non-self (Witzany 2020). Just like cells, viruses may be small but should not be considered stupid.

Accordingly, coordinated viral activities can now be properly placed in familiar engineering terms as it is a product of the purposeful use of information, its direct communication, and the deployment of available resources to yield products that satisfy specific functions. To support this activity, viruses demonstrate the flexible use of tools and modular genetic segments as 'plug-in cassettes' that can modify genetic code in concerted segments, permit divisions of labor, and deploy an extensive range of both viral and cellular resources. Contrary to what had been previously believed, these capacities are skillfully employed by viruses to attain their ends and commonly expressed as viral-cellular symbiotic co-engineering. Therefore, the concepts of natural engineering and their resulting niche constructions apply across living scales through the coordinated use of biological tools. This same reiterative pattern characterizes biofilms or termite mounds, or humans. Beneath their varied physical embodiments, the central role of symbiosis thrives from its first beginnings with the creation of the first competent cell or the crucial endosymbiotic merger that enabled eukaryotic life (Aanen and Eggleton 2017).

It can also be contended that viruses exhibit their own form of phenotype. Viral quasispecies are viral populations with extremely large numbers of variant genomes, termed mutant swarms or mutant clouds (Domingo and Perales 2019). As cellular co-engineers, quasispecies can now be reconceptualized as the viral means of exploring information space, with each quasispecies representing a variant viral phenotype (Wilke and Novella 2003). Villarreal and Witzany (2013) further argued that these viral quasispecies represent a collective consortium of RNA stem-loops as cooperative modules representing stem-loop societies that act as a viral social group. Correlatively, other viral elements can associate, including viroids, mobile genetic elements, and ribozymes that can leverage coordinated effects on stem-loop phenotypic variation.

It is not only viral phenotype that resembles cellular architecture. Viruses can be assessed as having senomic capacity, exerted through their nucleocapsid, akin to some of the functions of the cellular plasma membrane. It has been proposed that the senome model is also appropriate to viruses based on the complex sensory capacity of their arbitrium system that links to viral memory (Miller et al. 2022). Through this means, viruses link to cellular information space and the active cellular N-space Episenome to achieve their own form of participatory sense-awareness that energizes virocellular symbioses. Necessarily, cells must reciprocally link to viral information space for concordant measurement of environmental and cellular stresses and to gauge the extent of cellular resources that can be devoted to co-partnership.

Within this narrative, the virome is a vital intercessory among the three cellular domains. As such, viral actions extend well beyond pathogenesis, functioning as a significant partner in mutualized information management across the domains (Miller and Torday 2018). Accordingly, viruses must be directing some components of that symbiosis as co-engineering partners with cells as part of their complex reproductive strategies. And further, these must be conducted in non-random patterns toward generally mutualizing goals through the reciprocating use of viral and cellular tools (Roossinck 2008, Ryan 2016, 2019, Rossininck and Bazán 2017, Villarreal and Witzany 2021).

Among these viral and cellular tools, vesicles are among the most useful, utilizing sections of cell membranes for their formation or alternatively forming as lipid-membrane encapsulated inclusion bodies (Leeks et al. 2019). Viruses deploy vesicles to disperse within cells or external to them in collective infectious units either as an aggregate of virions or as a single composite virion with multiple genomes (polypoid virion). Both these vesicular forms can be considered examples of viral tool-making since these vesicles can achieve variable structural forms dependent on viral families and targeted cells. In the same manner, the viral capsid can be considered a type of viral tool since the dynamic interaction of viral genomes with their capsids enables viruses to engage with their target cells in multiple ways. In particular, and dependent on capsular structure, two types of non-infectious viral subforms can be produced after replication (Liu and Hu 2019). One of these forms is an aggregate of viral envelop proteins that can exist in quantities up to 100,000 times more plentiful than complete virions. The second type represents 'empty' virions with a capsid and envelop proteins but no viral genome within it. Surprisingly, this form is much more common than complete virions.

Viral replication requires cells and their various compartments, including their nucleus, cytoplasm, subcellular structures, or virus-induced rearranged membranes (Alvisi 2013). Thus, viruses use cells as their tools. However, cells use viruses to stimulate and supplement their immune status and as agents of information transfer among themselves. Consequently, viruses are also cellular tools. Extracellular vesicles serve both viruses and cells, representing a continuum of membranous particles that can derive from either or both (Nolte et al. 2016, Noble 2020). The exact origin of extracellular vesicles is unknown, but exosomes so closely resemble RNA viruses that enveloped viruses can be interpreted as a form of extracellular vesicle (Margolis and Sadovsky 2019).

Consequently, the prior long-standing conceptions of viral dynamics are challenged. It was previously assumed that the viral co-option of cellular machinery was non-contingent biological automatic templating. The contemporary view is that these actions express an overarching and critical planetary viral-cellular symbiosis with a wide range of contingent possibilities decided by either party or both (Villarreal 2005, 2007, Ryan 2009, 2016, 2019, Villarreal and Witzany 2021). As such, the virome significantly shapes every aspect of biology (Adachi 2020) through a cohering framework of viral-cellular partnerships and co-engineering. Outright pathogenesis is only one small aspect of that fuller narrative in which virocell metabolism and its encoded metabolic products mold aquatic and land-based environments, either directly or indirectly through giant viral intermediaries that enable mutualizing adaptations (Mizuno et al. 2019, Moniruzzaman et al. 2020).

In this co-respondent process, the virome, including NCLDVs, profoundly shapes the biogeochemistry of the planet by modulating and shaping microbial metabolic capabilities, infecting a wide range of eukaryotes (Endo et al. 2020). The virome represents a storehouse of potential microbial genes, including transposons that can fluidly exchange across the cellular domains, heavily influencing an extraordinary range of metabolic and evolutionary processes with abundant genetic transfers across the kingdoms (Dinsdale et al. 2008, Malik et al. 2017). Accordingly, it is now understood that viruses represent fundamental evolutionary drivers creating proteins with novel functions (Villarreal 2005, Baluška 2009, Forterre and Prangishvili 2009, Ryan 2009, Villarreal and Witzany 2010, Koonin and Wolf 2012, Cosby et al. 2021).

Naturally, the question of whether or not viruses are alive is pertinent. Using new bioinformatic techniques and analyzing protein folds in their genome rather than the viral genetic code itself, there is strong evidence that viruses should be considered alive. Their protein folds are ancient and can now be traced phylogenetically, revealing that viruses most likely originated from ancient cells and continue to share some cellular protein fold sequences in patterns resembling those of living organisms (Nasir and Caetano-Anollés 2015). However, viruses have also diverged significantly enough to represent an independent branch of life. Analysis indicates that viruses did not capture all their genetic material from cells, as previously argued, but instead represent independent creators of evolutionary novelty as spreaders of genetic diversity. Despite this evidence, there is a weighty countervailing body of opinion that strenuously argues that viruses are not alive. However, what is not in doubt is that viruses demonstrate many living faculties in cells, such as communication, cooperation, a sense of 'self', the purposeful use of information, and memory. Through these capabilities, cells can co-engineer to achieve both consensual and non-consensual outcomes. Deemed alive or not, it is their biological activity that matters more.

Forterre (2016) asserts that viruses should be considered alive specifically because they are inseparably involved in the living process. Furthermore, both viruses proper and the many varieties of capsid-less viral elements, such as transposable elements have been crucial to all major evolutionary transitions by facilitating the trading of genetic information and thereby driving both unicellular and multicellular novelty (Koonin 2016). For the moment, it is best to suppose that viruses occupy a contingent zone, best defined as "entities that are 'sufficiently alive to engineer' as a liminal form of life whose range of action is context-dependent" (Miller et al. 2022).

Accordingly, the contemporary view of viruses is no longer that of obligatory pathogens. Instead, they are mainly participants in viral symbiogenesis in which any competitive parasitism is balanced by intimate symbiotic partnerships with cells functioning through mutualism and commensalism (Villarreal 2007, Roossinck 2008, Ryan 2016, 2019, Rossinck and Bazán 2017). Hence, the outmoded traditional host-parasite model of viral behavior has to be discarded and replaced with a contemporary viewpoint that infectious dynamics, in its full range of manifestations, is a crucial driving force in evolution (Baluška 2009, Miller 2013, 2016). Undoubtedly, the transfer of genetic material as cell-cell communication of assessed information is the perpetual cardinal aspect of evolution (see Fig. 1, Chapter 7). The influence of the virome and sub-viral genetic particles extends into every feature of eukaryotic life as deeply embedded partners in our normal transcriptional state (Virgin 2014). As the greatest agent of genetic diversity on the planet and the dominant biological form, viruses are the central drivers of evolution, functioning primarily as symbionts rather than parasites. Viruses are full participants in the replacement and creation of new molecular "words" to be used as parts of the solutions biological problems. (Marijuán and Navarro 2022). In this perpetual narrative, viruses are tools of cells, and cells are tools of viruses, each serving as co-partners in co-engineered solutions to planetary stresses (Miller et al. 2021, 2022).

Viruses and cells should be viewed as tool kits for the survival of each by meeting environmental stresses at their own tempos and cooperating through continuous cellular-viral symbiogenesis (Villareal 2005, Ryan 2009, 2019, Miller and Torday 2018, Miller et al. 2021). The virome serves as an active agent of information transfer among cells supporting a reservoir of novel genes that move globally, permitting cells to successfully adapt and manipulate environments through niche constructions (Rohwer and Thurber 2009, Miller et al. 2021, 2022). Therefore, both biofilms and multicellular eukaryotes are the coordinated products of Natural Cellular Engineering, Natural Viral Engineering, and natural genomic editing, representing the transfer and coordination of information across scales. This partnership crystallizes in the intricately coordinated holobionic form as exquisite non-random viral-cellular engineering in which viral exchanges of all kinds and among all partners effect the fluid information transfers within the information management systems that support life's complexity.

References

Aanen, D.K. and Eggleton, P. 2017. Symbiogenesis: Beyond the endosymbiosis theory? Journal of Theoretical Biology 434: 99–103.

Abedon, S.T. 2020. Phage-Phage, Phage-Bacteria, and Phage-Environment Communication. pp. 23–70. *In*: Biocommunication of Phages. Cham: Springer International Publishing.

Adachi, A. 2020. Grand Challenge in Human/Animal Virology: Unseen, Smallest Replicative Entities Shape the Whole Globe. Frontiers in Microbiology 11.

Alvisi, G. 2013. Reprogramming the host: Modification of cell functions upon viral infection. World Journal of Virology 2(2): 16.

Bândea, C.I. 1983. A new theory on the origin and the nature of viruses. Journal of Theoretical Biology 105(4): 591–602.

Baluška, F. 2009. Cell-Cell Channels, Viruses, and Evolution. Annals of the New York Academy of Sciences 1178: 106–119.

Cheng, S., Wong, G.K.-S. and Melkonian, M. 2021. Giant DNA viruses make big strides in eukaryote evolution. Cell Host & Microbe 29(2): 152–154.

Chuong, E.B. 2013. Retroviruses facilitate the rapid evolution of the mammalian placenta. BioEssays 35(10): 853–861.

Claverie, J.-M. 2006. Viruses take center stage in cellular evolution. Genome Biology 7(6): 110.

Claverie, J.-M. and Abergel, C. 2016. Giant viruses: The difficult breaking of multiple epistemological barriers. Studies in History and Philosophy of Science Part C: Studies in History and Philosophy of Biological and Biomedical Sciences 59: 89–99.

Clokie, M.R.J., Millard, A.D., Letarov, A.V. and Heaphy, S. 2011. Phages in nature. Bacteriophage 1(1): 31–45.

Cosby, R.L., Judd, J., Zhang, R., Zhong, A., Garry, N., Pritham, E.J. and Feschotte, C. 2021. Recurrent evolution of vertebrate transcription factors by transposase capture. Science 371(6531): eabc6405.

de Castro, I.F., Volonté, L. and Risco, C. 2013. Virus factories: biogenesis and structural design. Cellular Microbiology 15(1): 24–34.

den Boon, J.A., Diaz, A. and Ahlquist, P. 2010. Cytoplasmic Viral Replication Complexes. Cell Host & Microbe 8(1): 77–85.

Díaz-Muñoz, S.L., Sanjuán, R. and West, S. 2017. Sociovirology: Conflict, Cooperation, and Communication among Viruses. Cell Host & Microbe 22(4): 437–441.

Dinsdale, E.A., Edwards, R.A., Hall, D., Angly, F., Breitbart, M., Brulc, J.M., … Rohwer, F. 2008. Functional metagenomic profiling of nine biomes. Nature 452(7187): 629–632.

Dolgin, E. 2019. The secret social lives of viruses. Nature 570(7761): 290–292.

Domingo-Calap, P., Segredo-Otero, E., Durán-Moreno, M. and Sanjuán, R. 2019. Social evolution of innate immunity evasion in a virus. Nature Microbiology 4(6): 1006–1013.

Domingo, E. and Perales, C. 2019. Viral quasispecies. PLOS Genetics 15(10): e1008271.

Endo, H., Blanc-Mathieu, R., Li, Y., Salazar, G., Henry, N., Labadie, K., ... Ogata, H. 2020. Biogeography of marine giant viruses reveals their interplay with eukaryotes and ecological functions. Nature Ecology & Evolution 4(12): 1639–1649.

Erez, Z., Steinberger-Levy, I., Shamir, M., Doron, S., Stokar-Avihail, A., Peleg, Y., ... Sorek, R. 2017. Communication between viruses guides lysis–lysogeny decisions. Nature 541(7638): 488–493.

Fischer, M.G. 2016. Giant viruses come of age. Current Opinion in Microbiology 31: 50–57.

Forterre, P. 2013. The virocell concept and environmental microbiology. The ISME Journal 7(2): 233–236.

Forterre, P. 2016. To be or not to be alive: How recent discoveries challenge the traditional definitions of viruses and life. Studies in History and Philosophy of Science Part C: Studies in History and Philosophy of Biological and Biomedical Sciences 59: 100–108.

Forterre, P. 2017. The Origin, Nature and Definition of Viruses (and Life): New Concepts and Controversies. pp. 15–26. *In*: P.J. Heidt, P.L. Ogra, M.S. Riddle and V. Rusch (eds.). Old Herborn University Seminar Monograph 31: Evolutionary Biology of The Virome, and Impacts In Human Health and Disease. Old Herborn University.

Forterre, P. and Prangishvili, D. 2009. The Great Billion-year War between Ribosome- and Capsid-encoding Organisms (Cells and Viruses) as the Major Source of Evolutionary Novelties. Annals of the New York Academy of Sciences, 1178: 65–77.

Holmfeldt, K., Nilsson, E., Simone, D., Lopez-Fernandez, M., Wu, X., de Bruijn, I., ... Dopson, M. 2021. The Fennoscandian Shield deep terrestrial virosphere suggests slow motion 'boom and burst' cycles. Communications Biology 4(1): 307.

Koonin, E.V. 2016. Viruses and mobile elements as drivers of evolutionary transitions. Philosophical Transactions of the Royal Society B: Biological Sciences 371(1701): 20150442.

Koonin, E.V. and Wolf, Y.I. 2012. Evolution of microbes and viruses: a paradigm shift in evolutionary biology? Frontiers in Cellular and Infection Microbiology 2(119).

Koonin, E.V. and Krupovic, M. 2017. Polintons, virophages and transpovirons: a tangled web linking viruses, transposons and immunity. Current Opinion in Virology 25: 7–15.

Leeks, A., Sanjuán, R. and West, S.A. 2019. The evolution of collective infectious units in viruses. Virus Research 265: 94–101.

Liu, K. and Hu, J. 2019. Secretion of empty or complete hepatitis B virions: envelopment of empty capsids versus mature nucleocapsids. Future Virology 14(2): 95–105.

Malik, S.S., Azem-e-Zahra, S., Kim, K.M., Caetano-Anollés, G. and Nasir, A. 2017. Do Viruses Exchange Genes across Superkingdoms of Life? Frontiers in Microbiology 8.

Margolis, L. and Sadovsky, Y. 2019. The biology of extracellular vesicles: The known unknowns. PLOS Biology 17(7): e3000363.

Marijuán, P.C. and Navarro, J. 2022. The biological information flow: From cell theory to a new evolutionary synthesis. Biosystems 213: 104631.

Miller, W.B. 2013. The Microcosm Within: Evolution and Extinction in the Hologenome. Boca Raton, Florida, United States: Universal Publishers.

Miller, W.B. 2016. Cognition, Information Fields and Hologenomic Entanglement: Evolution in Light and Shadow. Biology 5(2): 21.

Miller, W.B., Enguita, F.J. and Leitão, A.L. 2021. Non-Random Genome Editing and Natural Cellular Engineering in Cognition-Based Evolution. Cells 10(5): 1125.

Miller, W.B., Reber, A.S., Marshall, P. and Baluška, F. 2022. Viral-Cellular Natural Engineering in Cognition-Based Evolution. Communicative and Integrative Biology, Accepted in Press.

Miller, W.B. and Torday, J.S. 2018. Four domains: The fundamental unicell and Post-Darwinian Cognition-Based Evolution. Progress in Biophysics and Molecular Biology 140: 49–73.

Mirzaei, K.M., Xue, J., Costa, R., Ru, J., Schulz, S., Taranu, Z.E. and Deng, L. 2021. Challenges of Studying the Human Virome—Relevant Emerging Technologies. Trends in Microbiology 29(2): 171–181.

Mizuno, C.M., Guyomar, C., Roux, S., Lavigne, R., Rodriguez-Valera, F., Sullivan, M.B., … Krupovic, M. 2019. Numerous cultivated and uncultivated viruses encode ribosomal proteins. Nature Communications 10(1): 752.

Mohsen, M., Gomes, A., Vogel, M. and Bachmann, M. 2018. Interaction of Viral Capsid-Derived Virus-Like Particles (VLPs) with the Innate Immune System. Vaccines 6(3): 37.

Moniruzzaman, M., Martinez-Gutierrez, C.A., Weinheimer, A.R. and Aylward, F.O. 2020. Dynamic genome evolution and complex virocell metabolism of globally-distributed giant viruses. Nature Communications 11: 1710.

Moniruzzaman, M., Weinheimer, A.R., Martinez-Gutierrez, C.A. and Aylward, F.O. 2020. Widespread endogenization of giant viruses shapes genomes of green algae. Nature 588(7836): 141–145.

Nasir, A. and Caetano-Anollés, G. 2015. A phylogenomic data-driven exploration of viral origins and evolution. Science Advances 1(8): e1500527.

Netherton, C.L. and Wileman, T. 2011. Virus factories, double membrane vesicles and viroplasm generated in animal cells. Current Opinion in Virology 1(5): 381–387.

Niehl, A., Peña, E.J., Amari, K. and Heinlein, M. 2013. Microtubules in viral replication and transport. The Plant Journal 75(2): 290–308.

Noble, D. 2020. Exosomes, gemmules, pangenesis and Darwin. pp. 487–501. *In*: L. Edelstein, J. Smythies, P. Quesenberry and D. Noble (eds.). Exosomes: A Clinical Compendium. Cambridge, Massachusetts: Academic Press.

Nolte-'t Hoen, E., Cremer, T., Gallo, R.C. and Margolis, L.B. 2016. Extracellular vesicles and viruses: Are they close relatives? Proceedings of the National Academy of Sciences 113(33): 9155–9161.

Novoa, R.R., Calderita, G., Arranz, R., Fontana, J., Granzow, H. and Risco, C. 2005. Virus factories: associations of cell organelles for viral replication and morphogenesis. Biology of the Cell 97(2): 147–172.

Ofir, G. and Sorek, R. 2018. Contemporary Phage Biology: From Classic Models to New Insights. Cell 172(6): 1260–1270.

Patton, J.T., Silvestri, L.S., Tortorici, M.A., Vasquez-Del Carpio, R. and Taraporewala, Z.F. 2006. Rotavirus Genome Replication and Morphogenesis: Role of the Viroplasm. pp. 169–187. *In*: Reoviruses: Entry, Assembly and Morphogenesis. Springer Berlin Heidelberg.

Rohwer, F. and Thurber, R.V. 2009. Viruses manipulate the marine environment. Nature 459(7244): 207–212.

Roossinck, M.J. 2008. Symbiosis, Mutualism and Symbiogenesis. pp. 157–164. *In*: Plant Virus Evolution. Berlin, Heidelberg: Springer.

Roossinck, M.J. and Bazán, E.R. 2017. Symbiosis: Viruses as Intimate Partners. Annual Review of Virology 4: 123–139.

Ryan, F. 2009. Virolution. New York, NY, USA: HarperCollins Publishers.

Ryan, F. 2019. Virusphere: From common colds to Ebola epidemics—why we need the viruses that plague us. HarperCollins UK.

Ryan, F.P. 2016. Viral symbiosis and the holobiontic nature of the human genome. APMIS 124(1–2): 11–19.

Sanjuán, R. 2017. Collective Infectious Units in Viruses. Trends in Microbiology 25(5): 402–412.

Sanjuán, R. and Thoulouze, M.-I. 2019. Why viruses sometimes disperse in groups†. Virus Evolution 5(1): vez014.

Schoelz, J.E. and Leisner, S. 2017. Setting Up Shop: The Formation and Function of the Viral Factories of Cauliflower mosaic virus. Frontiers in Plant Science 8(1832).

Serganov, A. and Patel, D.J. 2007. Ribozymes, riboswitches and beyond: regulation of gene expression without proteins. Nature Reviews Genetics 8(10): 776–790.

Shapiro, J. 2016. Nothing in Evolution Makes Sense Except in the Light of Genomics: Read–Write Genome Evolution as an Active Biological Process. Biology 5(2): 27.

Thoulouze, M.-I. and Alcover, A. 2011. Can viruses form biofilms? Trends in Microbiology 19(6): 257–262.

Trinh, J.T., Székely, T., Shao, Q., Balázsi, G. and Zeng, L. 2017. Cell fate decisions emerge as phages cooperate or compete inside their host. Nature Communications 8: 14341.

Villarreal, L.P. 2005. Viruses And The Evolution Of Life. Washington, D.C.: Amer Society for Microbiology(ASM) Press.

Villarreal, L.P. 2007. Virus-host symbiosis mediated by persistence. Symbiosis 44: 1–9.

Villarreal, L.P. and Witzany, G. 2010. Viruses are essential agents within the roots and stem of the tree of life. Journal of Theoretical Biology 262(4): 698–710.

Villarreal, L.P. and Witzany, G. 2013. Rethinking quasispecies theory: From fittest type to cooperative consortia. World Journal of Biological Chemistry 4(4): 79.

Villarreal, L.P. and Witzany, G. 2021. Social Networking of Quasi-Species Consortia drive Virolution via Persistence. AIMS Microbiology 7(2): 138–162.

Virgin, H.W. 2014. The Virome in Mammalian Physiology and Disease. Cell 157(1): 142–150.

Wang, Q., Guan, Z., Pei, K., Wang, J., Liu, Z., Yin, P., ... Zou, T. 2018. Structural basis of the arbitrium peptide–AimR communication system in the phage lysis–lysogeny decision. Nature Microbiology 3(11): 1266–1273.

Wilke, C.O. and Novella, I.S. 2003. Phenotypic mixing and hiding may contribute to memory in viral quasispecies. BMC Microbiology 3: 11.

Witzany, G. 2020. What Does Communication of Phages Mean? pp. 1–22. *In*: Biocommunication of Phages. Springer Cham.

Zhang, Y.-Z., Chen, Y.-M., Wang, W., Qin, X.-C. and Holmes, E.C. 2019. Expanding the RNA Virosphere by Unbiased Metagenomics. Annual Review of Virology 6(1): 119–139.

Chapter 10

Holobionts: A Consensual 'We'

All multicellular macro-organisms are holobionts as an intimate partnership between eukaryotic body cells and a co-aligned microbiome. As in all holobionts, our human microbiome is a vast confederation of trillions of associated bacteria, archaea, viruses, and fungi in us and on us. Bacterial cells alone outnumber our eukaryotic body cells by at least 1.3 to 1 by some estimates (Sender et al. 2016, Gilbert et al. 2018). The influence of the microbiome is not merely a matter of sheer numbers. The human microbiome is exceptionally diverse, with constituents from each of the four domains (Prokaryota, Archaea, Eukaryota, Virome). Among bacterial participants in the human gut microbiome, an estimated 1,000 bacterial species comprise at least 2,000,000 unique genes (Gilbert et al. 2018). Thus, bacterial genes alone outnumber our human genes, numbering approximately 22,000, by 100 to 1 (Bäckhed et al. 2005, Gilbert et al. 2018).

Similarly, the human virome is colossal, estimated in the many trillions per individual, not even accounting for the fact that the great majority of that virome is yet unknown, representing a type of 'dark matter' unable to be adequately sequenced with current techniques (Liang and Bushman 2021). Just as with our cellular microbiome, our companion virome is capable of a wide range of interactions with target cells. Our prior conceptualization of viruses as obligatory pathogens no longer suffices and has yielded a contemporary viewpoint of viruses as potentially valuable cellular partners in symbiogenesis alongside prokaryotics and unicellular eukaryotes (Roossinck 2015, Ryan 2020). Consequently, the virome contributes to flexible cellular responses to environmental stresses.

Over the last several decades, research has confirmed that the microbiome of holobionts has a substantial role in biological development based on extensive mutual dependencies (Kelly et al. 2017, Stiemsma and Michels 2018). Through that partnering contribution, microbiomes significantly contribute to holobionic phenotypes (Lynch and Hsiao 2019). For example, and as will be discussed in greater detail in the next chapter, our microbiome is essential for the proper function of our human gut (Cho and Blaser 2012), brain and central nervous system (Cryan and Dinan 2012), immune system (Postler and Ghosh 2017), and the regulation of bone metabolism (Hernandez et al. 2016). Further, accumulating data connects our microbiome to many phenotypic states and disease processes (Huitzil et al. 2018), including satiety (Lee et al. 2020), obesity (John and Mullin 2016), inflammatory bowel disease (Miyoshi and Chang 2017), cancer (Vogtmann and Goedert 2016), and even neurobiological disorders such as schizophrenia and autism (Nguyen et al. 2018, Pulikkan et al. 2019).

Indeed, there is no aspect of our metabolism that does not have a co-dependent interrelationship with our partnering microbiome. In particular, the microbiome of holobionts is a critical determinant of our vital innate and adaptive immunity throughout all life-cycle stages (Hooper et al. 2012). Correspondingly, when this reciprocal balance is displaced, there are organism-wide consequences experienced as varying degrees of metabolic dysregulation or outright disease. For instance, even in childhood, significant gut dysbiosis links to metabolic syndrome and obesity (Carrizales-Sánchez et al. 2021), altered brain function, and dysfunctional behaviors (Heijtz et al. 2011). Even viral bacteriophages contribute. For example, gut bacteriophages are associated with improved executive function and memory in flies, mice, and humans (Mayneris-Perxachs et al. 2022).

As well as serving in a coordinated physiological role, microbial balance at all body sites is crucial to our protection from pathogens and is, therefore, necessary for our survival. For example, bifidobacteria in our gut microbiome protect us from enteropathogenic infections such as Salmonella or norovirus through the production of acetate and other short-chain fatty acids (Lian et al. 2022).

Given the depth of these intimately interconnected interrelationships, multicellular eukaryotic organisms like ourselves can no longer be regarded as unitary beings. Instead, we must consider ourselves a cellular super-assemblage comprised of constituents from each of the four domains that combine into a functional, seamless 'superorganism' (Gilbert et al. 2010, Miller 2013, 2016a,b, 2018, Gordon et al. 2013, Chiu and Gilbert 2015, Torday and Miller 2020).

Research has confirmed that no part of the body is sacrosanct with respect to microbial contact at any stage of life. All body sites have at least some trace elements of a co-existing microbial presence, though nowhere as voluminous as our gut microbiome. For example, the testes have a dedicated microbiome (Altmäe et al. 2019). The pancreas does, too (Geller et al. 2017). The pancreatic microbiome closely relates to pancreatic functionality, and its disruption can lead to overt disease. Pointedly, fungal migration from the gut lumen to the pancreatic duct can drive a localized dysbiosis that promotes tolerogenic reprogramming of the immune system and potential induction of pancreatic ductal carcinoma (Aykut et al. 2019).

Even the developing fetus is not exempt. Studies have demonstrated DNA from oral and intestinal bacteria in placental samples, suggesting that there is horizontal and vertical transmission of bacterial DNA between mother and fetus that resides as a low-mass placental microbiome (Zakis et al. 2022). Current evidence indicates that there is active microbial transfer and communication in the fetal stage, leading to fetal microbial colonization in utero that establishes a low-level fetal microbiome (Mishra et al. 2021). By the second trimester, fetal organs have diverse bacterial genera that can be propagated and cultured, having a significant influence on priming the fetal immune system. It is now established that the education of the human immune system begins in utero via T cell activation and memory development that derives in part from trace microbial exposures and maternal sources (Parker et al. 2021).

Irrespective of its full extent, this fledgling fetal microbiome participates within a developmental arc that will eventually become a full neonatal microbiome. Its influence is great enough that it functions as a developmental organ of its own as a "virtual organ within an organ" (O'Hara and Shanahan 2006, Miller 2016a). Therefore, the fetal and neonatal periods represent critical stages of microbial aggregation and combination. Moreover, that microbial contribution serves as a full co-partner in holobionic growth and development rather than the mere appendages as previously assumed (Miller 2016a,b).

Within this new framework, Gilbert (2014) has suggested a comprehensive "re-tell" of the human birth narrative. Birth is not simply another individual of the species but is the certifiable origin of a new community of ecological associations. Birth should be re-imagined as a complex re-elaboration of an intricate set of overlapping eukaryotic cellular and microbial symbiotic relationships. Consequently, the arc from conception through pregnancy, fetal development, and birth are all parts of an orchestrated, coordinated architecture for the co-development of interlocking niche constructions as the "passage from one set of symbiotic relationships to another" (Gilbert 2014). Therefore, the co-apted microbiome of every holobiont must now be examined in terms of a life-long developmental arc, beginning at conception but continuously extending across future generations, meeting environmental stresses through collaborative cellular responses to epigenetic modifications as adaptive cellular problem-solving (Miller et al. 2020a, Torday and Miller 2020). Thus, holobionic life stretches continuously through invisible threads, established prior to conception, nurtured ingeniously in the womb, and extending "beyond individual life itself, as either epigenetic shadow or privilege into the generations beyond" (Torday and Miller 2020).

The term holobiont was first introduced to a wide audience in 1990 by Lynn Margulis, resolutely asserting against fierce resistance that organisms should be viewed as host-microbe units (O'Malley 2017). In 1994, Richard Jefferson introduced a hologenome theory of evolution at a presentation at Cold Spring Harbor Laboratory, underscoring the co-evolving symbiotic nature of eukaryotic life (Miller 2013). That concept was based on genetic research on glucuronide metabolism by vertebrate-associated microbes, which led to a concept of a holobiont as a genomic scaffold. Years later, Eugene Rosenberg and Ilana Zilber-Rosenberg skillfully articulated an expanded hologenome concept of evolution, arguing the actual object of natural selection extends beyond any host macroorganism to encompass its associated symbiotic community (Rosenberg et al. 2009, Rosenberg and Zilber-Rosenberg 2011). This inclusive hypothesis represented a crucial stepping-stone though it remained fundamentally adherent to the synergism theories of Maynard Smith and Szathmáry, firmly ensconced within traditional Neodarwinism (Suki 2012). Its major theoretical difference from standard theory was a consequential shift of the focus of selection beyond the confines of the central genome to encompass a more pluralistic bandwidth of selective influences on holobionts including its associated microbiome.

Chiu and Gilbert (2015) regard holobionts as superorganisms representing an assemblage of multiple species of persistent symbionts structured through "reciprocal scaffolding". Burgeoning research confirms these essential reciprocations between eukaryotic body cells and the microbiome of holobionts. Furthermore, the range and intimacy of these associations argue that superorganisms are more than constellations of confederated mixed cellular and viral ecologies but a fully integrated entirety assessed as something other than a sum of parts (Miller 2013, 2016a, 2018, Miller et al. 2020a, Torday and Miller 2020). Multicellular eukaryotes are vast consortia of collaborative, co-linked, cooperative, co-dependent, and competitive ecologies merged so harmoniously that they function together as a single discrete, seamless entity. Living organisms cannot be understood except when they are appreciated as an "organized whole" (Ho 2008). Consequently, it is directly asserted that the ingrained concepts of "host" and "guest" that are universally used to distinguish eukaryotic body cells from microbial companions require a thorough reconsideration. For each of its cellular constituents, holobionic life is its best means of protecting individual self-integrity as self-identity and individual homeorhetic equipoise (Miller and Torday 2018, Miller et al. 2019). Therefore, the notion of the combination of eukaryotic 'us' and microbial 'other' should be reappraised as a consistently negotiated, consensual 'we' (Miller 2016a,b, Miller et al. 2019). When so considered, it is evident that such an entity can only exist through common pathways of information sharing and communication across all linked cellular ecologies.

The respect for individual cellular identities within holobionts is foundational, enforced via sensitive immunological rules to maintain functional co-alignment. Accordingly, it can be appreciated then that immunology is the bioactive expression of the maintenance of 'self' versus 'other', tasked to safeguard self-referential individuality. Consequently, immunological order must be maintained at all scales, not merely to protect from pathogens but to maintain crucial self-identity, which is the cornerstone of the compact underpinning holobionic life. Therefore, in holobionts, immunology must be highly attuned to deliberately govern the balance of the diverse constituents that comprise its collaborating viral-cellular ecologies (Miller 2013, Gilbert 2014, Miller et al. 2020a).

Our contemporary realization that all cells from all domains are cognitive agents is crucial to this conceptual realignment. The integrated, nested cellular ecologies of holobionts are all populated by self-referential participants. Hence, the conventional and seemingly convenient concepts of facultative and obligate microbial symbionts also require adjustment (Moya et al. 2008, Łukasik et al. 2013). Our perception of the defining borders between microbial commensalists, mutualists, symbionts, latency, and pathogens will have to shift to accommodate this actual reality, most notably concerning the pervasive role of the virome. The prior notions of specific microbes as either all 'good' or entirely 'bad' will need to be updated, acknowledging that for many strains, their range of actions can be either beneficial or pernicious depending on context (Round and Mazmanian 2009, Jochum and Stecher 2020). In some immunological scenarios, these 'pathobionts' can provide essential reciprocal functions with eukaryotic body cells; in others, they might exhibit a detrimental health effect.

The benefits conferred by a holobiont's microbiome extend beyond immediate health. Growing evidence indicates that an attendant microbiome grants environmental plasticity as a key driver of phenotypic adjustment to stress (Schneider and Meyer 2017). For instance, in the freshwater crustacean *Dalphnia magna,* cyanobacterial tolerance is related to a combination of eukaryotic central genotype and gut microbiome composition. Thus, this phenotype is both vertically and environmentally acquired, with the microbial contribution representing a significant driver of adaptation (Hendry 2016). This type of intercommunication is surprisingly flexible and robust. As one facet of these mechanisms, bacteria release extracellular vesicles that serve as a common language between eukaryotic body cells and their originating bacteria, mediated by the transport of bioactive molecules, including proteins, nucleic acids, lipids, and metabolites (Ñahui Palomino et al. 2021). Indeed, the extent of this transkingdom communication is astonishing. Plants and animal cells communicate and collaborate through extracellular vesicles (exosomes), transporting mRNA, miRNA, bioactive lipids, and proteins as a form of cellular interspecies communication (Lefebre and Lecuyer 2017, Leitão et al. 2020).

Incredibly, investigations confirm that there are measurable biological effects from edible plant-derived exosome-like nanoparticles (EPDENs) on mammalian cells that affect mammalian gene expression (Mu et al. 2014). Exosomes ingested from ginger, carrot, grape, and grapefruit can be absorbed by macrophages and intestinal stem cells and can influence mammalian cell expression and behavior, thereby suggesting that the phrase 'you are what you eat' is more than just a trope. Undoubtedly though, this exchange of exosome-like nanoparticles present in common fruits and vegetables is direct plant communication with mammalian cells. Such exosome-mediated interspecies communication is present in all taxonomic kingdoms, particularly common among bacteria and fungi, and further, serves as one means of parasite-parasite intercommunication.

Indeed, this same mechanism may be the pathway for many known but as yet unexplained interspecies gene transfers. For instance, the whitefly *Bemisia tabaci* has managed to swipe a plant gene, enabling it to neutralize a toxin that some plants produce to defend against insect pests (Xia et al. 2021). Consequently, the fabric of holobionic life is specifically designed to facilitate this type of trans-kingdom cross-talk, enabling holobionts to remain in constant reciprocation with their environment to promote its consistent and continuous assimilation.

Others have noted that the framework of language that is employed to describe the complexity of holobionic life captures some truths and obscures others (Morar and Bohannan 2019). Competing metaphors are commonly used to describe the nature of the association of microbes with eukaryotic cells, such as 'organ' or 'second genome' or 'symbiont' or 'community'. Instead, the framework of cellular engineering helps to better illuminate the complexity of these holobionic relationships. Engineering is the collaborative use of information and resources by self-referential participants that can be directed toward common ends. It is well-accepted that humans should seek to manipulate our microbiomes and that of plants and animals for our betterment through bioengineering. And certainly, if there were no appreciable engineering at the level of the microbiome itself, then we humans, with our still limited tools, would be unable to do any successful bioengineering to serve our own ends. Surely, our current successes merely leverage long-standing biological engineering pathways (Foo et al. 2017).

Life, in all its stages, is a constant negotiation that both enables and requires the trading of resources among ecological participants. When cellular ecologies, holobionts, or superorganism systems are observed, our natural tendency is to frame them from our human viewpoint. However, all such activities are necessarily the endpoints of a succession of self-directed cellular decisions which are not overtly apparent to our human observations. A cellular planet need not problem-solve in any manner that suits our human ideas. Holobionic demise is such an example. Those microbes that disseminate from their partnering eukaryotic cells post-mortem have experienced that association and have incorporated their influences. Indeed, holobionic demise is actually the birth of new communities that extend beyond our own life cycle as our necrobiome (Burcham et al. 2019). Some portion of a reciprocating association continues forward through future microbial generations, which will have eventual repercussions within other holobionts. Thus, the function of a holobiont can be interpreted as part of the support system for the perpetuation of each of the three cellular domains and the virome across evolutionary space-time (Miller and Torday 2018).

It has been asserted that all multicellular life is an environmental prediction (Miller et al. 2019, Miller et al. 2020a). This should not be controversial insofar as that is the defining feature of engineering, and multicellularity is cellular engineering. No matter its scale, engineering is always a form of prediction as a productive means of dealing with problems and stresses. When each cell in a tissue ecology reacts in its individual interest as it is self-interpreted, natural cellular engineering, niche construction, and holobionts can now be appreciated as differing manifestations of conjoining cellular predictions. Furthermore, and crucially, any engineering prediction is predicated on measurement. Therefore, the multicellular form is the cellular means for the collective measuring assessment of available information from environmental cues to be deployed as engineering activities. Directly then, holobionts as an engineering product must be viewed as a biological expression of composite cellular prediction (Miller and Torday 2018, Miller et al. 2020a,b). The constituent cells of a holobiont are predicting the best conjoint pathways towards their sustenance which is directly related to their inferential appraisal of their future environment. Toward that aim, environmental changes are sensed down to the single cell level and are encoded as memory through changes in protein conformation space (Stan et al. 2019).

From within this better understanding of multicellularity, holobionts can now be reconceptualized as concordant cellular solutions to cellular stresses. As this living entity is guided by self-referential cognition and dedicated to the preservation of self-identity, multicellularity can now be correctly appraised as a networked informational interactome in which information has multiple dimensions across scales. Consequently, holobionts are cellular multidimensional problem-solving, representing a conjoining information space empowered by the fluidity of its shared communications (Miller 2016b, 2018, Miller et al. 2020a, Torday and Miller 2020).

Accordingly, to sustain this level of shared information and its communication across scales and among many species, it is obliged that there is a templating reference informational platform. As all cells from all species are self-referential yet genetically dissimilar, with large amounts of contributory non-nuclear genetic material, gaining access to integrated information and its consonant measurement requires a shared information space to allow seamless coordination. Consequently, there must be a shared N-space attachment where each cellular Pervasive Information field (PIF) can co-align with others of like kind and permit the abundant intercellular and interspecies communications that are now known to occur. The previously described N-space Episenome is that unity, serving as the critical element of the information management system of the holobiont from the zygotic conception forward (Miller et al. 2020b). Thus, the N-space Episenome functions at the zygotic level as a heritable spatiotemporal templating informational matrix that continues governing the precise holobionic cellular compact across that holobiont's lifespan.

That same informational template serves as a platform for sorting ambiguous environmental cues and random events of all kinds that can impact any living entity. The presence of this informational platform resolves the long-standing dilemma of the exquisite coordination of epigenesis in holobionts when many epigenetic impacts might be random or neutral to the organism. In holobionts, comprised of 'knowing' cells, noise within the system must be separated from random events that might be significant and productively assessed. This process has been designated as the " harnessing of stochasticity", whereby some of the welter of random events is productively channeled. In such a process, Noble and Noble (2018) maintain that organisms " ..... must employ a comparator to find the solution that fits the challenge". They further assert that the "harnessing of stochastic and/or chaotic processes is essential to the ability of organisms to have agency and make choices" (Noble and Noble 2018). The N-space Episenome informational architecture is that reference 'comparator' serving as a standard through which holobionts can measure short-term stresses against longer-term environmental requirements. In this manner, the N-space Episenome provides the means for the productive constraint and channeling of stochasticity to become an effective tool for cellular problem-solving (Miller et al. 2020b). N-space architecture solves the conundrum of controlling trillions of eukaryotic body cells and obligatory companion microbes to enable unified and co-linked tissue ecologies. Despite constitutive differences from others, each participant must be in service to the total organism. It could only do so through congruent measurement.

Pertinently, there is no need for any explicit endpoint to any informational architecture. That living framework serves continuous cellular adaptation through consistent amendment to sustain cellular equipoise versus environmental stresses. One means of conceptualizing how holobionts are situated within this informational matrix is through the theoretical framework of Bohm and Hiley (1975). In their organizational scheme, biological entities organize along a 'super-system---system---sub-system' structure that permits the compatible overlapping of the superpositions of possibilities that every entity within a complex network experiences.

Within this frame, individual and shared information spaces connect across many levels through multiple means, including quantum non-locality, so that all can work together as an interlinked ensemble. When so conceptualized, none of the individual constituents can be properly evaluated outside the whole. Any attempt at a reduction that ignores the entire scope of the network constitutes a loss of meaning. Critically too, within this framework, just as in all biology, there is no specifically privileged level of causation since all represent a continuous interactome (Saetzler et al. 2011, Nicholson 2014, Noble 2015, Miller 2016b). Thus, individual cells and their information spaces as individual PIFs link together within multicellular organisms.

These PIFs conjoin and overlap at the level of multicellular tissue ecologies and then across scales to encompass the entire holobiont in a unitary holobionic information space as its N-space Episenome. Thus, this comprehensive information architecture represents the multicellular organism's attachment to the outward N-space information space-time matrix that comprises its relevant encompassing environmental context.

References

Altmäe, S., Franasiak, J.M. and Mändar, R. 2019. The seminal microbiome in health and disease. Nature Reviews Urology 16(12): 703–721.

Aykut, B., Pushalkar, S., Chen, R., Li, Q., Abengozar, R., Kim, J.I., ... Miller, G. 2019. The fungal mycobiome promotes pancreatic oncogenesis via activation of MBL. Nature 574(7777): 264–267.

Bäckhed, F., Ley, R.E., Sonnenburg, J.L., Peterson, D.A. and Gordon, J.I. 2005. Host-Bacterial Mutualism in the Human Intestine. Science 307(5717): 1915–1920.

Bohm, D.J. and Hiley, B.J. 1975. On the intuitive understanding of nonlocality as implied by quantum theory. Foundations of Physics 5: 93–109.

Burcham, Z.M., Pechal, J.L., Schmidt, C.J., Bose, J.L., Rosch, J.W., Benbow, M.E. and Jordan, H.R. 2019. Bacterial Community Succession, Transmigration, and Differential Gene Transcription in a Controlled Vertebrate Decomposition Model. Frontiers in Microbiology 10: 745.

Carrizales-Sánchez, A.K., García-Cayuela, T., Hernández-Brenes, C. and Senés-Guerrero, C. 2021. Gut microbiota associations with metabolic syndrome and relevance of its study in pediatric subjects. Gut Microbes 13(1): e1960135.

Chiu, L. and Gilbert, S.F. 2015. The Birth of the Holobiont: Multi-species Birthing Through Mutual Scaffolding and Niche Construction. Biosemiotics 8(2): 191–210.

Cho, I. and Blaser, M.J. 2012. The human microbiome: at the interface of health and disease. Nature Reviews Genetics 13(4): 260–270.

Cryan, J.F. and Dinan, T.G. 2012. Mind-altering microorganisms: the impact of the gut microbiota on brain and behaviour. Nature Reviews Neuroscience 13(10): 701–712.

Foo, J.L., Ling, H., Lee, Y.S. and Chang, M.W. 2017. Microbiome engineering: Current applications and its future. Biotechnology Journal 12(3): 1600099.

Geller, L.T., Barzily-Rokni, M., Danino, T., Jonas, O.H., Shental, N., Nejman, D., ... Straussman, R. 2017. Potential role of intratumor bacteria in mediating tumor resistance to the chemotherapeutic drug gemcitabine. Science 357(6356): 1156–1160.

Gilbert, J.A., Blaser, M.J., Caporaso, J.G., Jansson, J.K., Lynch, S.V and Knight, R. 2018. Current understanding of the human microbiome. Nature Medicine 24(4): 392–400.

Gilbert, S.F. 2014. Symbiosis as the way of eukaryotic life: The dependent co-origination of the body. Journal of Biosciences 39(2): 201–209.

Gilbert, S.F., McDonald, E., Boyle, N., Buttino, N., Gyi, L., Mai, M., ... Robinson, J. 2010. Symbiosis as a source of selectable epigenetic variation: taking the heat for the big guy. Philosophical Transactions of the Royal Society B: Biological Sciences 365(1540): 671–678.

Gordon, J., Knowlton, N., Relman, D.A., Rohwer, F. and Youle, M. 2013. Superorganisms and holobionts. Microbe 8: 152–153.

Heijtz, R.D., Wang, S., Anuar, F., Qian, Y., Björkholm, B., Samuelsson, A., ... Pettersson, S. 2011. Normal gut microbiota modulates brain development and behavior. Proceedings of the National Academy of Sciences 108(7): 3047–3052.

Hendry, A.P. 2016. Key Questions on the Role of Phenotypic Plasticity in Eco-Evolutionary Dynamics. Journal of Heredity 107(1): 25–41.

Hernandez, C.J., Guss, J.D., Luna, M. and Goldring, S.R. 2016. Links Between the Microbiome and Bone. Journal of Bone and Mineral Research 31(9): 1638–1646.

Ho, M.-W. 2008. The Rainbow and the Worm: The Physics of Organisms (3rd ed.). Singapore: World Scientific Publishing Company.

Hooper, L.V., Littman, D.R. and Macpherson, A.J. 2012. Interactions Between the Microbiota and the Immune System. Science 336(6086): 1268–1273.

Huitzil, S., Sandoval-Motta, S., Frank, A. and Aldana, M. 2018. Modeling the Role of the Microbiome in Evolution. Frontiers in Physiology 9(1836).

Jochum, L. and Stecher, B. 2020. Label or Concept—What Is a Pathobiont? Trends in Microbiology 28(10): 789–792.

John, G.K. and Mullin, G.E. 2016. The Gut Microbiome and Obesity. Current Oncology Reports 18(7): 45.

Kelly, J.R., Minuto, C., Cryan, J.F., Clarke, G. and Dinan, T.G. 2017. Cross Talk: The Microbiota and Neurodevelopmental Disorders. Frontiers in Neuroscience 11(490).

Lee, C.J., Sears, C.L. and Maruthur, N. 2020. Gut microbiome and its role in obesity and insulin resistance. Annals of the New York Academy of Sciences 1461(1): 37–52.

Lefebvre, F.A. and Lécuyer, E. 2017. Small Luggage for a Long Journey: Transfer of Vesicle-Enclosed Small RNA in Interspecies Communication. Frontiers in Microbiology 8(377).

Leitão, A.L., Costa, M.C., Gabriel, A.F. and Enguita, F.J. 2020. Interspecies Communication in Holobionts by Non-Coding RNA Exchange. International Journal of Molecular Sciences 21(7): 2333.

Lian, S., Liu, J., Wu, Y., Xia, P. and Zhu, G. 2022. Bacterial and Viral Co-Infection in the Intestine: Competition Scenario and Their Effect on Host Immunity. International Journal of Molecular Sciences 23(4): 2311.

Liang, G. and Bushman, F.D. 2021. The human virome: assembly, composition and host interactions. Nature Reviews Microbiology 19(8): 514–527.

Łukasik, P., van Asch, M., Guo, H., Ferrari, J. and Charles J. Godfray, H. 2013. Unrelated facultative endosymbionts protect aphids against a fungal pathogen. Ecology Letters 16(2): 214–218.

Lynch, J.B. and Hsiao, E.Y. 2019. Microbiomes as sources of emergent host phenotypes. Science 365(6460): 1405–1409.

Mayneris-Perxachs, J., Castells-Nobau, A., Arnoriaga-Rodríguez, M., Garre-Olmo, J., Puig, J., Ramos, R., … Fernández-Real, J.-M. 2022. Caudovirales bacteriophages are associated with improved executive function and memory in flies, mice, and humans. Cell Host & Microbe 30(3): 340–356.e8.

Miller, W.B. 2013. The Microcosm Within: Evolution and Extinction in the Hologenome. Boca Raton, Florida, United States: Universal Publishers.

Miller, W.B. 2016a. Cognition, Information Fields and Hologenomic Entanglement: Evolution in Light and Shadow. Biology 5(2): 21.

Miller, W.B. 2016b. The Eukaryotic Microbiome: Origins and Implications for Fetal and Neonatal Life. Frontiers in Pediatrics 4(96).

Miller, W.B. 2018. Biological information systems: Evolution as cognition-based information management. Progress in Biophysics and Molecular Biology 134: 1–26.

Miller, W.B., Baluška, F. and Torday, J.S. 2020a. Cellular senomic measurements in Cognition-Based Evolution. Progress in Biophysics and Molecular Biology 156: 20–33.

Miller, W.B. and Torday, J.S. 2018. Four domains: The fundamental unicell and Post-Darwinian Cognition-Based Evolution. Progress in Biophysics and Molecular Biology 140: 49–73.

Miller, W.B., Torday, J.S. and Baluška, F. 2019. Biological evolution as defense of "self". Progress in Biophysics and Molecular Biology 142: 54–74.

Miller, W.B., Torday, J.S. and Baluška, F. 2020b. The N-space Episenome unifies cellular information space-time within cognition-based evolution. Progress in Biophysics and Molecular Biology 150: 112–139.

Mishra, A., Lai, G.C., Yao, L.J., Aung, T.T., Shental, N., Rotter-Maskowitz, A., … Ginhoux, F. 2021. Microbial exposure during early human development primes fetal immune cells. Cell 184(13): 3394–3409.e20.

Miyoshi, J. and Chang, E.B. 2017. The gut microbiota and inflammatory bowel diseases. Translational Research 179: 38–48.

Morar, N. and Bohannan, B.J.M. 2019. The Conceptual Ecology of the Human Microbiome. The Quarterly Review of Biology 94(2): 149–175.

Moya, A., Peretó, J., Gil, R. and Latorre, A. 2008. Learning how to live together: genomic insights into prokaryote–animal symbioses. Nature Reviews Genetics 9(3): 218–229.

Mu, J., Zhuang, X., Wang, Q., Jiang, H., Deng, Z., Wang, B., … Zhang, H. 2014. Interspecies communication between plant and mouse gut host cells through edible plant derived exosome-like nanoparticles. Molecular Nutrition & Food Research 58(7): 1561–1573.

Ñahui Palomino, R.A., Vanpouille, C., Costantini, P.E. and Margolis, L. 2021. Microbiota–host communications: Bacterial extracellular vesicles as a common language. PLOS Pathogens 17(5): e1009508.

Nguyen, T.T., Kosciolek, T., Eyler, L.T., Knight, R. and Jeste, D.V. 2018. Overview and systematic review of studies of microbiome in schizophrenia and bipolar disorder. Journal of Psychiatric Research 99: 50–61.

Nicholson, D.J. 2014. The machine conception of the organism in development and evolution: A critical analysis. Studies in History and Philosophy of Science Part C: Studies in History and Philosophy of Biological and Biomedical Sciences 48: 162–174.

Noble, D. 2015. Evolution beyond neo-Darwinism: a new conceptual framework. Journal of Experimental Biology 218(8): 1273.

Noble, R. and Noble, D. 2018. Harnessing stochasticity: How do organisms make choices? Chaos: An Interdisciplinary Journal of Nonlinear Science 28(10): 106309.

O'Hara, A.M. and Shanahan, F. 2006. The gut flora as a forgotten organ. EMBO Reports 7(7): 688–693.

O'Malley, M.A. 2017. From endosymbiosis to holobionts: Evaluating a conceptual legacy. Journal of Theoretical Biology 434: 34–41.

Parker, E.L., Silverstein, R.B. and Mysorekar, I.U. 2021. Bacteria make T cell memories *in utero*. Cell 184(13): 3356–3357.

Postler, T.S. and Ghosh, S. 2017. Understanding the Holobiont: How Microbial Metabolites Affect Human Health and Shape the Immune System. Cell Metabolism 26(1): 110–130.

Pulikkan, J., Mazumder, A. and Grace, T. 2019. Role of the Gut Microbiome in Autism Spectrum Disorders. *In*: P.C. Guest (ed.). Reviews on Biomarker Studies in Psychiatric and Neurodegenerative Disorders (Vol. 1118, pp. 253–269). Cham, Switzerland: Springer.

Roossinck, M.J. 2015. Move Over, Bacteria! Viruses Make Their Mark as Mutualistic Microbial Symbionts. Journal of Virology 89(13): 6532–6535.

Rosenberg, E., Sharon, G. and Zilber-Rosenberg, I. 2009. The hologenome theory of evolution contains Lamarckian aspects within a Darwinian framework. Environmental Microbiology 11(12): 2959–2962.

Rosenberg, E. and Zilber-Rosenberg, I. 2011. Symbiosis and development: The hologenome concept. Birth Defects Research Part C: Embryo Today: Reviews 93(1): 56–66.

Round, J.L. and Mazmanian, S.K. 2009. The gut microbiota shapes intestinal immune responses during health and disease. Nature Reviews Immunology 9(5): 313–323.

Ryan, F. 2020. Viral Symbiosis in the Origins and Evolution of Life with a Particular Focus on the Placental Mammals. pp. 3–24. *In*: Symbiosis: Cellular, Molecular, Medical and Evolutionary Aspects. Cham, Switzerland: Springer.

Saetzler, K., Sonnenschein, C. and Soto, A.M. 2011. Systems biology beyond networks: Generating order from disorder through self-organization. Seminars in Cancer Biology 21(3): 165–174.

Schneider, R.F. and Meyer, A. 2017. How plasticity, genetic assimilation and cryptic genetic variation may contribute to adaptive radiations. Molecular Ecology 26(1): 330–350.

Sender, R., Fuchs, S. and Milo, R. 2016. Revised Estimates for the Number of Human and Bacteria Cells in the Body. PLOS Biology 14(8): e1002533.

Stan, R.C., Bhatt, D.K. and Camargo, M.M. 2020. Cellular Adaptation Relies on Regulatory Proteins Having Episodic Memory. BioEssays 42(1): 1900115.

Stiemsma, L.T. and Michels, K.B. 2018. The Role of the Microbiome in the Developmental Origins of Health and Disease. Pediatrics 141(4): e20172437.

Suki, B. 2012. The Major Transitions of Life from a Network Perspective. Frontiers in Physiology 3.

Torday, J. and Miller, W. 2020. Cellular-Molecular Mechanisms in Epigenetic Evolutionary Biology. Cham, Switzerland: Springer.

Vogtmann, E. and Goedert, J.J. 2016. Epidemiologic studies of the human microbiome and cancer. British Journal of Cancer 114(3): 237–242.

Xia, J., Guo, Z., Yang, Z., Han, H., Wang, S., Xu, H., ... Zhang, Y. 2021. Whitefly hijacks a plant detoxification gene that neutralizes plant toxins. Cell 184(7): 1693–1705.e17.

Zakis, D.R., Paulissen, E., Kornete, L., Kaan, A.M. (Marije), Nicu, E.A. and Zaura, E. 2022. The evidence for placental microbiome and its composition in healthy pregnancies: A systematic review. Journal of Reproductive Immunology 149: 103455.

Chapter 11

The Role of the Microbiome in the Evolution and Development of Holobionts

"Nature... tends to repeat the same organs in the same number and in the same relations, and varies to infinity only their form. In accordance with this principle I shall have to draw my conclusions, in determining the bones of the fish's skull, not from a consideration of their form, but from a consideration of their connections".

—Étienne Geoffroy St-Hilaire (1807)

A. Reciprocal dependencies

Scientists continue to debate the relative importance of the microbiome in eukaryotic life. Many express skepticism that multicellular eukaryotic life should be viewed as consistent co-development and co-evolution between eukaryotic cells and their co-existent microbiomes (Douglas and Werren 2016, Morris 2018). Indeed, many biologists insist that microbes and their hosts maintain strict separate, and individual identities as they evolve (Madhusoodanan 2019). This misconception is rooted in a fundamental misunderstanding of the exact nature of cellular life and the suite of cellular faculties that enable holobionts as the exclusive form of multicellular eukaryotic life.

All biological variations must be first expressed at the level of individual cells, no matter their initiating source (Shapiro 2011, Miller 2016a). Given that, any variation that might yield evolutionary development must be further expressed at the level of the organized cellular ecologies in macroorganisms. Consequently, each of these cellular ecologies is a product of natural cellular engineering. Crucially in holobionts, the constituents of these cellular ecologies are collectives of differentiated eukaryotic cell types and mixed microbial species. As these tissue ecologies all function effectively, evolutionary development in holobionts must represent a series of concordant cellular reciprocations among the myriad participants of those cellular tissue ecologies (Miller 2016a, 2018, Miller et al. 2019).

It is now well-accepted that phenotype is being continually adjusted by corresponding interactions between eukaryotic cells, a partnering microbiome, and the external environment (Miller 2016a). Given this requisite cellular inter-relationship, the actual purpose of phenotypes in holobionts can be recast from typical assumptions. Phenotype is the product of natural viral-cellular engineering whose purpose extends well beyond the immediate support of the macroorganism (Miller et al. 2019). Phenotype is the primary active means by which the aggregate holobionic cellular apparatus acquires environmental experiences to permit continuous organismal-environmental complementarity (Torday and Miller 2016a). Since phenotypes result from coordinated natural cellular engineering by intelligent, measuring cells, phenotypes are an engineering output as conjoint measured environmental prediction (Miller et al. 2020).

Inarguably, all engineering is a prediction. Significantly though, in cellular terms, engineering predictions are the summation of the actions of discrete self-referential cells. Accordingly, phenotype can be correctly characterized as best predictions by cells of collaborative cellular strategies directed toward the continuous defense of their individual self-identities and homeorhetic preferential equipoise (Miller 2016a, 2018). Notably, these concerted cellular actions are predictions toward sustaining environmental complementarity. For this to be achieved and maintained, all cells contributing to any given phenotype must coordinate through congruous aims. In holobionts, those participating cells consist of eukaryotic body cells and an obligate companion microbiome. Therefore, all must participate and assist across all tissue ecological levels in acquiring epigenetic impacts as relative signatures of current environmental conditions. Thus, holobionic environmental complementarity is sustained by both a wide palette of microbes and varied eukaryotic differentiated cells. (Miller 2016a, Torday and Miller 2016a,b).

Since phenotype is the product of ecological engineering that involves participants from each of the four domains (Prokaryota, Archaea, Eukaryota, Virome), evolutionary variation in holobionts is a concordant process involving eukaryotic cells and the constellation of their obligatory and facultative microbiome interrelating over an extended timespan. However, just as the macrorganism must sustain environmental complementarity, the individual participating cells seek the maintenance of their cellular self-integrity and individual homeorhetic equipoise (Miller et al. 2019). This specific biological duality offers a congruent pathway to reconciling the contentions between those who maintain that microbes in holobionts remain strictly self-directed and others that insist that the direct object of selection encompasses the entire holobiont as a unit with its constituent microbiome. Both means are simultaneously active at all times.

The multitude of diverse multispecies microbial participants that characterize holobionts support the efficient coordination of organism-wide adjustments to environmental stimuli. The holobionic unit can only thrive if the conjoining equipoise of all of its obligatory participants is attended. Further, this environmentally responsive goal can only be met when representatives of each of the four domains are embedded as an inseparable part of the entirety as an 'organized whole'(Miller et al. 2019). This is the best means for multicellular eukaryotes to remain most attuned to environmental fluxes and flexible in their responses (Miller and Torday 2018).

Voluminous research has delved into the extent of the reciprocating dependencies between our holobionts eukaryotic cells and their microbiome. Our microbial partners are essential for the proper function of our gut (Zmora et al. 2019), immune system (Dominguez-Bello et al. 2019), and brain and central nervous system, including behavioral disorders (Bell et al. 2019, Bastiaanssen et al. 2020). Given the depth of these intimately interconnected relationships, multicellular eukaryotic organisms, such as ourselves, can no longer be regarded as a unitary being. Instead, holobionts are a cellular super-assemblage comprised of constituents from each of the four domains that combine into a functional, seamless superorganism. (Gilbert et al. 2010, Miller 2013, 2016a, Gordon et al. 2013, Chiu and Gilbert 2015, Miller 2018, Torday and Miller 2020).

It is still controversial whether the developing fetus has a constituent microbiome. For example, bacterial DNA can be identified in the umbilical cord blood, amniotic fluid, meconium, and fetal membranes in healthy fetuses, presumably by translocation from maternal circulation or ascending from the vagina (Stinson et al. 2017). Although this would not constitute a microbiome, these circulating bacteria might be contributory to fetal health and development (Jiménez et al. 2005, Hummel et al. 2022). Moreover, the presence of a microbiome in the normal placenta is still contentious, although it, too, can harbor circulating microbes (de Goffau et al. 2019, Zakis et al. 2022).

The argument about a specifically dedicated fetal microbiome is truly moot since the development of the fetus is highly influenced by the maternal metabolism, which is directly related to her constituent microbiome. That fetal *in utero* experience impacts post-natal health over a lifetime (Jenmalm 2017). Indeed, each developmental stage of the macroorganic life cycle has its specific microbial contributions (Miller 2016b). As Gilbert (2014) has asserted about our human selves, the birthing narrative is best conceived as the origin of a new community of reciprocating symbiotic relationships between eukaryotic cells and microbial life. Each stage serves as a set of preamble conditions for the subsequent development and reproduction of the next. As a prime example of this life-sustaining co-dependency, Gilbert stresses, "Some material in the mother's milk is for the bacteria and not the infant" (Gilbert 2014). The microbiomes of the breast, nipple, and areola thrive on the sugars in breast milk. The newborn immune system reacts to those maternal sugars, reciprocally enhancing the reproduction of specific sets of bacteria. Similarly, the gut microbiome of a mammal cannot be fully understood outside the context of its eukaryotic companions. Amato et al. (2018) used metagenomic sequencing to study 18 species of wild non-human primates, demonstrating that holobiont phylogeny is a more substantial influence on gut microbial composition and function than dietary niches (Amato et al. 2018).

B. All development is co-development

The context of a multicellular eukaryotic architecture is one of reciprocating developmental mutualisms through shared ecological niche constructions between its eukaryotic body cells and its co-aligned microbiome (Chiu and Gilbert 2015).

The microbiome acts to help shape the immune status of the eukaryotic cellular ecologies and forms an essential component of the protection against pathogens (Chiu et al. 2017). These actions are the product of a reciprocating assessment of immunological status that depends on coordinated environmental sensing. Further, the microbiome functions as an interactive community to influence reproduction and developmental transitions (Shropshire and Bordenstein 2016, Jaspers et al. 2019). When considering the full extent of the influences of the microbiome in development, the proper end-point does not rest on whether or not every single organ or tissue ecology has its own dedicated and discrete microbiome. The entire body's microbiome offers its contributions. Research confirms the extensive influences of circulating metabolites from the gut microbiome on non-local phenotypes such as the cardiovascular system (Peng et al. 2018) and brain development and cognitive functions (Cryan and Dinan 2012). Even a single microbial enzyme can have significant phenotypic effects. As noted in Chapter 6, deletion of a single gene in a mouse gut microbe creates hydrolase enzyme-deficient mice that gain less weight and have lower levels of circulating lipids and cholesterol in their bloodstream and liver compared to normal controls (Yao et al. 2018). Thus, a single microbial genetic alteration can induce a preference for the metabolizing of fats rather than carbohydrates. However, the complexity of holobionic connections goes further. This same deficiency leads to documented changes in circadian rhythm and immune pathways affecting the gut and liver.

Bacteria are not the only microbes participating in this process. The mycobiome has its own direct and substantial influences as part of complex ecological partnerships (Enaud et al. 2018). The exploration of these complex interconnections in health and disease has only recently begun in depth (Krüger et al. 2019). Research confirms that both methanogenic and non-methanogenic halophilic Archaea are essential inhabitants of the human gut microbiome, specifically metabolizing hydrogen produced by anaerobic fermentation of carbohydrates to produce methane to transform heavy metals that can be toxic to human eukaryotic and bacterial cells (Nkamga et al. 2017).

The virome is a vital intercessory of communication between these cellular domains and has been shown to have its own reciprocating beneficial symbiotic influences on our human health through complex associations (Roossinck 2015, Roossinck and Bazán 2017, Miller and Torday 2018, Mirahmadizadeh et al. 2019). Recent research has revealed that bacteriophages that reside in the gut microbiome affect cellular microbial metabolism with downstream effects on neuronal gene expression and cognition (Blackmer-Raynolds and Sampson 2022). These viral influences can be surprisingly subtle and far-ranging. Studies on flies, mice, and humans indicate that higher levels of Caudovirales and Microviridae bacteriophages in gut bacteria are associated with improved cognition, verbal memory, and executive function (Mayneris-Perxachs et al. 2022).

Many well-detailed symbiotic relationships between eukaryotes and microbes are essential for survival. For example, coral relies on micro-algal endosymbionts for nearly all their energy requirements (Jaspers et al. 2019). Through mutualistic partnering and conjoint engineering with symbionts, three-dimensional calcium carbonate skeletons become part of entire reef ecosystems. In aphids, *Buchnera aphidicola* provides critical amino acids (Zhang et al. 2018). *Wigglesowrtha glossinidia* in the mid-gut of Tsetse flies enable the biosynthesis of B vitamins not provided through their diet (Bing et al. 2017). The Hawaiian bobtail squid Euprymna *scolopes* relies on environmentally acquired *Vibrio fischeri* employed for the morphogenesis of specialized cells needed for thermoluminescence (Montgomery and McFall-Ngai 1994).

Recent research on hydra, a member of Cnidaria that includes corals, jellyfish, and sea polyps, is uncovering how organisms and their microbiome reciprocally shape one another. In hydra, a large number of neuropeptides regulate the composition of its microbiome (Augustin et al. 2017). On hydra surfaces, a neuropeptide, NDA-1, inhibits the growth of either Gram-positive bacteria or specific Gram-negative *Curvibacter* sp. colonizers. Reciprocally, the hydra nerve net has sensory and ganglion neurons that intersperse among epithelial cells and maintain the temporal variation of the hydra microbiome, reciprocally supporting its metabolism and cooperating productively with its neuroimmune system. As a further example, any animal, such as an insect, that wants to digest cellulose must rely on its partnering gut microbiome to turn cellulose into digestible sugars. Both the innate animal cells and their microbiome are in continual reciprocity. The insect supplies the cellulose as bacterial nutrition, and the bacteria provide nutritional substrates to the insect (Bray 2009).

Studies in mice that directly apply to human physiology indicate that gut microbes communicate abundantly with intestinal epithelial cells, informing them how to better match dietary components. For example, microbes will actively promote lipid absorption, actively shifting genetic activity that regulates lipid metabolism and fatty acid oxidation (Lickwar et al. 2022).

A decade ago, Gilbert et al. (2010) contradicted then prevailing opinion that symbiotic relationships were too unstable to be of significance in evolution. Yet, the coral-algae symbiosis that evolved 240 million years ago continues apace to this day. Noting many such instances, Gilbert et al. (2010) asserted that "almost all development is co-development" (Gilbert et al. 2010. p. 673). Many years later, and with a growing body of ample supporting evidence, there is a fuller understanding of the extent of microbial contributions across the entire arc of holobionic life. From this accumulating research, and by placing it appropriately within a cellular framework, this prior dictum can now be appropriately rephrased: "All development is co-development".

C. All evolution is co-evolution

In the prokaryotic realm, there is no doubt that all evolution is co-evolution. The fluidity of horizontal gene transfer among microbes is sufficiently extensive that conclusive species definitions among Prokaryota and Archaea have proved elusive (Achtman and Wagner 2008). Furthermore, there is no doubt that microbes are accomplished and efficient engineers. They pursue aggregate aims among diverse types of microbial participants and deploy many types of sub-specializations to match environmental stresses by co-engineering biofilms (Kumar et al. 2017). Nonetheless, there has been little impetus to assume that these same cellular characteristics might account for the evolution of multicellular eukaryotes in which microbes are substantial participants. To a large degree, this relates to a general Neodarwinian belief that evolutionary development is primarily due to random genetic mutations and that natural selection has primacy in evolution (Koonin 2009, Pigliucci 2017). Therefore, even as the consensus shifts toward accepting extensive co-development between microbes and eukaryotic cells in holobionts, this adjustment has not translated into any particular support for the contention that the evolution of multicellular eukaryotes is substantially related to affiliated microbes. One reason for this disconnection is the difference in time frames between development and evolution. Exploring how microbial influences affect eukaryotic life-cycle development through reciprocating effects on the metabolome is much more readily interrogated than the more difficult task of proving co-evolution between eukaryotic cells and a companion microbiome over an evolutionary time scale.

Such an impasse can be overcome by placing the framework of evolution in terms of cellular-molecular engineering in which all cells, whether eukaryotic or microbial, are self-referential agents. All the domains are represented in holobionts either through a dedicated organ-based microbiome or via significant, indirect contributions from the total organism's microbiome. Jaspers et al. (2019) emphasize that microbes are an essential aspect of evolutionary innovation in multicellular eukaryotes, noting that the evolution of their nervous systems derives as much for the control of microbes as it does for sensory inputs, muscle control for movement, or coordination. Indeed, the primary means of defense against some intestinal pathogens is the production of specific antimicrobial peptides. However, these same peptides stimulate the mobilization of specific cytokines, such as interleukin-18, from nerve cells alongside intestinal barrier cells that protect overall gut barrier integrity and further participate in active overall immune surveillance (Flayer and Sokol 2020). Consequently, innate immunity in holobionts is directed just as much to the recognition of beneficial microbes as to combat pathogens (Amato et al. 2018). Thus, those partnerships between an obligatory microbiome and correspondent eukaryotic cells yielding vital metabolites and neurotransmitters, such as serotonin, depend on co-engineering and mutualizing niche constructions to successfully enact and maintain such intimate and enduring relationships.

Undoubtedly, the same reciprocating interactions between microbes and eukaryotic cells must have existed across evolutionary space-time. It is now recognized that at least 30% of mammalian metabolites have a bacterial origin, and 37% of human genes have homologs in Bacteria and Archaea with an additional substantial fraction originating in unicellular eukaryotes (McFall-Ngai et al. 2013). Consequently, Stuart Kauffman's 1995 dictum deserves reinforcement: All evolution is co-evolution. A direct line of reasoning supports that proposition. In holobionts whose tissue ecologies are multispecies assemblages, all engineering is co-engineering in such cellular ecologies. Accordingly, co-engineering enacts multicellularity, manifesting as ubiquitous biofilms or multicellular eukaryotes as holobionts (Miller 2016a,b). Unquestionably though, a lengthy longitudinal span of co-engineering is co-evolution in action.

There is no mystery why co-engineering defines planetary life. Co-engineering is information management as the principle cellular strategy to uphold self-integrity, defined as its preferred state of homeorhetic equipoise for each self-referential participant (Miller 2018, Miller et al. 2019). Multicellularity is the cellular means of enabling the shared assessment and validation of ambiguous information about the external environment (Miller et al. 2020). Multicellular conjoining measurement among the different species constituents of tissue ecologies is that best means. Thus, holobionic tissue ecologies should be regarded as products of co-engineering and niche constructions among all ecological constituents.

It is no longer controversial that viruses have played a crucial role in evolutionary development, participating in critical evolutionary innovations like the mammalian placenta (Ryan 2009, 2010, Koonin 2009, Baluška 2009, Villarreal and Witzany 2010). Transposons and transposable elements of viral origin are now recognized as having had a substantial diverse role in neurobiological development, including helping maintain cell identity and initiating cellular novelty. For example, the origin of vertebral myelin resulted from a retroviral insertion (Ghosh et al. 2022).

It is now well-accepted that microorganisms, in general, have had an essential role in eukaryotic evolution. Previously this was based on their pathogenic actions that yielded selective culling. However, a more pluralistic viewpoint now emphasizes symbioses (Sandoval-Motta et al. 2017, Hutizil et al. 2018). For instance, the vast majority of bacterial strains in termite guts are specific to termites, substantiating a long ongoing co-evolutionary relationship (Hongoh 2010). There is also abundant evidence that there has been long-standing co-evolution of hominids with their microbiome to support their immune systems and development (Madhusoodanan 2019). Furthermore, the range of those influences is astounding. For example, regeneration in *Dugesia japonica* planarians is partly due to an endogenous commensal bacterium, *Aquitalea* sp. FJL05, that participates in the regulation of axial and head patterning by modulating transcription genes (Williams et al. 2020).

Multiple lineages of bacterial taxa predominant in the primate gut arose via cospeciation with humans, chimpanzees, bonobos, and gorillas over the past 15 million years (Moeller et al. 2014, 2016). Necessarily, such contributions must express through individual cells and cellular tissue ecologies as products of cellular ecological co-engineering. These interconnecting influences are complex and profound across scales. For example, chemical cues in the gut microbiome of honeybees define social group membership in honey bee ecologies (Vernier et al. 2020). In reciprocation, social dynamics constrain the shape of organismal microbiomes. Studies in primates that acquire gut microbial conspecifics from direct social contact and shared environmental exposures show that social group composition shapes gut microbial communities (Perofsky et al. 2021). However, and crucially, the primary means by which mammals acquire their gut microbiota is through maternal vertical transmission rather than horizontal transfers. In contrast, the incidence of pathogens is more likely from horizontal spread (Moeller et al. 2018).

As noted previously, bacteria in the human body swap genes at a higher rate than usually seen, estimated to be 30% more frequent than from outside of the human body (Jeong et al. 2019). Furthermore, those genes are being transferred across body parts, no matter their ecological tissue of origin. Boto et al. (2019) have found that human and other animal microbiomes are 'hot spots' for horizontal gene transfer. These transfers extend across the cellular domains among Prokaryota, Archaea, and Eukaryota. Recently, the discovery of tiny protein-coding sequences within the human microbiome has provided evidence of thousands of previously cryptic conserved gene families participating in cross-talk between the eukaryotic architecture of holobionts and their microbial constituents (Mittleman and Burstein 2019). Thus, the physiological and metabolic effects of an animal's microbiome are closely aligned with the evolution of the holobiont (Brooks et al. 2016). Large multicellular eukaryotes can evolve more quickly and efficiently through gene acquisitions from an affiliated microbiome, which can respond faster than eukaryotic cells to shifts in environmental conditions (Soucy et al. 2015, Alberdi et al. 2016). In that regard, studies in scyphozoan jellyfish ("true jellies") *Chrysaora plocamia* and *Aurelia aurita* found that a core microbial community could be identified in each of their varied life cycle states, including the mother polyp, dormant cysts, or polyps excyted from podocysts. Each phase functions as a key driver of nutrient and mineral cycling (Lee et al. 2018). Moreover, more than half of that core microbiome was shared across all life cycle stages of both species, indicating that their microbiome has significantly contributed to their 650 million years of evolutionary success.

Certainly, these types of inter-relationships are not exceptional to the animal kingdom. Plant microbiomes are critical to their health, development, and growth (Berendsen et al. 2012, Müller et al. 2016). Nor is there doubt that microbial associations have been critical to plant evolution. The mutualistic cooperative pattern between plants and nitrogen-fixing bacteria at the level of root nodules is well documented (Martínez-Hidalgo and Hirsch 2017). Studies demonstrate that the leaf microbiome contributes to plant diversity and ecosystem productivity (Laforest-Lapointe et al. 2017). Further, molecular evidence indicates that the association between mycorrhizal fungi and green algae was fundamental to the evolution of land plants, beginning 700 million years ago (Turner et al. 2013). That specific association still exists among many plant species.

In assessing the sweep of influences of the microbiome on eukaryotic evolution, it is estimated that only a fraction of the total bacterial component has been evaluated, and the full extent of the mycobiome and virome has been little explored (Zárate et al. 2017, Kong and Morris 2017, Richard and Sokol 2019). Even our most advanced tools for evaluating a microbiome have acknowledged limitations, particularly in assessing the qualitative contributions of those fractions of the microbiome that represented in relatively small populations (Quince et al. 2017). In particular, less common or less abundant species are commonly overlooked, particularly within the mycobiome (Huseyin et al. 2017). Yet, a low presence does not imply inactivity. The engineering of multispecies tissue ecologies in holobionts must necessarily involve all of its constituents. The fact that their presence remains cryptic within the microbiome does not diminish the possibility of their considerable influence.

There is continued resistance to viewing epigenetic acquisitions as a source of enduring heritability (Deichmann 2016). A salient issue had been the long-standing presumption that since the target of epigenetic marks are somatic cells, there is no ready mechanism for their participation in transgenerational inheritance (Surani

2016). However, recent research demonstrates abundant communication between somatic cells and the germline via epigenetic microRNAs, which can be propagated over generations and even permanently fixed (Sharma 2015a,b, Sharma et al. 2018). All regulatory RNAs released from somatic cells can reach the epididymis, permitting their uptake by spermatozoa through LINE-1-encoded reverse transcriptase activity (Villarreal and Witzany 2018). Genome-invading RNA networks are the most consequential of these epigenetic impacts, consisting of regulatory RNAs derived from infectious events (Villarreal and Witzany 2019). These can be inserted into or deleted from a target genome. Thus, epigenetic impacts, especially those emanating from the microbial sphere, have been demonstrated to shape the embryonic landscape and ultimately drive ontogenic changes and evolutionary outcomes (Spadafora 2018). Furthermore, those microbes most closely aligned within a eukaryotic network are most likely to participate in these instances (Boto et al. 2019).

D. A new narrative for multicellular life

From that accumulating evidence, it is now evident that focusing on nuclear genomes to determine co-evolution is unnecessarily limiting. Alternatively, the depth and length of co-relationships can be better estimated through an organism's metabolic repertoire and community dynamics than through phylogenetic relationships (Bauer et al. 2015, Mendes-Soares and Chia 2017, Tripathy and Pradhan 2018). Tipathy and Pradhan (2018) have suggested that there are dual fundamental co-evolutionary driving factors in holobionts. These are the principles of symbiotic purposive association and mutualized forms of competition. It is this combination that leads to "cooperation of gainful evolution". Yet, another critical principle should be added to better understand this dyad. Genes are tools of intelligent cells, and genes from microbial partners are similarly constructively deployed (Miller 2016a, 2018, Miller and Torday 2018, Miller et al. 2019).

Annila and Baverstock (2014) regard genes as a means of specifying polypeptides with natural selection acting on the flow of cellular and organismal free energy consumption in patterns that contribute to phenotype. Importantly, when energy flows are considered predominant, all cellular systems are interconnected, and genes do not have hierarchical primacy. Accordingly, phenotypes emerge from the network in their entirety, including its microbiome, through multilevel reciprocations. Annila and Baverstock (2014) propose that this relationship accounts for "missing heritability", explaining why specific matching correlations between genetic variations and complex traits for most human diseases have yet to be found.

How, then does selection operate in the context of co-evolution between eukaryotic cells and their constituent microbiome across evolutionary space-time? Chiu and Gilbert (2015) argue that the relationship between eukaryotic cells and their companion microbiome in holobionts is an instance of reciprocal scaffolding, developmental mutualism, and ecological niche construction. As an example of this complex and intimate relationship, it is not yet clear whether eukaryotic immunity is for the benefit of the eukaryotic cellular architecture or to permit its productive co-habitation by mutualist microbes (Gilbert 2014).

Contemporary evidence indicates that the collective partnership among constituents in multicellular eukaryotes is not focused on the level of their central eukaryotic genomes. Jaspers et al. (2019, p. 83) offer a definition of metaorganisms as "closely integrated ecosystems-specialized environments with community members that have direct and indirect impacts on one another". This perspective specifically recognizes that companionate life is based on the trading of information and ecological resources. As crucial as genetic phylogenies might be in determining the extent of the evolutionary influence of the microbiome on evolution, co-engineering centers on trading information and resources which has no necessary genetic footprint and is better traced through shared metabolic pathways.

Recent research directly confirms this perspective. An extensive examination of the composition of the constituents of the human gut microbiome was conducted, exploring their evolutionary relationship with us. Scientists at the Max Planck Institute for Biology discovered that gut microbes share a parallel evolutionary history with humans (Suzuki et al. 2022). Over 60% of gut microbial strains matched their evolutionary histories with humans through a pattern of consistent co-diversification over hundreds of thousands of years. These strains are heavily dependent on other constitutents in the human gut microbiome, possessing smaller genomes and greater sensitivity to oxygen and temperature levels than other microbial strains that are less co-evolved. The latter have metabolisms that are more like free-living bacteria. In contrast, the highly co-evolved gut microbial strains partner with us so intimately that they behave as if they are part of our human genome.

The advantage of the holobionic form is that the whole organism, including its cohabiting microbiome, can explore the informational content of environmental cues, traditionally termed its 'fitness landscape' (Zilber-Rosenberg and Rosenberg 2008, Rosenberg and Zilber-Rosenberg 2016, Henry et al. 2019). To serve that end, phenotype as a combination of eukaryotic cells and contributing microbial partners is the holobiont's means of exploring the environment (Torday and Miller 2016a). Further, even external environmental conditions, including the planetary microbiome, are part of that interconnected context. For example, plant leaves harbor highly diverse microbial communities, termed the phyllosphere (Vacher et al. 2016). That phyllosphere is influenced by eukaryotic cells of the plant, the plant microbiome, and the external environment, including atmospheric conditions and the local planetary microbiome composition (Vacher et al. 2016). In turn, the holobiont reciprocally influences the planetary environment in which it resides. The result is a pattern of mutualized cues that can be purposed toward self-directed niche construction activities as part of 'nested ecosystems' (Wimberley 2009).

The microbiome of sponges exerts significant metabolic influences on their companion eukaryotic cells. Sponge cells are modulated by that interchange and further affect the biogeochemical cycling of key nutrients like carbon, nitrogen, and phosphorus in the external environment (Pita et al. 2018). Similarly, plant microbiomes are crucial to the structure and function of forest ecosystems (Mishra et al. 2020). In reciprocation, that outward ecosystem has its reverberating effect on plant microbial phenotypes.

In such circumstances, no ecological level can be sensibly evaluated outside of its myriad connections. In the case of microbiomes, each of these microbial networks can be considered a microbial hub in which different taxa assume specific roles (Agler et al. 2016). These microbial hubs act as mediators between relevant eukaryotic architecture and its constituent microbiome as major links to phenotypic form and function (van der Heijden and Hartmann 2016). Nevertheless, these interconnections, enacted in concert as they are, can be seen as forms of mutualizing niche construction and co-engineering. Eukaryotic cells interact with microbial hubs in consistent reciprocation, transmitting information among tissue ecologies directed toward individualized cellular problem-solving. Therefore, plant and animal evolution can only be properly assessed within a context where both eukaryotic and microbial constituents play complementary roles.

Where, then, is the epicenter of environmental filtering selection in holobionts? Selection has obvious effects on the macroorganic form at the level of holobionts functioning as an effective reproductive unit. Filtering selection at this level assures total organismal-environmental complementarity. However, the true success of a holobiont depends on cohesive cellular action where each participant is a self-directed agent (Miller 2016a). Multicellularity represents coordinate co-engineering that yields effective tissue ecologies. In all multicellular eukaryotic organisms, those co-engineering constituents are both eukaryotic and microbial. Thus, selection is acting to assure that the vital particulars of that cellular ecological co-engineering always remain in concert with environmental strictures. Importantly too, any macroorganic elaboration as a holobiont must pass through a unicellular zygotic stage as an obligatory recapitulation. In consequence, selection must also be acting at that level.

Therefore, the environment, as filtering selection, operates at all levels of a holobiont along concurrent tracts of differing cellular timelines, each with its own extent of holobionic influence in which no level has absolute primacy. Yet, at all cellular levels, filtering selection is always post-facto to the basic cellular engineering that enables united tissue ecologies within the macroorganic form and permits their further variation (Miller et al. 2020). Fortunately, all this reduces. Selection assures that those cellular faculties persist that permit productive concordant cellular engineering. Only in this manner can the individual cellular participants' individual homeorhetic requirements be continuously protected against environmental exigencies.

Any engineering is crucially dependent on credible measurement. In such circumstances, the object of evolutionary selection further clarifies. *The primary object of filtering selection is the measuring cell.* The enduring success of cells entirely depends on valid measurements that enable useful predictions. Cohabitant life evolves from this particular. As a filtering process, selection promotes the consistent assimilation of the environment to assure continuous organismal-environmental complementarity at all successive cellular ecological levels, all of which are based on cellular predictions. Consequently, selection favors those accurate cellular measurements that sustain the holobionic spectrum in continuous support of the individual self-identities of all ecological participants from each cellular domain.

Accumulating evidence indicates that the prior viewpoint that the microbiome of holobionts is merely confined to matters of health and disease can no longer be supported. Instead, data from quantitative genetics, evolutionary biology, and community ecology indicate that the holobionic microbiome plays a critical role in the development, phenotypic variations, and the evolution of holobionts (Miller 2016a, Dominguez-Bello et al. 2019, Torday and Miller 2020, Miller et al. 2020, Henry et al. 2021).

References

Achtman, M. and Wagner, M. 2008. Microbial diversity and the genetic nature of microbial species. Nature Reviews Microbiology 6(6): 431–440.

Agler, M.T., Ruhe, J., Kroll, S., Morhenn, C., Kim, S.-T., Weigel, D. and Kemen, E.M. 2016. Microbial Hub Taxa Link Host and Abiotic Factors to Plant Microbiome Variation. PLOS Biology 14(1): e1002352.

Alberdi, A., Aizpurua, O., Bohmann, K., Zepeda-Mendoza, M.L. and Gilbert, M.T.P. 2016. Do Vertebrate Gut Metagenomes Confer Rapid Ecological Adaptation? Trends in Ecology & Evolution 31(9): 689–699.

Amato, K.R., G. Sanders, J., Song, S.J., Nute, M., Metcalf, J.L., Thompson, L.R., … R. Leigh, S. 2018. Evolutionary trends in host physiology outweigh dietary niche in structuring primate gut microbiomes. The ISME Journal 13(3): 576–587.

Amato, K.R., G. Sanders, J., Song, S.J., Nute, M., Metcalf, J.L., Thompson, L.R., … R. Leigh, S. 2019. Evolutionary trends in host physiology outweigh dietary niche in structuring primate gut microbiomes. The ISME Journal 13(3): 576–587.

Annila, A. and Baverstock, K. 2014. Genes without prominence: a reappraisal of the foundations of biology. Journal of The Royal Society Interface 11(94): 20131017.

Augustin, R., Schröder, K., Murillo Rincón, A.P., Fraune, S., Anton-Erxleben, F., Herbst, E.-M., … Bosch, T.C.G. 2017. A secreted antibacterial neuropeptide shapes the microbiome of Hydra. Nature Communications 8: 698.

Baluška, F. 2009. Cell-Cell Channels, Viruses, and Evolution. Annals of the New York Academy of Sciences 1178: 106–119.

Bastiaanssen, T.F.S., Cussotto, S., Claesson, M.J., Clarke, G., Dinan, T.G. and Cryan, J.F. 2020. Gutted! Unraveling the Role of the Microbiome in Major Depressive Disorder. Harvard Review of Psychiatry 28(1): 26–39.

Bauer, E., Laczny, C.C., Magnusdottir, S., Wilmes, P. and Thiele, I. 2015. Phenotypic differentiation of gastrointestinal microbes is reflected in their encoded metabolic repertoires. Microbiome 3: 55.

Bell, J.S., Spencer, J.I., Yates, R.L., Yee, S.A., Jacobs, B.M. and DeLuca, G.C. 2019. Invited Review: From nose to gut—the role of the microbiome in neurological disease. Neuropathology and Applied Neurobiology 45(3): 195–215.

Berendsen, R.L., Pieterse, C.M.J. and Bakker, P.A.H.M. 2012. The rhizosphere microbiome and plant health. Trends in Plant Science 17(8): 478–486.

Bing, X., Attardo, G.M., Vigneron, A., Aksoy, E., Scolari, F., Malacrida, A., … Aksoy, S. 2017. Unravelling the relationship between the tsetse fly and its obligate symbiont Wigglesworthia : transcriptomic and metabolomic landscapes reveal highly integrated physiological networks. Proceedings of the Royal Society B: Biological Sciences 284(1857): 20170360.

Blackmer-Raynolds, L.D. and Sampson, T.R. 2022. The gut-brain axis goes viral. Cell Host & Microbe 30(3): 283–285.

Boto, L., Pineda, M. and Pineda, R. 2019. Potential impacts of horizontal gene transfer on human health and physiology and how anthropogenic activity can affect it. The FEBS Journal 286(20): 3959–3967.

Bray, D. 2009. Wetware: A Computer in Every Living Cell. Yale University Press.

Brooks, A.W., Kohl, K.D., Brucker, R.M., van Opstal, E.J. and Bordenstein, S.R. 2016. Phylosymbiosis: Relationships and Functional Effects of Microbial Communities across Host Evolutionary History. PLOS Biology 14(11): e2000225.

Chiu, L., Bazin, T., Truchetet, M.-E., Schaeverbeke, T., Delhaes, L. and Pradeu, T. 2017. Protective Microbiota: From Localized to Long-Reaching Co-Immunity. Frontiers in Immunology 8(1678).

Chiu, L. and Gilbert, S.F. 2015. The Birth of the Holobiont: Multi-species Birthing Through Mutual Scaffolding and Niche Construction. Biosemiotics 8(2): 191–210.

Cryan, J.F. and Dinan, T.G. 2012. Mind-altering microorganisms: the impact of the gut microbiota on brain and behaviour. Nature Reviews Neuroscience 13(10): 701–712.

de Goffau, M.C., Lager, S., Sovio, U., Gaccioli, F., Cook, E., Peacock, S.J., … Smith, G.C.S. 2019. Human placenta has no microbiome but can contain potential pathogens. Nature 572(7769): 329–334.

Deichmann, U. 2016. Epigenetics: The origins and evolution of a fashionable topic. Developmental Biology 416(1): 249–254.

Dominguez-Bello, M.G., Godoy-Vitorino, F., Knight, R. and Blaser, M.J. 2019. Role of the microbiome in human development. Gut 68(6): 1108–1114.

Douglas, A.E. and Werren, J.H. 2016. Holes in the Hologenome: Why Host-Microbe Symbioses Are Not Holobionts. MBio 7(2): e02099–15.

Enaud, R., Vandenborght, L.-E., Coron, N., Bazin, T., Prevel, R., Schaeverbeke, T., … Delhaes, L. 2018. The Mycobiome: A Neglected Component in the Microbiota-Gut-Brain Axis. Microorganisms 6(1): 22.

Flayer, C.H. and Sokol, C.L. 2020. Nerves of Steel: How the Gut Nervous System Promotes a Strong Barrier. Cell 180(1): 15–17.

Ghosh, T., Almeida, R.G., Zhao, C., Gonzale, M.G., Stott, K., Adams, I., … Franklin, R.J. 2022. A retroviral origin of vertebrate myelin. BioRxiv.

Gilbert, S.F. 2014. A holobiont birth narrative: the epigenetic transmission of the human microbiome. Frontiers in Genetics, 5.

Gilbert, S.F., McDonald, E., Boyle, N., Buttino, N., Gyi, L., Mai, M., … Robinson, J. 2010. Symbiosis as a source of selectable epigenetic variation: taking the heat for the big guy. Philosophical Transactions of the Royal Society B: Biological Sciences 365(1540): 671–678.

Gordon, J., Knowlton, N., Relman, D.A., Rohwer, F. and Youle, M. 2013. Superorganisms and holobionts. Microbe 8(4): 152–153.

Henry, L.P., Bruijning, M., Forsberg, S.K.G. and Ayroles, J.F. 2019. Can the microbiome influence host evolutionary trajectories? BioRxiv (700237).

Henry, L.P., Bruijning, M., Forsberg, S.K.G. and Ayroles, J.F. 2021. The microbiome extends host evolutionary potential. Nature Communications 12: 5141.

Hongoh, Y. 2010. Diversity and Genomes of Uncultured Microbial Symbionts in the Termite Gut. Bioscience, Biotechnology, and Biochemistry 74(6): 1145–1151.

Huitzil, S., Sandoval-Motta, S., Frank, A. and Aldana, M. 2018. Modeling the Role of the Microbiome in Evolution. Frontiers in Physiology 9(1836).

Hummel, G.L., Austin, K. and Cunningham-Hollinger, H.C. 2022. Comparing the maternal-fetal microbiome of humans and cattle: a translational assessment of the reproductive, placental, and fetal gut microbiomes. Biology of Reproduction, 107(2): 371–381.

Huseyin, C.E., O'Toole, P.W., Cotter, P.D. and Scanlan, P.D. 2017. Forgotten fungi—the gut mycobiome in human health and disease. FEMS Microbiology Reviews 41(4): 479–511.

Jaspers, C., Fraune, S., Arnold, A.E., Miller, D.J., Bosch, T.C.G. and Voolstra, C.R. 2019. Resolving structure and function of metaorganisms through a holistic framework combining reductionist and integrative approaches. Zoology 133: 81–87.

Jenmalm, M.C. 2017. The mother-offspring dyad: microbial transmission, immune interactions and allergy development. Journal of Internal Medicine 282(6): 484–495.

Jeong, H., Arif, B., Caetano-Anollés, G., Kim, K.M. and Nasir, A. 2019. Horizontal gene transfer in human-associated microorganisms inferred by phylogenetic reconstruction and reconciliation. Scientific Reports 9(5953).

Jiménez, E., Fernández, L., Marín, M.L., Martín, R., Odriozola, J.M., Nueno-Palop, C., … Rodríguez, J.M. 2005. Isolation of Commensal Bacteria from Umbilical Cord Blood of Healthy Neonates Born by Cesarean Section. Current Microbiology 51(4): 270–274.

Kauffman, S. 1995. At Home in the Universe: The Search for the Laws of Self-Organization and Complexity. New York, NY, United States: Oxford University Press.

Kong, H.H. and Morris, A. 2017. The emerging importance and challenges of the human mycobiome. Virulence 8(3): 310–312.

Koonin, E.V. 2009. Darwinian evolution in the light of genomics. Nucleic Acids Research 37(4): 1011–1034.

Krüger, W., Vielreicher, S., Kapitan, M., Jacobsen, I. and Niemiec, M. 2019. Fungal-Bacterial Interactions in Health and Disease. Pathogens 8(2): 70.

Kumar, A., Alam, A., Rani, M., Ehtesham, N.Z. and Hasnain, S.E. 2017. Biofilms: Survival and defense strategy for pathogens. International Journal of Medical Microbiology 307(8): 481–489.

Laforest-Lapointe, I., Paquette, A., Messier, C. and Kembel, S.W. 2017. Leaf bacterial diversity mediates plant diversity and ecosystem function relationships. Nature 546(7656): 145–147.

Lee, M.D., Kling, J.D., Araya, R. and Ceh, J. 2018. Jellyfish Life Stages Shape Associated Microbial Communities, While a Core Microbiome Is Maintained Across All. Frontiers in Microbiology 9: 1534.

Lickwar, C.R., Davison, J.M., Kelly, C., Mercado, G.P., Wen, J., Davis, B.R., ... Rawls, J.F. 2022. Transcriptional Integration of Distinct Microbial and Nutritional Signals by the Small Intestinal Epithelium. Cellular and Molecular Gastroenterology and Hepatology 14(2): 465–493.

Madhusoodanan, J. 2019. Do hosts and their microbes evolve as a unit? Proceedings of the National Academy of Sciences 116(29): 14391–14394.

Martínez-Hidalgo, P. and Hirsch, A.M. 2017. The Nodule Microbiome: N 2-Fixing Rhizobia Do Not Live Alone. Phytobiomes Journal 1(2): 70–82.

Mayneris-Perxachs, J., Castells-Nobau, A., Arnoriaga-Rodríguez, M., Garre-Olmo, J., Puig, J., Ramos, R., ... Fernández-Real, J.-M. 2022. Caudovirales bacteriophages are associated with improved executive function and memory in flies, mice, and humans. Cell Host & Microbe 30(3): 340–356.e8.

McFall-Ngai, M., Hadfield, M.G., Bosch, T.C.G., Carey, H.V., Domazet-Lošo, T., Douglas, A.E., ... Wernegreen, J.J. 2013. Animals in a bacterial world, a new imperative for the life sciences. Proceedings of the National Academy of Sciences 110(9): 3229–3236.

Mendes-Soares, H. and Chia, N. 2017. Community metabolic modeling approaches to understanding the gut microbiome: Bridging biochemistry and ecology. Free Radical Biology and Medicine 105: 102–109.

Miller, W.B. 2013. The Microcosm Within: Evolution and Extinction in the Hologenome. Boca Raton, Florida, United States: Universal Publishers.

Miller, W.B. 2016a. Cognition, Information Fields and Hologenomic Entanglement: Evolution in Light and Shadow. Biology 5(2): 21.

Miller, W.B. 2016b. The Eukaryotic Microbiome: Origins and Implications for Fetal and Neonatal Life. Frontiers in Pediatrics 4(96).

Miller, W.B. 2018. Biological information systems: Evolution as cognition-based information management. Progress in Biophysics and Molecular Biology 134: 1–26.

Miller, W.B., Baluška, F. and Torday, J.S. 2020. Cellular senomic measurements in Cognition-Based Evolution. Progress in Biophysics and Molecular Biology 156: 20–33.

Miller, W.B. and Torday, J.S. 2018. Four domains: The fundamental unicell and Post-Darwinian Cognition-Based Evolution. Progress in Biophysics and Molecular Biology 140: 49–73.

Miller, W.B., Torday, J.S. and Baluška, F. 2019. Biological evolution as defense of "self". Progress in Biophysics and Molecular Biology 142: 54–74.

Mirahmadizadeh, A., Yaghobi, R. and Soleimanian, S. 2019. Viral ecosystem: An epidemiological hypothesis. Reviews in Medical Virology 29(4): e2053.

Mishra, S., Hättenschwiler, S. and Yang, X. 2020. The plant microbiome: A missing link for the understanding of community dynamics and multifunctionality in forest ecosystems. Applied Soil Ecology 145: 103345.

Mittelman, K. and Burstein, D. 2019. Tiny Hidden Genes within Our Microbiome. Cell 178(5): 1034–1035.

Moeller, A.H., Caro-Quintero, A., Mjungu, D., Georgiev, A.V., Lonsdorf, E.V., Muller, M.N., ... Ochman, H. 2016. Cospeciation of gut microbiota with hominids. Science 353(6297): 380–382.

Moeller, A.H., Li, Y., Mpoudi Ngole, E., Ahuka-Mundeke, S., Lonsdorf, E.V., Pusey, A.E., ... Ochman, H. 2014. Rapid changes in the gut microbiome during human evolution. Proceedings of the National Academy of Sciences 111(46): 16431–16435.

Moeller, A.H., Suzuki, T.A., Phifer-Rixey, M. and Nachman, M.W. 2018. Transmission modes of the mammalian gut microbiota. Science 362(6413): 453–457.

Montgomery, M.K. and McFall-Ngai, M. 1994. Bacterial symbionts induce host organ morphogenesis during early postembryonic development of the squid Euprymna scolopes. Development 120(7): 1719–1729.

Morris, J.J. 2018. What is the hologenome concept of evolution? F1000Research 7: 1664.

Müller, D.B., Vogel, C., Bai, Y. and Vorholt, J.A. 2016. The Plant Microbiota: Systems-Level Insights and Perspectives. Annual Review of Genetics 50: 211–234.

Nkamga, V.D., Henrissat, B. and Drancourt, M. 2017. Archaea: Essential inhabitants of the human digestive microbiota. Human Microbiome Journal 3: 1–8.

Peng, J., Xiao, X., Hu, M. and Zhang, X. 2018. Interaction between gut microbiome and cardiovascular disease. Life Sciences 214: 153–157.

Perofsky, A.C., Ancel Meyers, L., Abondano, L.A., Di Fiore, A. and Lewis, R.J. 2021. Social groups constrain the spatiotemporal dynamics of wild sifaka gut microbiomes. Molecular Ecology 30(24): 6759–6775.

Pigliucci, M. 2017. Darwinism After the Modern Synthesis. pp. 89–103. *In*: R.G. Delisle (ed.). The Darwinian Tradition in Context. Cham: Springer.

Pita, L., Rix, L., Slaby, B.M., Franke, A. and Hentschel, U. 2018. The sponge holobiont in a changing ocean: from microbes to ecosystems. Microbiome 6(1): 46.

Quince, C., Walker, A.W., Simpson, J.T., Loman, N.J. and Segata, N. 2017. Shotgun metagenomics, from sampling to analysis. Nature Biotechnology 35(9): 833–844.

Richard, M.L. and Sokol, H. 2019. The gut mycobiota: insights into analysis, environmental interactions and role in gastrointestinal diseases. Nature Reviews Gastroenterology & Hepatology 16(6): 331–345.

Roossinck, M.J. 2015. Move Over, Bacteria! Viruses Make Their Mark as Mutualistic Microbial Symbionts. Journal of Virology 89(13): 6532–6535.

Roossinck, M.J. and Bazán, E.R. 2017. Symbiosis: Viruses as Intimate Partners. Annual Review of Virology 4: 123–139.

Rosenberg, E. and Zilber-Rosenberg, I. 2016. Microbes Drive Evolution of Animals and Plants: the Hologenome Concept. MBio 7(2): e01395.

Ryan, F. 2009. Virolution. New York, NY, USA: HarperCollins Publishers.

Ryan, F. 2010. You are half virus. New Scientist 205(2745): 32–35.

Sandoval-Motta, S., Aldana, M. and Frank, A. 2017. Evolving Ecosystems: Inheritance and Selection in the Light of the Microbiome. Archives of Medical Research 48(8): 780–789.

Shapiro, J.A. 2011. Evolution: A View from the 21st Century. Upper Saddle River, NJ: FT Press.

Sharma, A. 2015a. Systems genomics analysis centered on epigenetic inheritance supports development of a unified theory of biology. Journal of Experimental Biology.

Sharma, A. 2015b. Transgenerational epigenetic inheritance: resolving uncertainty and evolving biology. Biomolecular Concepts 6(2): 87–103.

Sharma, U., Sun, F., Conine, C.C., Reichholf, B., Kukreja, S., Herzog, V.A., ... Rando, O.J. 2018. Small RNAs Are Trafficked from the Epididymis to Developing Mammalian Sperm. Developmental Cell, 46(4): 481–494.e6.

Shropshire, J.D. and Bordenstein, S.R. 2016. Speciation by Symbiosis: the Microbiome and Behavior. MBio 7(2): e01785.

Soucy, S.M., Huang, J. and Gogarten, J.P. 2015. Horizontal gene transfer: building the web of life. Nature Reviews Genetics 16(8): 472–482.

Spadafora, C. 2018. The "evolutionary field" hypothesis. Non-Mendelian transgenerational inheritance mediates diversification and evolution. Progress in Biophysics and Molecular Biology 134: 27–37.

Stinson, L.F., Payne, M.S. and Keelan, J.A. 2017. Planting the seed: Origins, composition, and postnatal health significance of the fetal gastrointestinal microbiota. Critical Reviews in Microbiology 43(3): 352–369.

Surani, M.A. 2016. Breaking the germ line–soma barrier. Nature Reviews Molecular Cell Biology 17(3): 136.

Suzuki, T.A., Fitzstevens, J.L., Schmidt, V.T., Enav, H., Huus, K.E., Mbong Ngwese, M., Grie'hammer, A., Pfleiderer, A., Adegbite, B.R., Zinsou, J.F. and Esen, M. 2022. Codiversification of gut microbiota with humans. Science 377(6612): 1328–1332.

Torday, J. and Miller, W. 2016a. Phenotype as Agent for Epigenetic Inheritance. Biology 5(3): 30.

Torday, J. and Miller, W. 2016b. The Unicellular State as a Point Source in a Quantum Biological System. Biology 5(2): 25.

Torday, J. and Miller, W. 2020. Cellular-Molecular Mechanisms in Epigenetic Evolutionary Biology Cham, Switzerland: Springer.

Tripathy, A. and Pradhan, R.K. 2018. Symbiotic Interactions, Law of Purposive Association and the+/+ Nature of all Co-evolution. Indian Journal of Science and Technology 11: 43.

Turner, T.R., James, E.K. and Poole, P.S. 2013. The plant microbiome. Genome Biology 14(6): 209.

Vacher, C., Hampe, A., Porté, A.J., Sauer, U., Compant, S. and Morris, C.E. 2016. The Phyllosphere: Microbial Jungle at the Plant–Climate Interface. Annual Review of Ecology, Evolution, and Systematics 47: 1–24.

van der Heijden, M.G.A. and Hartmann, M. 2016. Networking in the Plant Microbiome. PLOS Biology 14(2): e1002378.

Vernier, C.L., Chin, I.M., Adu-Oppong, B., Krupp, J.J., Levine, J., Dantas, G. and Ben-Shahar, Y. 2020. The gut microbiome defines social group membership in honey bee colonies. Science Advances 6(42).

Villarreal, L.P. and Witzany, G. 2010. Viruses are essential agents within the roots and stem of the tree of life. Journal of Theoretical Biology 262(4): 698–710.

Villarreal, L.P. and Witzany, G. 2018. Editorial: Genome Invading RNA Networks. Frontiers in Microbiology 9(581).

Villarreal, L.P. and Witzany, G. 2019. That is life: communicating RNA networks from viruses and cells in continuous interaction. Annals of the New York Academy of Sciences 1447(1): 5–20.

Williams, K.B., Bischof, J., Lee, F.J., Miller, K.A., LaPalme, J.V., Wolfe, B.E. and Levin, M. 2020. Regulation of axial and head patterning during planarian regeneration by a commensal bacterium. Mechanisms of Development 163: 103614.

Wimberley, E.T. 2009. Nested Ecology: The Place of Humans in the Ecological Hierarchy. Baltimore: Johns Hopkins University Press.

Yao, L., Seaton, S.C., Ndousse-Fetter, S., Adhikari, A.A., DiBenedetto, N., Mina, A.I., … Devlin, A.S. 2018. A selective gut bacterial bile salt hydrolase alters host metabolism. ELife 7: e37182.

Zakis, D.R., Paulissen, E., Kornete, L., Kaan, A.M. (Marije), Nicu, E.A. and Zaura, E. 2022. The evidence for placental microbiome and its composition in healthy pregnancies: A systematic review. Journal of Reproductive Immunology 149: 103455.

Zárate, S., Taboada, B., Yocupicio-Monroy, M. and Arias, C.F. 2017. Human Virome. Archives of Medical Research 48(8): 701–716.

Zhang, Y., Su, X., Harris, A., Caraballo-Ortiz, M.A., Ren, Z. and Zhong, Y. 2018. Genetic Structure of the Bacterial Endosymbiont Buchnera aphidicola from Its Host Aphid Schlechtendalia chinensis and Evolutionary Implications. Current Microbiology 75(3): 309–315.

Zilber-Rosenberg, I. and Rosenberg, E. 2008. Role of microorganisms in the evolution of animals and plants: the hologenome theory of evolution. FEMS Microbiology Reviews 32(5): 723–735.

Zmora, N., Suez, J. and Elinav, E. 2019. You are what you eat: diet, health and the gut microbiota. Nature Reviews Gastroenterology & Hepatology 16(1): 35–56.

CHAPTER 12

Four Domains and the Primacy of the Unicellular State

"From the paramecium to the human race, all life forms are meticulously organized, sophisticated aggregates of evolving microbial life. Far from leaving microorganisms behind on an evolutionary 'ladder,' we are both surrounded by them and composed of them."

Lynn Margulis and Dorian Sagan. Microcosmos, 1997

"The market demands that a trader follows all of her rules, every day, and every moment. Many just cannot thrive in this unbending universe."

Mike Bellafiore, trader

There are four domains among living systems: the cellular domains of Prokaryota, Archaea, Eukaryota, and the Virome, the latter constituted by viruses and associated subviral particles (Hug et al. 2016). Despite its wide diversity of observable macroscopic forms, Eukaryota is monophyletic with a lineage that extends back to an archaeal phylum 'Lokiarchaeota' and, therefore, can be correctly represented as one of these four major domains (Spang et al. 2015).

The date of the origin of the living cell can only be approximated. However, using multiple lines of evidence, the appearance of the 'last universal common ancestor' (LUCA) protocell extends back at least 3.7 billion years based on carbon isotope signatures in Eoarchean (Sousa 2021). Using new molecular clock analysis and isotope geochemistry, Betts et al. (2018) argue that ancestral cellular life appeared 3.9 billion years ago. Based on microfossils in seabed sediments, an estimate of first life has been extended, possibly back 4.28 billion years (Papineau et al. 2022). Eukaryota is regarded as the youngest of the domains, but even so, dates back 1.5–1.8 billion years. Moreover, its core genes derive from even older Archaea. Many eukaryotic attributes have archaeal origins, such as DNA folding patterns, genomic compaction, and deployment of histone proteins (Mattiroli et al. 2017). Further, homologs of eukaryotic actin and tubulin critical to cytoskeletal function are present in archeal ancestors (Yutin and Koonin 2012).

The virome is similarly ancient, with origins that remain highly debated, centering on whether the virosphere predates cellular life or is derivative (Forterre and Prangishvili 2009, Holmes 2011). Undoubtedly though, the virosphere has been a co-companion of cellular life for billions of years. Recently, the virosphere has undergone a reordering, establishing new subdivisions of increasingly diverse subviral particles at one end and giant viruses at the other. Many believe that giant viruses should be regarded as a separate domain of life in the new order, Megavirales (Colson et al. 2012). However, irrespective of these classifications, the virome represents the vital intercessory among the cellular domains (Ryan 2009, Miller and Torday 2018). Indeed, evolution can only be understood through the continuous, active interchange of the entire virome, including capsidless selfish viroids and small genetic elements with the three cellular domains. (Koonin and Dolja 2014, Villarreal and Ryan 2018).

Two related issues are not contentious. First, all cells of any origin are characterized by basal self-referential cognition (Shapiro 2007, 2011, Ford 2004, 2009; 2017, Baluška and Mancuso 2009, Trewavas and Baluška 2011, Bechtel 2014, Lyon, 2015, Baluška and Levin 2016, Miller 2016, 2018, Koseska and Bastiaens 2017, Keijzer 2017, Vallverdú et al. 2018, Reber 2019, Miller et al. 2020a, Baluška et al. 2021). Secondly, all multicellular life, whether expressed as biofilms or holobionts, is sustained through collective associations among these four domains (Miller 2013, 2016, 2018, Richter and King 2013, Cai et al. 2015, Torday and Miller 2016a, Miller and Torday 2018).

All cellular life has been definitely defined by cooperative patterns for billions of years. Microfossils that closely match contemporary microbial mats and stromatolites have been dated at least as far back as 3.5 billion years (Noffke et al. 2013). Notably, single, widely dispersed cells cannot leave a fossil trace. Any cell-based microfossils constitute empirical evidence of collective life since our estimate of the origin of life hinges on evidence of collective cellular action that is sufficiently robust to leave a fossil trace. Thus, communal life is ancient. Furthermore, highly complex multicellularity is also vastly old. Recent research indicates that fungi may be a billion years old, 500 million years older than previously thought (Loron et al. 2019). Accordingly, collaborative multicellular metazoan life originated much earlier than initially estimated and may actually be older still since fossil evidence of that period is so scarce. Consequently, for billions of years, biological and evolutionary development has been the continuous narrative of the perpetual entanglement and interchanges among the four permanent domains.

Pertinently, though less widely appreciated, all biology is always centered at the level of self-reference experienced at the individual cellular level. All biological expression of all kinds necessarily emanates from cellular processes since all bioactive processes are deployed within cells. Therefore, although multicellular life is a collaborative exercise, the epicenter of action always materializes from its individual cellular architectural participants (Miller 2016, 2018, Torday and Miller 2016a, Miller and Torday 2018, Miller et al. 2020a,b). Derivative to this requisite, the collaborative multicellular form of life must serve to protect each of its self-aware cognitive constituents that comprise its organized cellular tissue ecologies. It follows that both unicellular and multicellular planetary evolution must now be construed as always dedicated to the perpetuation of the four domains. Only these four basic forms of life are perpetual.

In comparison, species holobionic life forms are fleeting, lasting a few millions of years compared to billions for the cellular domains. Indeed, it can now be recognized that continued cellular evolutionary success in the face of a constant barrage of environmental stresses is due to its mandate to engage in a collaborative holobionic form. The self-reinforcing, cognitive cellular framework that permits holobionic architecture encourages the interdomain transfers essential to continuous cellular assimilation of the external environment (Miller 2018).

As previously emphasized, multicellular life depends on a fundamental physical linkage implicit in the self-referential state. Triadic energy-information-communication impels an obliged work channel that is productively deployed as collaborative engineering. That collective engineering effort is sustained by identifiable self-reinforcing, elemental cellular rules: cooperation, collaboration, co-dependence, and, typically, mutualizing competition. However, there are other consequential stipulations that consistently impact the dynamics of multicellular life and permit it to flourish over evolutionary space-time as a self-organizing informational interactome.

First among these is the basic principle of reiteration. From a structural perspective, all collective life, from level to level, is self-similar patterning, likened to Russian nesting dolls (Agnati et al. 2009). All of the basic rules of life are reiterated in self-similar patterns, level by level. In collective life, these rules first express at the cellular-molecular level, are re-enacted within biofilms, serve as the governing rules in the localized tissue ecologies of holobionts, and then re-express idiosyncratically among macro holobionts as their societal rules and norms. Across all biology, consistent patterning is evident as reiterative self-consistent fractal or mosaic formulations underlying the observable self-consistency in the way living systems interact (Agnati et al. 2009). For all biology, the same elements differentially assemble to assume distinct, separate biological outcomes, i.e., as reiterating fractals or mosaics. Although emergent properties might arise within these specific morphologically varied structures, the underlying fundamental structural patterning is the same. Thus, a 'Principle of Biological Attraction' can be identified: "the assembly of basic biological units into a more complex system is navigated by an inherent drive for spontaneous attraction and merging of lower order complementary biological units/systems to generate higher organization level units/systems" (Agnati et al. 2009, p. 553). This biological force can be considered analogous to gravitational pull, with each organism creating its own attractional field that draws others toward it as a type of intercellular glue.

All of these biological phenomena must also have another dimension. Any physical manifestation of biological attractional effects must have its mirroring representation in information space-time. Consequently, the reiterative architecture of subsystems as individual cellular Pervasive Information Fields, systems-level collective multicellular N-space Episenomes, and its further collective reiteration as super-systems of co-engineering holobionts, and then on to encompassing N-space are all underscored through the fabric of information (Miller et al. 2019, 2020b). This overlap is frankly mandated. Self-reference, which defines the living state, is information-dependent. Further, self-reference is self-organization. Accordingly, the intercellular 'glue' that binds energetic or bioactive exchanges as physical phenomena is the biological expression of those consistent rules that permit the emergence of forms from the N-space information space-time matrix that governs living systems. Consequently, the basis of biological attraction can now be productively understood as the inherent cellular impulse to share self-referential information to validate environmental cues to achieve mutualized preferential states.

From the above, it is clear that self-referential information space-time is a scale-free informational architecture having its own intrinsic attractional effect. Holobionic multicellularity is driven by the collaborative assessment and deployment of cellular information through the productive coexistence of representatives of the four domains as mutualizing co-participants in a complex multi-layered system. Notably, though, each cell is its own information management system. Each self-referential cell seeks the highest levels of informational validity in the face of ambiguous environmental information, attempting to maximize effective information (*EI) to achieve energy efficiency and preferential homeorhetic balance (Miller 2018). Cells know their informational security is best achieved within the intimately entwined holobionic complexities that epitomize the multicellular eukaryotic form. Since every cellular deployment of its resources is always a prediction, it can now be appreciated that the collective multicellular, multi-domain holobionic form is the cellular solution to winnowing information space-time to permit a more manageable range of predictions. That improvement in information quality permits a minimization of variational free energy in support of its living systems enabling the effective restriction to a limited number of states through active inference (Friston et al. 2006). Consequently, the four domains work collaboratively as a living spatio-temporal informational matrix in which the patterning of information space supports maximal predictability in space-time.

As noted previously, the natural cellular engineering and its attendant niche constructions that enable multicellular eukaryotic life are products of an underlying information cycle. Indeed, this is the practical physical expression of the attractional effects of information. The information cycle triggers a correlative self-reinforcing engineering cycle in the self-referential frame. (Miller and Torday 2018). Its yield is holobionts.

The Principle of Optimality is an additional governing biological tenet (Igamberdiev and Shklovskiy-Kordi 2017). Beginning from a specific state, the initiating decisions that are made to solve the first problems become the framework of the optimal future path. Accordingly, the optimal path is the one in which all the remaining decisions remain in conformity with the basic rules established with the first decisions. Thus, solutions to any resulting problems must also be co-respondent with solutions applied to the initial problem. In other words, straying from first principles will ultimately and inevitably lead to sub-optimal results. Notably, this principle, which has been directly applied to computer programming, equally adapts to problem-solving organisms despite the differences in how they use information. Applying the Principle of Optimality to biological systems does much to explain the success of the holobionic form of life. The cellular search for an optimal path necessarily extends through a continuous reiteration of those initiating, elemental first principles that began the path to multicellularity and the mutual associations that enabled holobionts. This fundamental principle helps us understand why the basic cellular rules that permitted the first iterations of multicellularity have remained in force for billions of years.

With this background, the reason for the obligatory recapitulation of all eukaryotic multicellular organisms through a unicellular zygotic stage clarifies. The zygotic stage resets the cell to conform to the rules that must govern the entire life cycle of the newly conceived holobiont, thereby preventing any damaging skew. The egg and sperm have each experienced a long series of environmental impacts prior to their joining together for fetal conception. Each has acquired various epigenetic markers that can alter subsequent embryogenesis. It may not matter much in just one generation, but it could lead to fatal deviations over many successive ones. The zygotic state assures a reset to a fundamental launching baseline that permits essential multi-scalar resonances, enabling consistent cellular re-elaborations as holobionts. In essence, a set of cellular 'first principles' must be reinforced reiteratively to counteract obligatory drift that would occur within a biological system that requires continuous adaptations, especially in cell-cell communication (Miller and Torday 2018).

Among these crucial resonances that require resetting, reciprocation dominates biologic action, mitigating the perpetual balance among the four domains that have successfully co-existed on this planet over eons. Pertinently, the essence of cooperation is reciprocation. Undoubtedly then, the continued success of life obliges an intermittent re-centering phase to assure that the first governing principles are being perpetually respected to ensure that essential reciprocations continue to operate. The unicellular zygotic stage serves as the continuous focal point for rebalancing the eukaryotic architectural form to prevent an untoward drift from basic reiterating principles. It can be considered akin to the required rebooting of all computer systems to maintain optimal function. Absent that, a gradual deteriorating divergence would inevitably occur from cumulative short-term environmental impacts (Torday and Miller 2016a,b, Miller 2016, 2018). Therefore, the zygotic stage resets the critical rebalancing of long-term versus short-term requisites for continued living success, deconvoluting complex, interconnected physiological pathways back to their unicellular origins in conformity with established first principles (Torday and Miller 2016a).

In this regard, the unicellular zygote has a crucial role in the adjudication of epigenetic impacts. A remarkable degree of remodeling occurs just after eukaryotic fertilization, which restores totipotency within the zygotic form (Eckersley-Maslin et al. 2018). This process involves DNA demethylation of some epigenetic marks and not others, chromatin remodeling, genome spatial reorganization, and a range of transcriptional alterations. This orchestrated process is now known as the maternal-to-zygotic transition, elaborating within the unicellular form and continuing across early embryogenesis (Eckersley-Maslin et al. 2018). Its purpose is to assure that the re-elaborated holobionic form is both an adequate 'present' and a long-term dynamical forecast of future environmental trends.

Statistical studies in human finance help to understand the principle behind an obligatory unicellular recapitulation in multicellular holobionic life. Research reveals that financial trading success is not directly correlated with the highest forms of intelligence. Often, those individuals are overconfident in their trades and engage in a pattern of excessive over-commitment (Lowenstein 2000). This impulse is why hedge funds often collapse, and less staid unleveraged mutual funds rarely do. The same pattern is exhibited in basic biological systems, confirmed by research on bacterial populations in fitness landscapes. Those experiments demonstrate a Tortoise-Hare adaptive fitness pattern (Nahum et al. 2015). Bacterial migration patterns reveal that those populations that initially start crossing a fitness landscape fastest ultimately do worse than initial slow movers. The initial fast movers are the quickest to adapt to local conditions but invariably meet an ensuing set of conditions to which they are less robustly equipped to deal with these new pressures since they over-adapted to the earliest stresses (Nahum et al. 2015). They become trapped within high amplitude fitness 'hills' from overly rapid adaptations. The fastest initial movers become increasingly less flexible as they move over the entire fitness landscape than the initially slow migrators. This group had more time to explore potential adaptations and, by doing so, ultimately make better choices. In effect, the rapid migrators skewed from optimal long-term patterns of adaptation by over-emphasizing short-term stimuli.

This pattern is the same snare that afflicts traders that test potentially profitable financial strategies through back-testing within narrow time frames, leading to backtest-overfitting. The derived results might look good on paper, but the nature of essentially chaotic inputs is that they do not follow the same pattern of progression in the future. When any model is developed that deliberately seeks to optimize returns over a confined sample period, it suffers from being inadequately flexible to meet the next unpredictable series of progressions (Bailey et al. 2014). Those trading systems that are too optimally fit within any sample back-test period have been shown to significantly fail as future price action unfolds, inadvertently leading to a systematic elevation of risk (Lopez de Prado 2013). What has been overlooked in the testing is the non-uniform distribution of random price data and its impact on future price fluctuations. The reassessment and re-centering functions of the unicellular zygote are precisely designed to deal with the requirement to balance short-term adjustments versus long-term adaptations within the context of a stream of chaotic and only partially predictable future inputs.

Experienced traders learn to accommodate this form of systematic risk by bet-hedging, taking the biggest risks off the table, and making smaller commitments. Bet-hedging is also how the unicellular zygote behaves, removing outlier epigenetic marks (Miller and Torday 2018). Consequently, when information is profoundly linked to survival, and epigenetic environmental cues are valuable information, there will have to be a robust system for their productive deployment. Although the adjudication of heritable expression or suppression of epigenetic marks centers within the zygotic unicell and its immediate embryological after-states, it should not be overlooked that this biological process reverberates across the integrity of the entire organism's information management system (Miller et al. 2019). Even suppressed epigenetic marks represent vital environmental cues as potential solutions to future stresses. Unexpunged but suppressed epigenetic marks constitute stored information as a form of environmental memory. In some circumstances, that suppressed epigenetic capacity can become its future.

With this background, we can now see that the traditional concept of phenotypes has to be critically rethought. Macroorganic phenotype serves a purpose beyond metabolic or physical support of the holobiont. Phenotype is the primary active means by which the aggregate holobionic cellular apparatus acquires epigenetic environmental experiences to permit its continuous organismal-environmental complementarity (Torday and Miller 2016a, Miller and Torday 2018). Since phenotypes result from coordinated natural cellular engineering by intelligent, measuring cells, it follows that phenotypes are conjoint measured environmental predictions (Miller et al. 2020a). In a framework of cellular cognition and co-engineering, phenotypes are cellular products representing forecasts and anticipation, just as in human engineering (Miller et al. 2020a,b). Significantly though, all phenotypic outputs are the aggregate of individual cellular actions to maintain individual states of preferred homeorhesis (Miller et al. 2019, 2020a). Therefore, the responses to epigenetic modulations are not inadvertent or random. Instead, phenotypic adjustments and evolutionary variations are continuous cellular problem-solving through collaborative natural engineering and mutualized niche construction (Miller 2018, Miller and Torday 2018, Miller et al. 2020a, Torday and Miller 2020a). As De Loof (2015) asserts, all biology is problem-solving. There need be no design, simply conjoining aims. Phenotypic adaptation is an emergent property of consensual cellular problem-solving. And further, it can also now be understood that phenotypes at all scales are collaborative predictions.

Notably, the holobionic microbiome is a purposeful co-respondent within phenotypic outcomes, serving as another cellular means of exploring the external environment to continuously secure successful environmental assimiliation. The direct purpose of phenotype is to extend outward to 'taste' the environment and return its informational cues to the unicellular zygote. Certainly, then, as reciprocating co-participants in both that outer environment and within companionate holobionts, the microbiome establishes independent and co-dependent linkages in this continuous process of constant reciprocation. Multicellular eukaryotic life thrives by having these two co-dependent and corresponding linked moieties, which necessarily contain constituents of each of the four domains, each responding to environmental stresses somewhat differently and, more importantly, along differing time scales. In this manner, the unicellular zygote receives a revenue stream of environmental information, diversified by source, measured impact, and time frame.

Any success within this complex framework is clearly dependent on fluid cell-cell communication and the further participation of the virome. Thus, the full extent and varieties of transkingdom genetic information and communication cross-talk described in the previous chapter are no longer surprising. On the contrary, it represents one more iterative manifestation of the essential interconnectivity among the four domains.

One other issue requires emphasis. Variations are as much the products of constraints as liberties. Constraints are present at every level and reinforced through the principle of biological reiteration. At the most basic self-referential level, the simple reception of information is work done across an energy gradient, becoming an obliged communication to some other self-referential participant. This binding linkage represents a constitutive constraint because that inescapable communication inevitably initiates a reiterative work channel (Deacon 2011, Miller and Torday 2018). Further, this same constraint institutes a first-level reciprocation: "every constraint is information in a referential frame" (Miller and Torday 2018).

At the scale of entire holobionts, all reciprocations and reiterations are expressions of underlying restrictions within nested, self-similar patterns. These are, in turn, subject to the constraining activities of the continuous reinforcement of elemental cellular rules through unicellular zygotic recapitulation. The remodeling within the maternal-zygote transition, the zygote itself, and early embryogenesis each exert substantial limitations on epigenetic imprinting and derivatively constrain the impact of immediate environmental fluxes gleaned through the varieties of holobionic phenotype. Continuous selection is unquestionably a constraint. Even the obligatory halving of genomic endowments through meiosis is another consistent source of reinforced constraint that consistently canalizes holobionic metabolism, physiology, and phenotype within longer-term boundaries and parameters (Torday and Miller 2020a).

These same types of constraints reverberate across all living scales. Stem cell research reveals that clonal lineages of cells distribute within a defined set of boundary conditions that limit cellular variety (Chang et al. 2008). Discrete restrictions are placed on the clonal transcriptome, the heterogeneity of gene expression, and other aspects of clonal reproduction that deliberately reinforce conforming 'metastable states' (Chang et al. 2008). Therefore, even within clonal lineages, there is a consistent biologic attraction effect that imposes its particular form of constraint so that any cellular outliers inevitably move toward the median status of the lineage. However, among all the various forms of biological constraint that control living processes, the paramount one is the constellation of immunological pathways that preserve cellular self-integrity.

A fresh narrative spills from this fundamental re-analysis of the unicellular stage in holobionts and the constraints it places on evolutionary development. It is not accidental that holobionts are the exclusive endpoints of eukaryotic multicellularity. Each holobiont is a unique combination of the four domains, and all cellular participants are self-referential entities. Consequently, every holobiont with its varieties of expressive phenotypes is equipped to explore the environment slightly differently, enabling its improved collective assessment in cellular terms. Within those holobionts, the necessary maintenance of environmental complementarity is implemented at successive self-referential cellular levels, coordinated by congruous aims. Therefore, it is clear that all biology and evolution is a narrative of the continual defense of instantiated cellular self-reference and its requisites, dependent just as powerfully on limitations and boundaries as upon any freedom of action.

Stephen Hawkings is widely accredited as indicating that "Intelligence is the ability to adapt to change". If that is judged to be an appropriate benchmark, then our level of human intelligence extending no farther back than 200,000 to 300,000 years is barely tested (Miller et al. 2019). By contrast, the unicellular domains have demonstrated uninterrupted successful, intelligent adaptation for billions of years (Miller and Torday 2018). Indeed, over 99% of species that have ever lived are now extinct (Jablonski 2004). The planet has ever been a cellular realm (McFall-Ngai et al. 2013). Macro-organic species come and go in our biological system. Indeed, they are mere elaborations of elemental cellular requirements. Instead, evolution is a timeless continuum (Torday and Miller 2020b). Only the four domains are perennial, and their unflagging perpetuation is evolution's ceaseless narrative.

References

Agnati, L.F., Baluška, F., Barlow, P.W. and Guidolin, D. 2009. Mosaic, self-similarity logic and biological attraction principles. Communicative & Integrative Biology 2(6): 552–563.

Bailey, D.H., Borwein, J.M., Lopez de Prado, M. and Zhu, Q.J. 2015. The Probability of Back-Test Over-Fitting. Journal of Computational Finance, Forthcoming.

Baluška, F. and Levin, M. 2016. On Having No Head: Cognition throughout Biological Systems. Frontiers in Psychology, 7.

Baluška, F. and Mancuso, S. 2009. Deep evolutionary origins of neurobiology: Turning the essence of "neural" upside-down. Communicative & Integrative Biology 2(1): 60–65.

Baluška, F., Miller, W.B. and Reber, A.S. 2021. Biomolecular Basis of Cellular Consciousness via Subcellular Nanobrains. International Journal of Molecular Sciences 22(5): 2545.

Bechtel, W. 2014. Cognitive Biology: SurprisingModel Organismsfor Cognitive Science. Proceedings of the Annual Meeting of the Cognitive Science Society, 36.

Betts, H.C., Puttick, M.N., Clark, J.W., Williams, T.A., Donoghue, P.C.J. and Pisani, D. 2018. Integrated genomic and fossil evidence illuminates life's early evolution and eukaryote origin. Nature Ecology & Evolution 2(10): 1556–1562.

Cai, X., Wang, X., Patel, S. and Clapham, D.E. 2015. Insights into the early evolution of animal calcium signaling machinery: A unicellular point of view. Cell Calcium 57(3): 166–173.

Chang, H.H., Hemberg, M., Barahona, M., Ingber, D.E. and Huang, S. 2008. Transcriptome-wide noise controls lineage choice in mammalian progenitor cells. Nature 453(7194): 544–547.

Colson, P., De Lamballerie, X., Fournous, G. and Raoult, D. (2012). Reclassification of giant viruses composing a fourth domain of life in the new order Megavirales. Intervirology 55(5): 321–332.

De Loof, A. 2015. Organic and Cultural Evolution can be Seamlessly Integrated Using the Principles of Communication and Problem-Solving: The Foundations for an Extended Evolutionary Synthesis (EES) as Outlined in the Mega-Evolution Concept. Life: The Excitement of Biology 2(4): 247–269.

Deacon, T.W. 2011. Incomplete Nature: How Mind Emerged from Matter. New York City, NY, USA: W. W. Norton & Company.

Eckersley-Maslin, M.A., Alda-Catalinas, C. and Reik, W. 2018. Dynamics of the epigenetic landscape during the maternal-to-zygotic transition. Nature Reviews Molecular Cell Biology 19(7): 436–450.

Ford., B.J. 2004. Are Cells Ingenious? Microscope 52(3/4): 135–144.

Ford, B.J. 2009. On Intelligence in Cells: The Case for Whole Cell Biology. Interdisciplinary Science Reviews, 34(4), 350–365.

Ford, B.J. 2017. Cellular intelligence: microphenomenology and the realities of being. Progress in Biophysics and Molecular Biology 131: 273–287.

Forterre, P. and Prangishvili, D. 2009. The origin of viruses. Research in Microbiology 160(7): 466–472.

Friston, K., Kilner, J. and Harrison, L. 2006. A free energy principle for the brain. Trends in Cognitive Sciences 13(7): 293–301.

Holmes, E.C. 2011. What Does Virus Evolution Tell Us about Virus Origins? Journal of Virology 85(11): 5247–5251.

Hug, L.A., Baker, B.J., Anantharaman, K., Brown, C.T., Probst, A.J., Castelle, C.J., … Banfield, J.F. 2016. A new view of the tree of life. Nature Microbiology 1(5): 16048.

Igamberdiev, A.U. and Shklovskiy-Kordi, N.E. 2017. The quantum basis of spatiotemporality in perception and consciousness. Progress in Biophysics and Molecular Biology 130: 15–25.

Jablonski, D. 2004. Extinction: past and present. Nature 427(6975): 589.

Keijzer, F.A. 2017. Evolutionary convergence and biologically embodied cognition. Interface Focus 7(3): 20160123.

Koonin, E.V. and Dolja, V.V. 2014. Virus World as an Evolutionary Network of Viruses and Capsidless Selfish Elements. Microbiology and Molecular Biology Reviews 78(2): 278–303.

Koseska, A. and Bastiaens, P.I. 2017. Cell signaling as a cognitive process. The EMBO Journal 36(5): 568–582.

Lopez de Prado, M. 2013. The Probability of Back-Test Over-Fitting. SSRN Electronic Journal.

Loron, C.C., François, C., Rainbird, R.H., Turner, E.C., Borensztajn, S. and Javaux, E.J. 2019. Early fungi from the Proterozoic era in Arctic Canada. Nature 570(7760): 232–235.

Lowenstein, R. 2000. When Genius Failed: The Rise and Fall of Long-Term Capital Management. New York, NY, United States: Random House.

Lyon, P. 2015. The cognitive cell: bacterial behavior reconsidered. Frontiers in Microbiology 6(264).

Margulis, L. and Sagan, D. 1997. Microcosmos: Four billion years of microbial evolution. Berkley and Los Angeles, CA: Univ of California Press.

Mattiroli, F., Bhattacharyya, S., Dyer, P.N., White, A.E., Sandman, K., Burkhart, B.W., … Luger, K. 2017. Structure of histone-based chromatin in Archaea. Science 357(6351): 609–612.

McFall-Ngai, M., Hadfield, M.G., Bosch, T.C.G., Carey, H.V., Domazet-Lošo, T., Douglas, A.E., … Wernegreen, J.J. 2013. Animals in a bacterial world, a new imperative for the life sciences. Proceedings of the National Academy of Sciences 110(9): 3229–3236.

Miller, W.B. 2013. The Microcosm Within: Evolution and Extinction in the Hologenome. Boca Raton, Florida, United States: Universal Publishers.

Miller, W.B. 2016. Cognition, Information Fields and Hologenomic Entanglement: Evolution in Light and Shadow. Biology 5(2): 21.

Miller, W.B. 2018. Biological information systems: Evolution as cognition-based information management. Progress in Biophysics and Molecular Biology 134: 1–26.

Miller, W.B., Baluška, F. and Torday, J.S. 2020a. Cellular senomic measurements in Cognition-Based Evolution. Progress in Biophysics and Molecular Biology 156: 20–33.

Miller, W.B. and Torday, J.S. 2018. Four domains: The fundamental unicell and Post-Darwinian Cognition-Based Evolution. Progress in Biophysics and Molecular Biology 140: 49–73.

Miller, W.B., Torday, J.S. and Baluška, F. 2019. Biological evolution as defense of "self". Progress in Biophysics and Molecular Biology 142: 54–74.

Miller, W.B., Torday, J.S. and Baluška, F. 2020b. The N-space Episenome unifies cellular information space-time within cognition-based evolution. Progress in Biophysics and Molecular Biology 150: 112–139.

Nahum, J.R., Godfrey-Smith, P., Harding, B.N., Marcus, J.H., Carlson-Stevermer, J. and Kerr, B. 2015. A tortoise–hare pattern seen in adapting structured and unstructured populations suggests a rugged fitness landscape in bacteria. Proceedings of the National Academy of Sciences 112(24): 7530–7535.

Noffke, N., Christian, D., Wacey, D. and Hazen, R.M. 2013. Microbially Induced Sedimentary Structures Recording an Ancient Ecosystem in the ca. 3.48 Billion-Year-Old Dresser Formation, Pilbara, Western Australia. Astrobiology 13(12): 1103–1124.

Papineau, D., She, Z., Dodd, M. S., Iacoviello, F., Slack, J.F., Hauri, E., … Little, C.T.S. 2022. Metabolically diverse primordial microbial communities in Earth's oldest seafloor-hydrothermal jasper. Science Advances 8(15).

Reber, A.S. 2019. The First Minds: Caterpillars, Karyotes, and Consciousness (Illustrate). New York, NY, USA: Oxford University Press.

Richter, D.J. and King, N. 2013. The Genomic and Cellular Foundations of Animal Origins. Annual Review of Genetics 47: 509–537.

Ryan, F. 2009. Virolution. New York, NY, USA: HarperCollins Publishers.

Shapiro, J.A. 2007. Bacteria are small but not stupid: cognition, natural genetic engineering and socio-bacteriology. Studies in History and Philosophy of Science Part C: Studies in History and Philosophy of Biological and Biomedical Sciences, 38(4): 807–819.

Shapiro, J.A. 2011. Evolution: A View from the 21st Century. Upper Saddle River, NJ: FT Press.

Sousa, C. 2021. Origin of Life. The American Biology Teacher 83(2): 76–79.

Spang, A., Saw, J.H., Jørgensen, S.L., Zaremba-Niedzwiedzka, K., Martijn, J., Lind, A.E., … Ettema, T.J.G. 2015. Complex archaea that bridge the gap between prokaryotes and eukaryotes. Nature 521(7551): 173–179.

Torday, J. and Miller, W. 2016a. The Unicellular State as a Point Source in a Quantum Biological System. Biology 5(2): 25.

Torday, J.S. and Miller, W.B. 2016b. Biologic relativity: Who is the observer and what is observed? Progress in Biophysics and Molecular Biology 121(1): 29–34.

Torday, J. and Miller, W. 2020a. Cellular-Molecular Mechanisms in Epigenetic Evolutionary Biology. Cham, Switzerland: Springer.

Torday, J.S. and Miller, W.B. 2020b. Evolution as a timeless continuum. *In*: J. Guex, J.S. Torday and W.B.J. Miller (eds.). Morphogenesis, Environmental Stress and Reverse Evolution. Cham, Switzerland: Springer.

Trewavas, A.J. and Baluška, F. 2011. The ubiquity of consciousness. EMBO Reports 12(12): 1221–1225.

Vallverdú, J., Castro, O., Mayne, R., Talanov, M., Levin, M., Baluška, F., … Adamatzky, A. 2018. Slime mould: The fundamental mechanisms of biological cognition. Biosystems 165: 57–70.

Villarreal, L.P. and Ryan, F. 2018. Viruses in the Origin of Life and Its Subsequent Diversification. *In*: Handbook of Astrobiology. Boca Raton, Florida, United States: CRC Press.

Yutin, N. and Koonin, E.V. 2012. Archaeal origin of tubulin. Biology Direct 7: 10.

Chapter 13

Speciation

The impetus for holobionic life is that every constituent, directly or indirectly, participates in continuous shared protection (Miller et al. 2019). Macroscopic phenotypes arise as cellular collaborations expressed through concordant natural viral-cellular engineering and ecological niche construction (Miller 2018, Miller et al. 2019, 2020 a,b). Therefore, phenotype represents coordinated information-based cellular problem-solving to support the self-referential homeorhetic equipoise of each constituent within the united cellular ecologies that comprise holobionts (Torday and Miller 2016). Consequently, as all cellular actions are the product of information assessment, all the particulars of multicelluarity should properly be considered cellular information management strategies, linking holobionic tissue ecologies into a seamless set of coordinated ecological informational interactomes (Miller et al. 2019, 2020a, Torday and Miller 2020).

In holobionts, phenotype serves as a means of accumulating epigenetic and environmental experiences to assure continuous organismal-environmental complementary in the face of environmental stresses (Torday and Miller 2016, Miller 2018). Certainly, in practical biological terms, the protection of self-identity is primarily based on immunological reactions to the stream of epigenetic encounters that impact all holobionts, including infectious agents. In CBE, speciation results from triggered incompatibilities within cellular recognition systems from these epigenetic exchanges sufficient to yield reproductive segregation (Miller 2013). These events are primarily grounded by immunological reactions triggered by infectious interchanges at a variety of levels. Such incompatibilities permit the divergence of a reproductive population from within an originating single holobionic population cohort. In this manner, two populations diverge due to an infectious incursion triggering an immunological response great enough in the infected to stimulate reproductive separation from the uninfected or those who were infected and did not have that consequential reaction. Therefore, cellular recognition system incompatibilities reside within a framework of infectious disease dynamics and their linked immunological outcomes. In this framework, intermittent immunological events or a series of linked events can be sufficient to trigger reproductive isolation between two sub-populations of a prior consonant species population.

To solve the puzzling problem of speciation, it is necessary to shift the epicenter of speciation theory away from genomic divergences towards immunological ones that decouple a previous concordant pattern of self-hood. By this means, speciation is placed into informational and cellular ecological terms. When this is done, the preservation of cellular self-identity becomes the primary biological driver of speciation, representing the same impetus that drives multicellularity. It is proposed that infectious epigenetic inputs can achieve epidemic spread intermittently, provoking an immunological reaction sufficient to yield two separate breeding populations as a population-specific speciation event. In such an instance, one population has been subject to an epigenetic adjustment, and the other has not. If the difference stimulates an immunological incompatibility, then reproductive isolation might ensue. In such circumstances, immunology rules. In this model, speciation events can be induced either by a single triggering epigenetic event, or a linked immunological combination of lesser impacts sufficient to yield reproductive isolation. As a further derivative, a triggering event might lead to permissive interspecific hybridization (the crossing of two species from the same genus) made possible in this instance by a transient lowering of immunological barriers from immunocompromise.

Once these two reproductively-isolated populations are newly established, and even though they look identical morphologically at first, each can drift towards separate developmental and phenotypic manifestations through now independent lineal real-time adaptations to environmental stresses over evolutionary space-time. By this means, the origin of species can be considered similar to the effects of breeding as Darwin had first envisioned based on animal husbandry but is now placed on a natural, intermittent basis.

In their comprehensive book 'Speciation', Coyne and Orr (2004) acknowledged that a singular problem in biology has been the failure to enact a uniform definition of a species. In fact, they detail eight differing ones. To illustrate this still modern quandary, they quote Brookfield: " The essence of the "species problem" is the fact that, while many different authorities have very different ideas of what species are, there is no set of experiments or observations that can be imagined that can resolve which of these views is the right one. This being so, the "species problem" is not a scientific problem at all, merely one about choosing and consistently applying a convention about how we use a word. So, we should settle on our favorite definition, use it, and get on with the science" (2002, p. 107).

Noting this controversy, decades ago, the influential Theodosius Dobzhansky (1935) proposed "a species is a group of individuals fully fertile inter se, but barred from interbreeding with other similar groups by it physiological properties (producing either incompatibility of parents, or sterility of hybrids, or both)" (as cited in Coyne and Orr, 2004, p. 28). Consequently, over time, the Biological Species Concept (BSC) has emerged, defining species as groups of interbreeding individuals reproductively isolated from other groups (Bobay and Ochman 2017). This definition has been generally adopted. However, a substantial blurring of the lines occurred over time. For example, Templeton (1998) expanded the species definition to include species as 'cohesion mechanisms' characterized by recognition compatibilities or ecological niches that contribute to mate recognition factors influencing pre-mating patterns and isolating mechanisms that could affect discriminative mating. Based on such factors, the capricious concept of 'incipient' species has gained traction though having no formal definition (Charron et al. 2014).

Therefore, when it comes to the issue of species, considerable confusion reigns. As Haider (2018) notes, the variety of species concepts each has limitations, and no universally accepted species definition exists, so any of these might be used depending on context.

Although there is a high level of general agreement that chromosomal incompatibilities largely govern reproductive isolation and eventual speciation (Charron et al. 2014), the problem in conventional Neodarwinism is how to get there in any practical sense. A vast literature attempts to devise theoretical means of speciation by the long-term accumulation of random genetic mutations through genetic drift between segregated populations, or by various selective means based on discriminative mate selection, sexual selection, or local adaptations (Gavrilets 2003). Most of these theoretical mechanisms require allopatry (a population splits into two geographically isolated populations) or peripatric speciation that posits that new species form from the non-geographical isolation of smaller populations sufficient to impede genetic exchanges with the dominant population (Coyne and Orr 2004). Although highly controversial, others claim examples of sympatric speciation in which two distinct species derive from a single ancestral lineage occupying the same geographic location (Fitzpatrick et al. 2008). The problem with each is that there is no realistic means for achieving reproductive isolation since there is no mechanism for inducing reproductive incompatibilities simply through isolation. That would merely lead to selective breeding, and we know from our own breeding efforts that speciation is not effected by that means.

There are quite a few knotty problems in speciation theory that require resolutions. For instance, it is not clear that there are any genuine species among bacteria despite the common usage of that term due to the extreme fluidity of genetic flow among bacterial strains (Fraser et al. 2009). It is similar among Archaea (Rosselló-Móra and Amann 2015). Further, even the examination of genomic sets among eukaryotic multicellular organisms to determine species differences is fraught as there is no true gold standard for that genomic assessment (Huang 2018). For example, recent taxonomic studies based on coalescent-based phylogenetic reconstructions and landscape genetic analyses indicate that the spatial distribution of genetic variations does not match current taxonomic designations, and to such a degree that they are incompatible with current models of ecological/environmental speciation (Tonzo et al. 2019).

Although continuous divergent adaptations governed by natural selection are a mainstay of allopatric speciation theory, a recent analysis of 50,000 species concluded that speciation is essentially random and that speciation and adaptation are two different processes (Hedges et al. 2015). Noting this, it has been suggested that all that is required for speciation is a mere 2 million years of accumulated random mutations. Although most of these are supposed neutral, this accumulation is theoretically sufficient to lead to speciation through a gradual buildup of incompatibilities. However, if that were true, all species would eventually be substantially altered by aggregated random genetic errors. If the average species duration is 6 million years, no species could conceivably resist speciation events. However, horseshoe crabs have existed essentially unchanged for over 400 million years, crocodiles for 240 million years, and the frilled shark for over 80 million years.

In the 21st century, a more nuanced view of genes has developed. Rather than a fixed heritable system akin to computer code, it is increasingly regarded as a flexible common language of communication between cognitive cellular constituencies deployed across the three cellular domains (Prokaryota, Archaea, Eukaryota). (Witzany 2014, Miller and Torday 2018, Miller et al. 2020a). Consequently, the genome is not passive nor a highly protected compartment merely subject to random mutational influences (Monroe et al. 2022). Instead, it is a substantive, active, and reciprocating cellular constituent in biological and evolutionary development as a major mode of information transfer and a vital means of cell-cell communication (Torday and Miller 2020).

The virome is a primary intermediary in problem-solving among the three cellular domains (Miller and Torday 2018). In the past, our appraisal of viruses was confined to their well-acknowledged role in infectious disease dynamics based on a traditional host/pathogen model. Consequently, its selective influences were conceptually centered within a conventional framework of Neodarwinian competitive selection and its culling impact on populations and stimulating disease-resistant immunocompetence (Williams and Nesse 1991). Yet, overt infectious pathogenesis represents only one aspect of a much larger viral dynamic, including the attempted or actual horizontal and vertical transfer of genetic material of viral or cellular origin among the cellular domains. Whether welcome or not, these genetic transfers constitute an exchange of information with an evolutionary consequence through the information cycle.

In the past, infectious interchanges were regarded as medical issues. Their only acknowledged range of outcomes was individual infectious pathogenesis, epidemic infection, symbiosis, parasitism, or infectious latency. But, based on extensive accumulated research, a further outcome from infectious events has been revealed that correlates direct genetic transfers as epigenetic incursions with the capacity to effect the generation of evolutionary shifts (Baluška 2009, Miller 2013). Thus, viruses, sub-viruses, bacteria, and fungi are not solely agents of pathogenesis. In addition, they share a partnership position in the evolution of genomes with the potential for a range of interactions that could yield either immediate or eventual heritable expression (Forterre 2006). Indeed, viruses, retroviruses, and retroelements can be viewed as operating within the crucial role of a collective archive of memory of community-wide genetic information (Goldenfeld and Woese 2007). Furthermore, even if a viral incursion is latent, any acquired genetic information becomes 'on call' for use in confronting future environmental stresses.

Accordingly, the perpetual process of infectious genetic interchange can now be deemed a vital aspect of the evolutionary process. High levels of retroviral and viral sequences in modern genomes accumulated over evolutionary space-time have been documented. For instance, there has been an enormous viral contribution to the human genome, now known to be as much as 50%, with 9% due to invasion by endogenous retroviruses (Ryan 2009, 2010). These exchanges are by no means necessarily problematic. This universality of horizontal genetic incursions and their assimilation as continuous information exchange ensures on-going organismal-environmental complementarity in the face of environmental shifts. All such actions impact an organism's attachment to information space-time, are sustained by the cellular information cycle and help form the informational architectures of cells that can eventually produce holobionts. As a result, contemporary evidence attests that infectious dynamics are a crucial driving force for evolution (Baluška 2009, Witzany, 2011, Miller 2013, 2016). Its impact need not be random or gradual (Boto 2010, Shapiro 2017). Transposable elements are now known factors in episodic intervals of rapid speciation and evolutionary transitions. Bursts of transposable element activity and horizontal transfer of retroviral inclusions energize rapid adaptive radiations (Oliver and Greene 2012). Therefore, the transfer of genetic material as cell-cell communication is a perpetual evolutionary feature across biological cycles. Incurred infectious agents can be active promotors, repressors, or latent residents in cells, forming a critical aspect of its current and potential information system, ultimately yielding both biological expression and shifts in evolutionary development.

Research has demonstrated that the LTR class I endogenous retrovirus (ERV) retroelements, distant relatives of HIV, have considerably impacted the transcriptional network of human tumor suppressor protein p53 (Wang et al. 2007). This protein complex is a master gene regulator crucial for primate differentiation. There is accumulating evidence that retroelements can significantly shape the regulatory network of transcription factors in a species-specific manner (Oliver and Greene 2012). The adaptive value of retrotransposon activation is a known modulator of environmental stresses and has contributed heavily to the functional regulatory machinery of the cell (Shapiro 2011). Furthermore, eukaryotic development is strongly dependent upon viral impacts and their incorporations, which can be co-opted towards holobionic development (Witzany 2011). As had been previously noted, researchers now believe that vertebrate myelin originated when retroviral-derived elements inserted in the genome of ancient vertebrates at key positions, triggering a massive expression of that signature basic myelin protein (Ghosh et al. 2022). Although it is our bias to consider these infectious elements as invading parasites, Ryan (2009) uses the term "symbionts" for such retroelements in acknowledgment of their often beneficial roles in evolutionary development (Ryan 2009).

The summation of these highly varied influences underscores an increasing awareness of genomes as products of a stream of complex, differing inputs. There is no need for those influences to be initially manifest. Genomes are continuously active products of insertions and deletions, duplications, and rearrangements. Large-scale evolutionary events can be triggered by an accumulation of earlier latent epigenetic inputs (Merhej and Raoult 2012). Merhej and Raoult (2012) note: "Genomes are collections of genes with different evolutionary histories that cannot be represented by a single tree of life (TOL). A forest, a network or a rhizome of life may be more accurate to represent evolutionary relationships among species". Crucial to this understanding is recognizing the genome as a flexible palette consistently altered by a stream of consequential epigenetic impacts representing information to cells.

In this regard, TEs have been previously categorized as intracellular parasites whose cell-cell transfer is considered a form of genetic information transfer (Hua-Van et al. 2011). Notably, TEs have been found in eukaryotic viruses and can be transmitted horizontally to eukaryotes in an infectious pattern of spread. Boeke and Stoye (1997) have asserted that retroviruses are a highly evolved component of a phylogenetic tree that comprises these invasive retroelements. Therefore, infectious retroviruses and retroelements share ancestry and effectively engage with eukaryotic cellular genomes. Further, it is now known that non-retroviral RNA has invaded the genome of many mammalian species (Horie et al. 2010, Feschotte and Gilbert 2012). Indeed, evolutionary history is now regarded as a series of sequential cross-species jumps of many viral families, especially those coded in RNA, with substantial alteration of entire metabolic pathways (Monier et al. 2009, Geoghegan et al. 2017).

Consequently, it is now felt that retroelements represent a record of previous infectious impacts on genomes, and further, these have had a significant evolutionary role (Witzany 2011). Retroelements participate in adaptation to environmental stresses, phenotypic and metabolic variation, and contribute to the regulatory machinery of cellular life (Shapiro 2011). Therefore, TEs are a propulsive evolutionary force as a source of biological novelty and eukaryotic metabolic innovation (Brandt et al. 2005, Schaack et al. 2010, Bennetzen and Wang 2014, Soucy et al. 2015). An analysis of 759 plant, fungal, and animal genomes identified a plethora of horizontal TE transfer events in eukaryotic species, implicating ticks and viruses as vectors of infectious transfer of LINE-1 and Bovine-B TEs, as the most abundant TE families in mammals. These are now accredited as critical drivers in evolution. Contrary to the previous view of viruses as merely eukaryotic parasites, research confirms that the essential characteristic of the virome is symbiosis among prokaryotes and eukaryotes. The outcomes can be either highly cooperative or intermittently competitive within a larger overarching narrative of mutualizing symbiogenesis (Grasis 2017, Villarreal and Ryan 2018).

The consequence of the infectious transmission of retroelements may even account for the first instance of planetary speciation. It is now believed that the origination of the ancestral eukaryotic cell was related to instances of retroelement invasion within preexisting cell lines triggering the increased genetic plasticity and networks that characterize Eukaryota (Lee et al. 2018). Retroelements are rare in bacteria. Further, DNA repair in eukaryotes significantly enhances the efficiency of retrotransposition and the influence of retroelements on cells, permitting additional cellular complexities.

Importantly, there is no requirement for these genetic transfers of retroelements as infectious interchanges to be isolated, small events. Instances of massive transfers among bdelloid rotifers have been documented (Eyres et al. 2015, Gladyshev et. al. 2008). Massive intermittent bursts of TE proliferation are now believed to have caused radical genomic rebuilding, triggering gross episodes of domestication, polyploidy, potential hybridizations, and shifting mate selection patterns (Belyayev 2014). TEs are also considered responsible for episodes of rapid genomic remodeling that can result in new gene regulatory networks (Rebollo et al. 2010). Experiments with natural and artificial speciation by interspecific hybridization confirm this. TEs and other means of genome restructuring can rapidly activate and generate sufficient genomic alterations so that novel species can quickly emerge (Shapiro 2022).

Some scientists insist that the genome concept requires a thorough recalculus. Genomes are products of massive genetic bombardments and successful incursions by mobile genetic elements (Brosius 1999). In that case, both adaptation and speciation could transpire when usual epigenetic constraints are relaxed either by cellular consensus or by an exogenous trigger (Rebollo et al. 2010). Notably, immunosuppression distinctly characterizes many human disease states, directly leading to a significant increase in the incidence of infectious events. In such instances, genomic viral incursions would be included among those events. Given our massive viral genomic contingent, humans have not been spared.

This exact viral-genome interaction has even been observed in real-time despite our limited historical window of observation. Exogenous retroviruses from the viral subfamily *Orthoretrovirinae* have invaded the Koala genome in the last two decades (Greenwood et al. 2018). This Koala retrovirus has become a permanent element in the genomes of a substantial population of Koalas. Exogenous retroviruses can become endogenous genomic participants capable of horizontal transmission among individuals. Significantly, endogenized retroviruses can become the source of further exogenous retroviruses. Accordingly, Koala retrovirus serves as an example of a distinct genomic invasion becoming heritable, infectious, and responsible for immune suppression and neoplasia. (Kinney and Pye 2016).This exact constellation of viral effects is one mode of how a speciation event could arise, either by triggering immunological incompatibilities that lead to separate reproductive sub-populations or by the induction of immune incompetence, permitting opportunistic interspecific hybridization when it might not otherwise have occurred.

Furthermore, it is also now known that the supposedly inviolable Weismann barrier is not complete as once believed. Small RNAs and potentially other genetic particles can be trafficked in mammalian sperm during post-testicular maturation in the epididymis, confirming that soma-germline RNA transfers occur, likely via vesicular transport from the epididymis to maturing sperm (Sharma et al. 2018).

Genetic transfers to germ lines could occur through multiple means. Retroviral incursions can extend directly into genomes. For example, such events accounted for the origin of syncytiotrophoblastic tissues and the mammalian placenta (Black et al. 2010). Undoubtedly, these instances occur more often than their achieving expression. Instead, they can reside quietly within the genome for long periods. Thereafter, any of these might be triggered during genetic recombinations, the accumulation of other retroelements, or bursts of TE activity. As an additional potential outcome, these genetic incursions might activate a fertility barrier based on the induction or triggering of specific incompatibility genes leading to cytonuclear incompatibility as nuclear-mitochondrial conflict, as has been identified in three yeast species (Chuo et al. 2010).

Since the microbiome is a crucial player in holobionts and participates in biological expression, symbiont disruption or amplification could similarly trigger cytoplasmic incompatibilities (Richardson et al. 2019) through indirect induction of a fertility barrier by sexual discrimination at multiple levels (Sharon et al. 2010). The introduction of select symbionts can lead to hybrid sterility, lethality, and induction of altered behaviors, including differential mate selection (Shropshire and Bordenstein 2016). For example, a newly discovered bacteria, *Cardinium* sp, causes various reproductive alterations in arthropods, including cytoplasmic incompatibility and feminization of genetic males. The endosymbiont *Wolbachia* also induces cytoplasmic incompatibilities in arthropods (Martinez et al. 2021). In addition, *Wolbachia* affects sperm, causing embryonic mortality when interbreeding with *Wolbachia*-free females. Consequently, there is no longer any doubt that microorganisms play a fundamental role not only in eukaryotic metabolism but also as evolutionary drivers (Sandoval-Motta, et al. 2017).

Given that speciation is an immunological event, it is reasonable to entertain the possibility of intermittent episodes of individual or widespread breakdowns of immunological capabilities allowing 'permissive hybridization' beyond instances triggered by microbial symbionts (Miller 2013). Our current understanding of immunological reactions in human diseases indicates that there are permissive states for infection. Consequently, certain ecological conditions might amplify the potential for generally enforced immunological rules to be breached, e.g., during catastrophic environmental conditions. In such rare intermittent circumstances, no matter their exact etiology, there might be bursts of speciation, thereby accounting for the documented case of massive TE expansion in many eukaryotic genomes, including humans.

Recent studies in genome dynamics have revealed surprising conclusions about evolutionary processes. Studies indicate that structural genomic variants are more important than single nucleotide polymorphisms (SNPs) that have long been supposed to account for the genetic drift that leads to speciation (Wellenreuther et al. 2019). Structural variations affect many more bases than SNPs. Furthermore, the mainstay presumption of Neodarwinism has been that species must naturally become more genetically diverse over time and that diversity should be increased in larger species populations. However, based on a meticulous trawling through five million DNA barcodes collected from 100,000 animal species, that belief was entirely contradicted. Instead, genetic diversity is about the same over time among the various species, and its extent was irrespective of population size when measured at the mitochondrial level (Stoeckle and Thaler 2018). Further, this lack of variation was most pronounced among so-called 'neutral' mutations, meaning that they are largely irrelevant in evolutionary terms.

The concept that drift is not a valid mechanism for speciation and that speciation occurs abruptly as a reordering and isolating event is supported by recent experimental evidence. Researchers at Yale and Harvard analyzed diverse genomic datasets for patterns of expression of genetic variation and mutation. They made inferences about genetic architecture to determine the role of stabilizing selection and mutation in shaping genetic expression, finding that gene expression evolves in a "domain of phenotype space well fit by the House of Cards model" (Hodgins-Davis et al. 2015). Mutations that are large enough to be considered consequential and effectively shuffle the deck explain evolutionary processes better than a model of small mutational accumulations. Large mutations would initiate a cascade of changes in genetic expression as a 'house-of-cards'. This conforms with the concepts of triggers and criticalities as a set of evolutionary interactions between the immune system, genetic incursions, and the genome proposed by Miller in 2013. Accordingly, the anchoring belief of evolution and speciation through accumulated minor random genetic mutations is decisively undercut.

There are two notable strengths in placing speciation squarely in the realm of infectious disease dynamics. One of these is obvious, and the other less so. First, as any random genetic mutation from replication error must arise in a single individual, it is formidably improbable that such an event could lead to speciation.

Any advantageous mutation that is expressed by a change in any such individual will be gradually vitiated through the serial dilutions of meiotic reproduction. The same can be argued about the long-term gradual accumulations of generally neutral mutations since there is no chance of a critical reproductive barrier unless a further triggering event ensues. Conversely, the nature of infectious spread assures that there is a foothold reproductive population right at the point of origination (Miller 2013). Only an infectious trigger offers a practical means for speciation to both arise and be sustained within a population since this is the only means for multiple reproducing individuals to be simultaneously affected. Indeed, there is an applicable model in biofims. Antibiotic-resistance plasmid transfers in biofilms are not one-off events between two bacterial cells. Instead, this is a colonial-wide phenomenon yielding a new phenotype. Similarly, a population-wide infectious event could trigger the sudden development of two different sub-populations within multicellular eukaryotes, resulting in the infected and the uninfected or infection-resistant with differing phenotypic capacities and at least some degree of reproductive incompatibility. It is from this juncture that selection might propel overt speciation.

Further yet, infectious dynamics provides a pathway by which speciation could even be considered a sexually transmitted event in which a sperm-mediated infectious agent or symbiont leads to selective mating incompatibilities. Indeed, when reconsidered in this manner, speciation can be seen as a natural defense system sustaining self-identity against sexually transmitted assaults. By providing a reproductive barrier, a new species' ground state is begun with a new immunological alphabet in the continuous defense of self-identity. The integrity of the original species is thus maintained.

Secondly, in an infectious model of speciation, various infectious agents can gain experience that might one day lead to successful endogenization. This action might proceed akin to viral quasispecies with rounds of epidemic spread, matching to our most recent experience with SARS- CoV-2. Although initially repulsed immunologically, eventually, a variant finds a way to reach beyond those defenses. It is entirely probable that over the course of many thousands of years, some infectious agents enter somatic cells and then insinuate into the genome or the germ line to ensure their further reproduction. In this way, there may be repeated cycles of attempted infiltration, ultimately leading to successful replication in the reproductive apparatus of some sub-population of infected individuals. This event leads to segregated immunocompatibility that effectively becomes a reproductively separate species. Indeed, studies in six baboon species (genus Papio) demonstrated that substantial genotypic and phenotypic differences can accumulate despite continued reproductive flow so that species can look alike but be genetically diverging (Rogers et al. 2019). All that would be lacking is the correct infectious episode to effect a direct speciation event through induced reproductive isolation.

There has been one aspect of communication among organisms that has been very generally misunderstood within the conventional Darwinian frame. Among the most significant forms of biocommunication is lateral genetic transfer. Viral infection is the chief intermediary. All such infectious incursions, whether rebuffed or successful, are attempts at biocommunication at the cellular level. All infectious processes, no matter their overt outcome, are aspects of this attempt at the horizontal transfer of information.

Among infectious agents, pathogenesis is only one outcome among many. The result of any infectious interchange is dependent upon its target of opportunity. A broad range of potential infectious outcomes can follow from any infectious incident ranging from individual infection, to epidemics, parasitism, mutualism, symbiosis, latency, epigenetic adjustments with the potential for evolutionary shifts, or the direct incursion of genetic material into the cytoplasm or genome. In this latter circumstance, speciation becomes one intermittent outcome, dependent on that infecting agent, becoming incorporated into cellular life through viral/subviral endogenization. Yet, in all such cases, any of these events becomes part of the cellular information cycle, ultimately contributing to natural cellular engineering and niche construction. Infectious dynamics is one way by which self-aware biological entities communicate, collaborate, and compete (Miller 2013). When placed into this framework, alliances and mergers among jostling living entities become the common currency of all biology. Infection, especially viral, is one example of this overarching principle. Infection may kill, but alternatively, it may also lead to productive symbioses, thereby becoming a driving force in evolutionary complexity (Baluška 2009).

From the preceding, pathogenesis can now be understood as one manifestation of a larger arena of information transfer whose range of influences encompasses an entire suite of actions. The final common denominator of all these phenomena is that they result from coordinated cellular processes that ultimately link to infectious dynamics and natural viral-cellular engineering. Infectious genetic transfers between domains were once discounted. However, newer research indicates that viruses have served over evolutionary space-time as continuous intermediaries among the domains, including trans-domain genetic cross-talk, such as bacteriophage infection of eukaryotic cells (Gill and Brinkman 2011). It is for this reason that there are a large number of universal protein domains (Malik et al. 2017).

Within this context, a further necessary linkage can be identified. Immunological rules must always apply. Thus, immunological concerns are paramount among all the domains and holobionts. When self-reference is the base, this is not surprising. Immunology is directed toward the defense of self-integrity. Further, immunology is also the linchpin of reproduction. Successful reproduction depends upon self-similar recognition through immunological compatibility. Natural selection filters for immunological competence as the chief litmus of successfully meeting environmental injunctions (Miller 2013). Since immunological factors determine reproductive success more than access to particular mates or any other macroscopic metric, immunological distinctions define the operating characteristics of our biologic system rather than the traditional measures of fitness. Therefore, selection is a derivative function of the immunological enforcement of self-referential identity, which is itself dependent on cellular information systems. Immunology is just another form of information/communication as an aspect of cognition that guides cellular decision-making within any information matrix (Tauber 2000). This perspective yields a productive reduction concerning speciation. Reproductive isolation is an immunological epiphenomenon that occurs through the triggering of immunologically-grounded incompatibilities, either chromosomal, cytoplasmic, cytonuclear mitochondrial, or gametic.

Thus, it is no longer surprising that epigenetic impacts or the transfers of transposons or small RNAs could be triggers of hybrid dysfunctions that yield effective reproductive isolation between populations (Michalak 2009). Once embedded, subsequent triggers may yield greater hybrid genetic dysfunctions as a more significant divergence of chromosomal incompatibilities, disrupting patterns of DNA methylation and imprinting, activation of transcriptional pathways, or the mobilization of TEs (Michalak 2009). Adaptive introgression or allopolyploidization might follow that would lead to instantaneous speciation (Abbott et al. 2013).

Extending this mechanistic path, Stindl (2014) offers a Telomeric Sync model of speciation, insisting that speciation requires the simultaneous emergence of differentiating features defining a new reproductive population. This requirement could be met through population-wide chromosomal rearrangements due to stress-induced telomere (the caps of eukaryotic chromosomes) shortening, yielding the saltatory emergence of differing chromosomal complements. Telomeric erosion leads to a deterioration of chromosomal integrity, predisposing to transposon-mediated genomic re-patterning in the germline of many individuals simultaneously and establishing a new breeding population with a distinct chromosomal profile.

From the preceding, it is plain that the issue of speciation specifically reduces to interactions among comparative recognition systems. Consequently, a species definition can be placed in an informational context. A species is a group of organisms that access N-space in a similar manner as that partitioning of N-space that can be interrogated by their concordant senomes. Since development and morphology in holobionts are strictly under the purview of their N-space Episenomes, it follows that those phenotypes that we use to conveniently identify species must be mirrored by a similar concordance in N-space. The same applies to the cellular senome that reciprocates with the genome and epigenome. When so considered, speciation can now be placed within a framework of parallel information assessment and utilization rather than phenotypic or genomic descriptions.

It is now understood that the microbiome of holobionts is not a loosely co-adapted appendage. Instead, a holobiont represents a seamlessly coordinated informational interactome that includes myriads of non-eukaryotic cellular constituents and a co-existent and participating virome. Consequently, there is an entailing further requirement for a shared informational platform for the coherent measured assessment by the entire assembly of participants. Accordingly, the occurrence of a new species must represent a shift of that consensual informational apparatus. Further too, that biological shift must be accompanied by its mirroring within its relevant N-space architecture that serves as a morphogenetic template for the next zygotic eukaryotic re-elaboration. In other words, speciation must be reflected by its concordance with N-space. Therefore, a species is better defined through its holobionic partnerships whose concurrence, in biological terms, is dominated by immunological factors and cell-cell signaling pathways as biological manifestations of cellular recognition systems.

Therefore, it is no longer sufficient to regard a species as adequately defined as a set of individuals with common traits and phenotypes. Instead, species are reproductively isolated sets of self-similar individuals with concordant holobionic partners co-existing within a common information-measuring platform. This new concept reduces satisfactorily. A *species is the biological expression of a conforming information set*. Consequently, the most accurate conception of a species might be that it represents a 'cluster in sequence space' rather than the traditional metrics (Thaler 2021). How might that be qualified? A species informational aggregate would be best defined based on the boundaries of its distinct N-space Episenome.

The subject of speciation has provoked an immense outpouring of thought for over 150 years. Despite those efforts, only a few conclusions are sustainable. Accumulated scientific data indicate that speciation results from recognition-system incompatibilities sufficient to effect reproductive isolation. No matter the explicit trigger, that event must be decisive enough to disrupt previously concordant cell-cell signaling pathways or provoke immunologically-based reproductive conflicts. After all, at the most basic conceptual level, reproductive isolation is the direct rejection of one or another agent of reproduction between the two parties. Even genic incompatibilities can only be effectuated through downstream signaling processes.

Furthermore, evidence now shows that infectious dynamics provide the most robust and flexible pathway to effect that end. Therefore, the potential triggers are manifold. Yet, any of these must enact as recognition-system incompatibilities that displace normative patterns of natural cellular engineering among self-referential intelligent measuring cells. Crucially, in multicellularity, the self-integrity of those cells is sustained through their mutualized attachment to a concordant information space as its unique N-space Episenome. Consequently, although speciation is a biological event, it is the manifest expression of a discontinuity in the mutualizing exchange of information, representing the division of a previously concordant information space-time species matrix into two separate self-identifying information sequence clusters. In sum, speciation is a breakdown of previously co-aligned N-space Episenomes. Clearly, speciation is an artifact of the consistent, continuous cellular drive to maintain self-integrity as opposed to any specific evolutionary end as macroorganic forms. Plainly then, speciation is an alternative expression of information management as an epiphenomenon of variation sufficient to result in recognition system incompatibilities incurred in the continuous defense of individual and collective self-identity.

References

Abbott, R., Albach, D., Ansell, S., Arntzen, J.W., Baird, S.J.E., Bierne, N., ... Zinner, D. 2013. Hybridization and speciation. Journal of Evolutionary Biology 26(2): 229–246.

Baluška, F. 2009. Cell-Cell Channels, Viruses, and Evolution. Annals of the New York Academy of Sciences 1178(1): 106–119.

Belyayev, A. 2014. Bursts of transposable elements as an evolutionary driving force. Journal of Evolutionary Biology 27(12): 2573–2584.

Bennetzen, J.L. and Wang, H. 2014. The Contributions of Transposable Elements to the Structure, Function, and Evolution of Plant Genomes. Annual Review of Plant Biology 65: 505–530.

Black, S.G., Arnaud, F., Palmarini, M. and Spencer, T.E. 2010. Endogenous Retroviruses in Trophoblast Differentiation and Placental Development. American Journal of Reproductive Immunology 64(4): 255–264.

Bobay, L.-M. and Ochman, H. 2017. Biological Species Are Universal across Life's Domains. Genome Biology and Evolution 9(3): 491–501.

Boeke, J.D. and Stoye, J.P. 1997. Retrotransposons, endogenous retroviruses, and the evolution of retroelements. *In*: J.M. Coffin, S.H. Hughes and H.E. Varmus (eds.). Retroviruses. Cold Spring Harbor, NY: Cold Spring Harbor Laboratory Press.

Boto, L. 2010. Horizontal gene transfer in evolution: facts and challenges. Proceedings of the Royal Society B: Biological Sciences 277(1683): 819–827.

Brandt, J., Schrauth, S., Veith, A.-M., Froschauer, A., Haneke, T., Schultheis, C., … Volff, J.-N. 2005. Transposable elements as a source of genetic innovation: expression and evolution of a family of retrotransposon-derived neogenes in mammals. Gene 345(1): 101–111.

Brookfield, J. 2002. Genes, Categories and Species, The Evolutionary and Cognitive Causes of the Species Problem. J. HEY. Oxford University Press. 2001. 217 pages. ISBN 0 19 514477 5. Price£ 34.95.(hardback). Genetics Research 79(1): 107–108.

Brosius, J. 1999. Genomes were forged by massive bombardments with retroelements and retrosequences. Genetica 107(1–3): 209–238.

Charron, G., Leducq, J.-B. and Landry, C.R. 2014. Chromosomal variation segregates within incipient species and correlates with reproductive isolation. Molecular Ecology 23(17): 4362–4372.

Chou, J.-Y., Hung, Y.-S., Lin, K.-H., Lee, H.-Y. and Leu, J.-Y. 2010. Multiple Molecular Mechanisms Cause Reproductive Isolation between Three Yeast Species. PLoS Biology 8(7): e1000432.

Coyne, J.A. and Orr, H.A. 2004. Speciation (Vol. 37). Sunderland, MA: Sinauer Associates.

Dobzhansky, T. 1935. A Critique of the Species Concept in Biology. Philosophy of Science 2(3): 344–355.

Eyres, I., Boschetti, C., Crisp, A., Smith, T.P., Fontaneto, D., Tunnacliffe, A. and Barraclough, T.G. 2015. Horizontal gene transfer in bdelloid rotifers is ancient, ongoing and more frequent in species from desiccating habitats. BMC Biology 13: 90.

Feschotte, C. and Gilbert, C. 2012. Endogenous viruses: insights into viral evolution and impact on host biology. Nature Reviews Genetics 13(4): 283–296.

Fitzpatrick, B.M., Fordyce, J.A. and Gavrilets, S. 2008. What, if anything, is sympatric speciation? Journal of Evolutionary Biology 21(6): 1452–1459.

Forterre, P. 2006. The origin of viruses and their possible roles in major evolutionary transitions. Virus Research 117(1): 5–16.

Fraser, C., Alm, E.J., Polz, M.F., Spratt, B.G. and Hanage, W.P. 2009. The Bacterial Species Challenge: Making Sense of Genetic and Ecological Diversity. Science 323(5915): 741–746.

Gavrilets, S. 2003. Perspective: Models of Speciation: What Have We Learned In 40 Years? Evolution 57(10): 2197–2215.

Geoghegan, J.L., Duchêne, S. and Holmes, E.C. 2017. Comparative analysis estimates the relative frequencies of co-divergence and cross-species transmission within viral families. PLOS Pathogens 13(2): e1006215.

Ghosh, T., Almeida, R.G., Zhao, C., Gonzale, M.G., Stott, K., Adams, I., … Franklin, R.J. 2022. A retroviral origin of vertebrate myelin. BioRxiv.

Gill, E.E. and Brinkman, F.S.L. 2011. The proportional lack of archaeal pathogens: Do viruses/phages hold the key? BioEssays 33(4): 248–254.

Gladyshev, E.A., Meselson, M. and Arkhipova, I.R. 2008. Massive Horizontal Gene Transfer in Bdelloid Rotifers. Science 320(5880): 1210–1213.

Goldenfeld, N. and Woese, C. 2007. Biology's next revolution. Nature 445(7126): 369.

Grasis, J.A. 2017. The Intra-Dependence of Viruses and the Holobiont. Frontiers in Immunology 8(1501).

Greenwood, A.D., Ishida, Y., O'Brien, S.P., Roca, A.L. and Eiden, M.V. 2018. Transmission, Evolution, and Endogenization: Lessons Learned from Recent Retroviral Invasions. Microbiology and Molecular Biology Reviews 82(1): e00044–17.

Haider, N. 2018. A Brief Review on Species Concepts with Emphasis on Plants. LS: International Journal of Life Sciences 7(3): 115–125.

Hedges, S.B., Marin, J., Suleski, M., Paymer, M. and Kumar, S. 2015. Tree of Life Reveals Clock-Like Speciation and Diversification. Molecular Biology and Evolution 32(4): 835–845.

Hodgins-Davis, A., Rice, D.P. and Townsend, J.P. 2015. Gene Expression Evolves under a House-of-Cards Model of Stabilizing Selection. Molecular Biology and Evolution 32(8): 2130–2140.

Horie, M., Honda, T., Suzuki, Y., Kobayashi, Y., Daito, T., Oshida, T., ... Tomonaga, K. 2010. Endogenous non-retroviral RNA virus elements in mammalian genomes. Nature 463(7277): 84–87.

Hua-Van, A., Le Rouzic, A., Boutin, T.S., Filée, J. and Capy, P. 2011. The struggle for life of the genome's selfish architects. Biology Direct 6: 19.

Huang, J.-P. 2018. What Have Been and What Can Be Delimited as Species Using Molecular Data Under the Multi-Species Coalescent Model? A Case Study Using Hercules beetles (Dynastes; Dynastidae). Insect Systematics and Diversity 2(2): 3.

Kinney, M.E. and Pye, G.W. 2016. Koala Retrovirus: A Review. Journal of Zoo and Wildlife Medicine 47(2): 387–396.

Lee, G., Sherer, N.A., Kim, N.H., Rajic, E., Kaur, D., Urriola, N., ... Kuhlman, T.E. 2018. Testing the retroelement invasion hypothesis for the emergence of the ancestral eukaryotic cell. Proceedings of the National Academy of Sciences 115(49): 12465–12470.

Malik, S.S., Azem-e-Zahra, S., Kim, K.M., Caetano-Anollés, G. and Nasir, A. 2017. Do Viruses Exchange Genes across Superkingdoms of Life? Frontiers in Microbiology 8.

Martinez, J., Klasson, L., Welch, J.J. and Jiggins, F.M. 2021. Life and Death of Selfish Genes: Comparative Genomics Reveals the Dynamic Evolution of Cytoplasmic Incompatibility. Molecular Biology and Evolution 38(1): 2–15.

Merhej, V. and Raoult, D. 2012. Rhizome of life, catastrophes, sequence exchanges, gene creations, and giant viruses: how microbial genomics challenges Darwin. Frontiers in Cellular and Infection Microbiology 2(113).

Michalak, P. 2009. Epigenetic, transposon and small RNA determinants of hybrid dysfunctions. Heredity 102(1): 45–50.

Miller, W.B. 2013. The Microcosm Within: Evolution and Extinction in the Hologenome. Boca Raton, Florida, United States: Universal Publishers.

Miller, W.B. 2016. Cognition, Information Fields and Hologenomic Entanglement: Evolution in Light and Shadow. Biology 5(2): 21.

Miller, W.B. 2018. Biological information systems: Evolution as cognition-based information management. Progress in Biophysics and Molecular Biology 134: 1–26.

Miller, W.B., Baluška, F. and Torday, J.S. 2020a. Cellular senomic measurements in Cognition-Based Evolution. Progress in Biophysics and Molecular Biology 156: 20–33.

Miller, W.B. & Torday, J. S. (2018). The Origins of Biological Deception: Ambiguous Information and Human Behavior. *Minding Nature*, *11*(3).

Miller, W.B., Torday, J.S. and Baluška, F. 2019. Biological evolution as defense of "self". Progress in Biophysics and Molecular Biology 142: 54–74.

Miller, W.B., Torday, J.S. and Baluška, F. 2020b. The N-space Episenome unifies cellular information space-time within cognition-based evolution. Progress in Biophysics and Molecular Biology 150: 112–139.

Monier, A., Pagarete, A., de Vargas, C., Allen, M.J., Read, B., Claverie, J.-M. and Ogata, H. 2009. Horizontal gene transfer of an entire metabolic pathway between a eukaryotic alga and its DNA virus. Genome Research 19(8): 1441–1449.

Monroe, J.G., Srikant, T., Carbonell-Bejerano, P., Becker, C., Lensink, M., Exposito-Alonso, M., ... Weigel, D. 2022. Mutation bias reflects natural selection in Arabidopsis thaliana. Nature 602(7895): 101–105.

Oliver, K.R. and Greene, W.K. 2012. Transposable elements and viruses as factors in adaptation and evolution: an expansion and strengthening of the TE-Thrust hypothesis. Ecology and Evolution 2(11): 2912–2933.

Rebollo, R., Horard, B., Hubert, B. and Vieira, C. 2010. Jumping genes and epigenetics: Towards new species. Gene 454(1–2): 1–7.

Richardson, K.M., Griffin, P.C., Lee, S.F., Ross, P.A., Endersby-Harshman, N.M., Schiffer, M. and Hoffmann, A.A. 2019. A Wolbachia infection from Drosophila that causes cytoplasmic incompatibility despite low prevalence and densities in males. Heredity 122(4): 428–440.

Rogers, J., Raveendran, M., Harris, R.A., Mailund, T., Leppälä, K., Athanasiadis, G., ... Worley, K.C. 2019. The comparative genomics and complex population history of Papio baboons. Science Advances 5(1): eaau6947.

Rosselló-Móra, R. and Amann, R. 2015. Past and future species definitions for Bacteria and Archaea. Systematic and Applied Microbiology 38(4): 209–216.

Ryan, F. 2009. Virolution. New York, NY, USA: HarperCollins Publishers.

Ryan, F. 2010. You are half virus. New Scientist 205(2745): 32–35.

Sandoval-Motta, S., Aldana, M. and Frank, A. 2017. Evolving Ecosystems: Inheritance and Selection in the Light of the Microbiome. Archives of Medical Research 48(8): 780–789.

Schaack, S., Gilbert, C. and Feschotte, C. 2010. Promiscuous DNA: horizontal transfer of transposable elements and why it matters for eukaryotic evolution. Trends in Ecology & Evolution 25(9): 537–546.

Shapiro, J.A. 2011. Evolution: A View from the 21st Century. Upper Saddle River, NJ: FT Press.

Shapiro, J.A. 2017. Exploring the read-write genome: mobile DNA and mammalian adaptation. Critical Reviews in Biochemistry and Molecular Biology 52(1): 1–17.

Shapiro, J.A. 2022. What we have learned about evolutionary genome change in the past 7 decades. Biosystems 215–216, 104669.

Sharma, U., Sun, F., Conine, C.C., Reichholf, B., Kukreja, S., Herzog, V.A., ... Rando, O.J. 2018. Small RNAs Are Trafficked from the Epididymis to Developing Mammalian Sperm. Developmental Cell, 46(4): 481–494.e6.

Sharon, G., Segal, D., Ringo, J.M., Hefetz, A., Zilber-Rosenberg, I. and Rosenberg, E. 2010. Commensal bacteria play a role in mating preference of Drosophila melanogaster. Proceedings of the National Academy of Sciences, 107(46): 20051–20056.

Shropshire, J.D. and Bordenstein, S.R. 2016. Speciation by Symbiosis: the Microbiome and Behavior. MBio 7(2).

Soucy, S.M., Huang, J. and Gogarten, J.P. 2015. Horizontal gene transfer: building the web of life. Nature Reviews Genetics 16(8): 472–482.

Stindl, R. 2014. The telomeric sync model of speciation: species-wide telomere erosion triggers cycles of transposon-mediated genomic rearrangements, which underlie the saltatory appearance of nonadaptive characters. Naturwissenschaften 101(3): 163–186.

Stoeckle, M.Y. and Thaler, D.S. 2018. Why should mitochondria define species? BioRxiv 276717.

Tauber, A.I. 2000. Moving beyond the immune self? Seminars in Immunology 12(3): 241–248.

Templeton, A.R. 1998. Human Races: A Genetic and Evolutionary Perspective. American Anthropologist 100(3): 632–650.

Thaler, D.S. 2021. Is Global Microbial Biodiversity Increasing, Decreasing, or Staying the Same? Frontiers in Ecology and Evolution 9(565649).

Tonzo, V., Papadopoulou, A. and Ortego, J. 2019. Genomic data reveal deep genetic structure but no support for current taxonomic designation in a grasshopper species complex. Molecular Ecology 28(17): 3869–3886.

Torday, J.S. and Miller, W.B. 2016. Phenotype as Agent for Epigenetic Inheritance. Biology 5(3): 30.

Torday, J.s. and Miller, W.B. 2020. Cellular-Molecular Mechanisms in Epigenetic Evolutionary Biology Cham, Switzerland: Springer.

Villarreal, L.P. and Ryan, F. 2018. Viruses in the Origin of Life and Its Subsequent Diversification. In: Handbook of Astrobiology. Boca Raton, Florida, United States: CRC Press.

Wang, T., Zeng, J., Lowe, C.B., Sellers, R.G., Salama, S.R., Yang, M., ... Haussler, D. 2007. Species-specific endogenous retroviruses shape the transcriptional network of the human tumor suppressor protein p53. Proceedings of the National Academy of Sciences 104(47): 18613–18618.

Wellenreuther, M., Mérot, C., Berdan, E. and Bernatchez, L. 2019. Going beyond SNPs: The role of structural genomic variants in adaptive evolution and species diversification. Molecular Ecology 28(6): 1203–1209.

Williams, G.C. and Nesse, R.M. 1991. The Dawn of Darwinian Medicine. The Quarterly Review of Biology 66(1): 1–22.

Witzany, G. 2011. Natural genome editing from a biocommunicative perspective. Biosemiotics 4(3): 349–368.

Witzany, G. 2014. Language and Communication as Universal Requirements for Life. pp. 349–369. *In*: V.M. Kolb (ed.). Astrobiology: An Evolutionary Approach. Boca Raton, Florida, United States: CRC Press.

CHAPTER 14

Sexual Reproduction and its Impact on Evolutionary Variation

It is believed that life relied exclusively on asexual reproduction until two billion years ago (Otto 2008). On first consideration, it can be assumed that asexual reproduction has an inherent advantage. It requires no costly competition to find a mate to produce offspring. Noting this, the central justification of the primacy of sexual reproduction in Neodarwinism has been that meiotic division produces greater levels of variation in organisms than nonsexual reproduction (Crow 1994). Sexual reproduction produces visible small variations in immediate offspring, and consequently, there is an anchoring belief that sex must confer an evolutionary advantage of enhanced genetic diversity given its dominance among 99.99% of eukaryotes (Otto 2008). Conversely, to justify its higher costs, sexual reproduction must be associated with a significant increase in the fitness of successive generations of offspring to balance the biological equation. Although this enduring dogma is generally accredited to the German evolutionary biologist August Weismann who articulated this perspective in 1889, that idea was first advanced in 1883 by an American, W.K. Brooks (Moseley 1886). Nonetheless, Weissman is generally recognized with that insight and, consequently, has been considered nearly as influential an evolutionary biologist as Darwin.

Despite near-universal acceptance, the validity of this assumption can be rightfully questioned based on new insights into cellular life in our contemporary era. Cognition-Based Evolution offers an opportunity to reappraise the puzzle of sexual reproduction in a new frame based on cellular problem-solving. When placed in that context, species can be appraised as combinations of cellular constituencies in discrete harmony, particularly concerning immunological competence. Accordingly, from that cellular perspective, any species with its particular and unique combination of linked viral-cellular ecologies has successfully met environmental exigencies. Hence, having survived continuous selective filtering, the perpetuation of that productive combinations of cellular ecologies would be a biological priority. Following that line of reasoning, sexual reproduction should serve as a means of species conservation rather than any attempt to generate biological novelty. The extraordinary longevity of certain species serves as confirmation of this atypical stance. Exceptionally long-lived species with highly conserved morphologies such as horseshoe crabs or sharks indicate that sexual reproduction serves dual purposes: maintaining species cellular homeorhetic balance while permitting sufficient genomic plasticity so that species continuity is maintained by flexible responses to the experienced succession of environmental stresses.

Given that sexual reproduction proceeds via meiosis, it is surprising that the concept that sexual reproduction produces increased genetic variation has gained such thorough currency. There could be no better means of long-term genomic averaging, particularly in a double allele system, than through the continuous meiotic haploid divisions that sexual reproduction imposes. Meiosis is essentially a mathematical halving. While it is true that any meiotic division of the genetic complements of the two parental genomes results in comparative variation in their immediate offspring, when that process is correctly placed in the context of large populations, it expressly becomes a reinforcing mechanism for the continuous averaging of a species genetic complement. Tens or hundreds of thousands or even millions of averaging episodes in continuous succession is a consistent method of achieving a long-term reversion to the mean, undercutting the chances for the persistence of outliers (Miller and Torday 2018). This phenomenon of long-term averaging is readily observable across biology at all scales, including among non-sexually reproducing prokaryotes. For instance, in *Escherichia coli*, frequent recombination is the steady keel that resists the imposition of outlier random genetic variations by assuring that these remain at a low frequency (Dixit et al. 2017).

Although clonal lineages consistently demonstrate a stable total genomic content, they still demonstrate sufficient variability to generally meet environmental challenges. Contrary to superficial expectations, asexual clonal reproduction yields higher rates of genetic variability at twice the efficiency of sexual reproduction (Gorelick and Heng 2011). Gorelick and Heng (2011) note that sex decreases most genetic variation, supporting their assertion through a range of evidence from ecology, paleontology, population genetics, and cancer research. Consequently, sexual reproduction is designed to produce species stasis. A genome is an entire information package rather than a simple collection of genes. Linking species to information systems focuses evolution on the accumulation, creation, and deployment of information rather than a genetic paradigm (Crkvenjakov and Heng 2021).

Further, species are protected at the phenotypic level through sexual reproduction. Epigenetic adjustments of the genomic complement provide limited plasticity to meet environmental stresses and cues to counterbalance the genome-conserving effects of sexual reproduction. However, even this additional epigenetic plasticity is still meant to serve species stability since, through their cumulative impact, the constituent cells of the species can better maintain their homeorhetic equipoise within a context of consistent environmental flux. In this manner, smaller-scale targeted epigenetic variations support the 'pitch' of the species within its natural boundaries (Miller 2013, Torday and Miller 2020). However, it is important to note that even in the instance of epigenetic shifts, sexual reproduction expressly serves to limit their impact too, since they are also being averaged and modulated by dual means – consistent meiotic division and the obligatory unicellular zygotic stage in which those marks are distinctly adjudicated (Torday and Miller 2016). Clearly, then, sexual reproduction is the cellular solution to prevent potentially damaging genetic drift, functioning as a steady keel over evolutionary space-time.

Indeed, there is scientific proof that this is the case, coming from an unexpected source. DNA matching is a powerful tool for identifying relatives. Inadvertently, that same tool has given precise data about the cumulative effects of meiotic division on the human genome measured through the centiMorgan scale (Krimsky 2021). The centiMorgan (cM) is a unit of genetic measurement that gauges the genetic distance between generations based on the length of identical segments of DNA that two people share due to descent from their common ancestor. Generally, people have 6,800 cMs, and any child inherits half from each parent. So a child should match roughly 3,400 cMs of their parent's DNA. Naturally, the more halving, the less the genetic segments match. A grandparent might match 1,700 cMs, for example. Reaching the fifth generation, only 200–340 cMs match. Surprisingly, this small fraction is still enough to establish ancestry; nonetheless, all fifth-generation progeny have thoroughly mixed patterns of ancestral DNA of a still entirely stable human genome.

In a biological system based on information and its valid assessment, sexual reproduction is more than a biologically material event. Sexual reproduction is also an advantaged means of sustaining concordant holobionic partnerships that must be mirrored in N-space. That platform serves multicellular information concordance, and assuring its validity among trillions of cells must be a priority. Indeed, in CBE, genes serve the information management systems of cells. Not only genomic integrity is being re-centered by the obligatory unicellular stage of sexual reproduction. The Pervasive Information Field (PIF) of the cell is also being reset. (Miller 2016). Absent this type of constraint, the critical information tableau of the individual cells as an intrinsic part of the information architecture of the holobiont would degrade. If so, its species identity would become increasingly chaotic. Further, the adjudication of epigenetic marks to meet both short-term environmental exigencies and long-term trends depends on its correlative information space attachment. Although that information field is not material in and of itself, it nonetheless has physicality since all information yields overt biological expression. Maintenance of that crucial information field is as necessary as the genome as it forms the crucial aspect of species morphogenesis and developmental patterning (Miller et al. 2020).

Thus, information content and its validity equate with form in biology, and the mandated recapitulation of sexual reproduction must also serve to protect the integrity of species-specific information management. Therefore, sexual reproduction is more than a means toward reiterative phenotypic expression and rationalizing epigenetic experiences. Sexual reproduction and meiotic averaging are the best means of re-centering the individual PIFs of the cellular constituents of a holobiont. Sexual reproduction also links to a constraining re-centering of the organismal N-space Episenome that governs species identity as an information sub-set and bandwidth of thermodynamic injunctions.

Chief among the self-limiting variations permitted by sexual reproduction is the reinforcement of immunological competence in varied environments. Compared to asexual reproduction, meiotic division is an enhanced mechanism for achieving greater immunological integrity for offspring based on the cumulative immunological experiences of the epigenetic impacts that two different parents contribute. For example, among algae, facultative sexual reproduction is an effective antiviral strategy (Morin 2008). The tiny alga *Emiliania huxleyi* plays a substantial role in ocean ecologies, forming intermittent massive blooms. Viral infections can cause algal bloom collapse, especially due to EhVs, a group of giant phycodnaviruses (Brussaard 2004). Research indicates that *E. huxleyi* will shift from a diploid blooming cell to a morphologically different motile flagellated haploid form with enhanced immune protection from invasive EhV. The haploid cell still gets infected but is resistant to the deleterious effects of EhV and does not undergo cell lysis (Mordecai et al. 2017). This remarkable alternative cellular state is deemed a haplococcolithovirocell. These haploid forms can subsequently mate and reconstitute diploid blooming cells. In this instance, facultative sexual reproduction has been shown to represent an anti-pathogen defense strategy that can be used by rapidly reproducing asexual clonal algal populations.

Research indicates that sexual reproduction provides useful variations while protecting species integrity, specifically relating to disease resistance. A study examined the chitinase A gene in 32 different species of evening primroses (Onagraceae) (Hersh-Green et al. 2012). Chitinase A controls chitinase production, an enzyme that plants can deploy to resist some pathogen-induced diseases such as the fungal disease, powdery mildew. Some primrose species can reproduce asexually, so a direct comparison could be made for the extent of chitinase A expression between the two forms of reproduction. Chitinase A expression was found to be substantially greater in those primrose species that reproduce sexually. The enzyme itself was not changed, just its level of expression. That degree of variation is limited, but sufficient to retain species vigor without compromising species self-identity, further suggesting that sexual reproduction exists to sustain species within boundaries (Miller 2013).

In certain instances, monoculturing techniques have become the mainstay for some consumer agricultural products. Banana cloning to enhance banana output has led to the Cavendish subgroup becoming the lead export banana. That clonal strain is sterile and highly susceptible to Black Leaf Streak and other fungal and viral diseases, representing a substantial impediment to the banana industry (Jones 2007).

Over the decades, many theories about the advantages of sexual reproduction have been advanced. The 'Red Queen Principle was introduced by Van Valen (1973), positing that predator and prey species needed to keep pace with one another in a continuous evolutionary arms race, maintaining co-evolutionary ecological balance. This same narrative has also been applied to explain the evolutionary narrative of the microbial realm (Brussaard 2004). Presuming a perpetual stand-off, the 'Red Queen' references the race between the Red Queen and Alice in Lewis Carroll's 'Through the Looking-Glass' (1871). Despite running furiously, the pair always end up in the same place. In theory, the benefit of sexual reproduction is the preservation of those particular genes that are actively engaged in this constant arms race between hosts and their rapidly reproducing and evolving pathogens and parasites. A long-term study of snails (Potamopyrgus antipodarum) capable of both sexual and asexual reproduction demonstrated asexual clonal populations were more susceptible to parasite infection than sexually reproducing ones (Jokela et al. 2009). Within a 7–10 year period, most clonal populations dwindled over time, even the ones that were initially the most infection-resistant. Conversely, the sexual populations were not significantly perturbed throughout that time frame, offering results in conformity with the Red Queen hypothesis. A further experiment with the nematode Caenorhabditis elegans stressed by exposure to the bacterial pathogen *Serratia marcescens* found that obligatory asexual populations were driven to extinction. In contrast, sexual populations persisted, demonstrating co-evolutionary adaptations with the pathogen (Morran et al. 2011).

Sexual selection theory has been the subject of lavish inquiry by evolutionary biologists. In general, this has been directed toward finding a mechanism for achieving sexual isolation as a means of speciation. Consequently, there has been extensive study of mating cues in animals, including pre-copulatory mechanisms such as sensory cues like bird songs, chemical attractants (pheromones), physical displays, or ritualized courtship (Pélissié et al. 2014). Others have concentrated on investigating post-copulatory mechanisms, including theories of sperm competition or cryptic female choice that might involve sperm preferences. However, significant problems loom. Most animal species are polyandrous and sexually indiscriminate, with a wide variety of partners. Even we humans are no different in that regard. Indeed, promiscuity is the norm, not fidelity. Most sexual activity across the animal kingdom is primarily a product of chance and availability. In 1948, Bateman published data on fruit flies that strongly influenced the subsequent decades of research, proposing that males produce abundant sperm and thereby, are likely to mate promiscously, whereas female produce far fewer, larger, and more 'expensive' eggs and are likely to be very discriminating by selecting only one fit partner. Despite the considerable currency this theory gained, careful analysis has revealed that it is flawed (Tang-Martínez 2016). Sexual selection is highly pliant and defies conventional theory. Further complicating any simple narrative of this sort, masturbation is prevalent in animals, from mammals to insects, and same-sex activity is also ubiquitous (Natterson-Horowitz and Bowers 2012).

Further yet, deception is rampant in sexual matters. Consequently, the dominating post-Victorian concept that mating rights go to the fittest individuals within species is not grounded in the available evidence. Stealthy mating is routine, and so are multiple matings, even among harem females who remain surprisingly promiscuous (Garg et al. 2012). In the Darwinian world, the dominant male fights and wins the rights to a docile female or has won her by dint of dominant sexual displays. That makes for fascinating nature documentaries but is not any evolutionary reality. In the fuller chronicle, there is a more complex dynamic at play. It is a planet filled with stealth, abundant non-specific sexual cues, and high levels of indiscriminate promiscuity.

Consequently, there is no mystery why biology preserves stability instead of any direct impulse towards variation and greater genetic diversity. The point of all the embedded constraints within biological systems is to explicitly maintain the ordered whole that defines that species' successful perpetuation against stresses. Since an act of speciation is a variation from a previous centrality of cellular harmony, then speciation would also be resisted. All holobionts represent a form of negotiated compact. If there is a substantial displacement, then the 'ordered whole' is challenged. All biological mechanisms direct toward protecting species' self-integrity because it represents a successful reiterating manifestation of core cellular functions. Limited variations are accommodated because they serve to protect that core. As the aim of evolution is the continuous cellular endogenization of environmental cues, a speciation event is not necessarily an improvement in that process. Not all variants are advantageous (Otto 2008).

Naturally, some variants are desirable as they are a means of adjusting to environmental stresses. How do modular species cope with this requirement? Clonal species such as corals and seagrasses are ancient and have achieved their balance of minor variations around a central genomic core. Through a system of somatic shifts in asexual reproduction, minor changes are accommodated that mirror soma-germline transfers in sexually reproducing species, once considered insurmountable (Reusch et al. 2021). Indeed, somatic evolution is embedded in both reproductive systems, seen in many clonal species among select plants, fungi, and animals, but also evinced within sexual reproduction.

A commonplace justification for sexual reproduction has been a persistent assertion that it confers selective advantages compared to asexually reproducing organisms (Barton and Charlesworth 1998). In theory, greater numbers of deleterious mutations should accumulate through obligate selfing than through sexual reproduction, leading to declining fitness, higher rates of extinction, or lack of speciation. Yet, bdelloid rotifers are 'ancient asexuals' with a history that extends back at least 25 million years based on fossil evidence, an exceptional species age (Poinar and Ricci 1992). Further, they demonstrate genetic diversity, allowing them to be divided into separate species based on genetic sequencing and divergent morphologies such as jaw shape (Fontaneto et al. 2007).

Assuredly, sexual reproduction is a dominating reality in our biological system. From that fact, placed within the proper framework, sexual reproduction persists precisely because it is a barrier to speciation (Crkvenjakov and Heng 2021). Typically, speciation is supposed to relate directly to genes. In sexual reproduction, slight variations obviously emerge when two individuals produce the next generation. In the context of populations, however, any significant variations are suppressed. Notably, this effect is not confined to genes but similarly involves critical membrane functions and cytoplasmic factors. Crucially, the essential information matrix upon which biological action relies is also being conserved as its own expression of the cellular compact that retains the balanced harmony of the linked cellular ecologies of holobionts.

Necessarily then, sexual reproduction should be regarded as a biological lock and key mechanism as one further critical constraint in a biological system governed by infectious dynamics and other potentially chaotic inputs. Certainly, fertilization by sexual reproduction is a tightly controlled genetic exchange which should be viewed as a proscribed infectious genetic cell-cell transfer. In a microbial world, there are many avenues for microbial mischief during the complex dynamics of fertilization. Sexual reproduction must be an advantaged means for these myriad potentials to be rigorously controlled. The continual mix and match of the great variety of immunological factors such as histocompatibility loci serves to sustain the totality of cellular and organismal defenses that are always under assault, thereby preserving species survivability, boundaries, and self-integrity. A species represents a time-tested set of collective cellular solutions to cellular problems. Sexual reproduction is the cellular means of preserving that advantaged solution.

References

Barton, N.H. and Charlesworth, B. 1998. Why sex and recombination?. Science 281(5385): 1986–1990.

Bateman, A.J. 1948. Intra-sexual selection in Drosophila. Heredity 2(3): 349–368.

Brussaard, C.P.D. 2004. Viral Control of Phytoplankton Populations-a Review1. The Journal of Eukaryotic Microbiology 51(2): 125–138.

Carroll, L. (2010; 1871). Through the looking glass and what Alice found there. London: Penguin.

Crkvenjakov, R. and Heng, H.H. 2021. Further illusions: On key evolutionary mechanisms that could never fit with Modern Synthesis. Progress in Biophysics and Molecular Biology 169–170: 3–11.

Crow, J.F. 1994. Advantages of sexual reproduction. Developmental Genetics 15(3): 205–213.

Dixit, P.D., Pang, T.Y. and Maslov, S. 2017. Recombination-Driven Genome Evolution and Stability of Bacterial Species. Genetics 207(1): 281–295.

Fontaneto, D., Herniou, E.A., Boschetti, C., Caprioli, M., Melone, G., Ricci, C. and Barraclough, T.G. 2007. Independently Evolving Species in Asexual Bdelloid Rotifers. PLoS Biology 5(4): e87.

Garg, K.M., Chattopadhyay, B., DOSS D, P. S., AK, V. K., Kandula, S. and Ramakrishnan, U. 2012. Promiscuous mating in the harem-roosting fruit bat, Cynopterus sphinx. Molecular Ecology 21(16): 4093–4105.

Gorelick, R. and Heng, H.H.Q. 2011. Sex Reduces Genetic Variation: A Multidisciplinary Review. Evolution: International Journal of Organic Evolution 65(4): 1088–1098.

Hersch-Green, E.I., Myburg, H. and Johnson, M.T.J. 2012. Adaptive molecular evolution of a defence gene in sexual but not functionally asexual evening primroses. Journal of Evolutionary Biology 25(8): 1576–1586.

Jokela, J., Dybdahl, M.F. and Lively, C.M. 2009. The Maintenance of Sex, Clonal Dynamics, and Host-Parasite Coevolution in a Mixed Population of Sexual and Asexual Snails. The American Naturalist 174(S1): S43–S53.

Jones, D.R. 2007. Disease And Pest Constraints To Banana Production. Acta Horticulturae (828): 21–36.

Krimsky, S. 2021. Understanding DNA Ancestry. Cambridge, England: Cambridge University Press.

Miller, W.B. 2013. The Microcosm Within: Evolution and Extinction in the Hologenome. Boca Raton, Florida, United States: Universal Publishers.

Miller, W.B. 2016. Cognition, Information Fields and Hologenomic Entanglement: Evolution in Light and Shadow. Biology 5(2): 21.

Miller, W.B. and Torday, J.S. 2018. Four domains: The fundamental unicell and Post-Darwinian Cognition-Based Evolution. Progress in Biophysics and Molecular Biology 140: 49–73.

Miller, W.B., Torday, J.S. and Baluška, F. 2020. The N-space Episenome unifies cellular information space-time within cognition-based evolution. Progress in Biophysics and Molecular Biology 150: 112–139.

Mordecai, G., Verret, F., Highfield, A. and Schroeder, D. 2017. Schrödinger's Cheshire Cat: Are Haploid Emiliania huxleyi Cells Resistant to Viral Infection or Not? Viruses 9(3): 51.

Morin, P.J. 2008. Sex as an algal antiviral strategy. Proceedings of the National Academy of Sciences 105(41), 15639–15640.

Morran, L.T., Schmidt, O.G., Gelarden, I.A., Parrish, R.C. and Lively, C.M. 2011. Running with the Red Queen: Host-Parasite Coevolution Selects for Biparental Sex. Science 333(6039): 216–218.

Moseley, H.N. 1886. Dr. August Weismann on the Importance of Sexual Reproduction for the theory of Selection 1. Nature 34(887): 629–632.

Natterson-Horowitz, B. and Bowers, K. 2012. Zoobiquity: What animals can teach us about health and the Science of Healing. Knopf, New York: Penquin Random House

Otto, S.P. 2008. Sexual Reproduction and the Evolution of Sex. Nature Education 1(1): 182.

Pélissié, B., Jarne, P., Sarda, V. and David, P. 2014. Disentangling Precopulatory And Postcopulatory Sexual Selection In Polyandrous Species. Evolution 68(5): 1320–1331.

Poinar, G.O. and Ricci, C. 1992. Bdelloid rotifers in Dominican amber: Evidence for parthenogenetic continuity. Experientia 48(4): 408–410.

Reusch, T.B.H., Baums, I.B. and Werner, B. 2021. Evolution via somatic genetic variation in modular species. Trends in Ecology & Evolution 36(12): 1083–1092.

Tang-Martínez, Z. 2016. Rethinking Bateman's Principles: Challenging Persistent Myths of Sexually Reluctant Females and Promiscuous Males. The Journal of Sex Research, 53(4–5): 532–559.

Torday, J. and Miller, W. 2016. The Unicellular State as a Point Source in a Quantum Biological System. Biology 5(2): 25.

Torday, J. and Miller, W. 2020. Cellular-Molecular Mechanisms in Epigenetic Evolutionary Biology. Cham, Switzerland: Springer.

Valen, L. Van. 1973. A new evolutionary law. Evolution Theory 1: 1–30.

Chapter 15

Extinction

Although extinctions are typically studied as a separate discipline within evolutionary biology, there is no point in considering the continuous process of species extinctions disembodied from a coherent theory of evolution. Extinction is as much a part of evolution as speciation. Placing extinction within that more comprehensive framework, both extinction and speciation can be viewed as separable outcomes emerging from the same biological interplay; speciation and extinction are both epiphenomena of the eternal dynamic of infectious interactions among the four domains.

Extinction theory was formally inaugurated in 1799 by the French naturalist Georges Cuvier (1769–1932). Until then, naturalists believed that any unusual fossil creatures that were uncovered were simply uncommon examples of existing species (Wiens and Worsley 2016). Cuvier's examination of the various strata of rock in the Paris basin convinced him that each had its distinct distribution of mammal fauna, with the oldest quite distant from living species. These fossils represented extinct species just as had been believed by classical Greek scholars, and Cuvier proposed a theory of catastrophism to account for them (Racki 2015).

Surprisingly, most books on evolution contain little or no information about extinctions. For example, Kirschner and Gerhart's (2005) book on evolution, 'The Plausibility of Life', contains no reference to extinctions. Shapiro (2011), in a brilliant text, 'Evolution: A View from the 21st Century', includes two sparse pages mentioning extinctions, focusing on the rapid diversifications that follow rather than their causes. In their comprehensive text, 'Speciation', Coyne and Orr (2004) barely touch on extinctions, only noting that reinforcement patterns and sexual selection might pertain. Accordingly, the causes of extinction are not of particular interest to most evolutionary theorists. However, a few scientists have concentrated on the field, offering many potential triggers.

In a meticulously researched 2006 book, 'Extinction: How Life on Earth Nearly Ended 250 Million Years Ago', paleobiologist Douglas Erwin catalogs multiple possible causes of extinctions, concentrating on the catastrophic mass extinction of the end Permian when 95% of all life was extinguished (Erwin 2006). Potential candidate triggers include extraterrestrial impacts from meteors or asteroids, and climatic shifts from supervolcanism, glaciation cycles, or oxygen depletion.

Curiously, however, infectious disease dynamics, the most consequential of all biological processes, is not even mentioned.

At least five prior episodes of mass extinction with losses of 75% of all life forms are known since the Cambrian and are generally traced to periods of geological upheaval (Cowie et al. 2022). Nonetheless, extinctions are always ongoing, with mass extinction events seen as episodes of extreme amplitude compared to the acknowledged background rate. Although subject to a widely disputed range, the general estimate of species extinctions is approximately one species extinction per million species years (Lamkin and Miller 2016). A further calculation suggests that the average species exists between five and ten million years before extinction (Lawton and May 1995). Those steady state extinctions have been attributed to various causes, including loss of habitat, lack of genetic diversity leading to inbreeding, inadequate adaptation, disease, and invasive species, including humans. It deserves emphasis that the 'Big Five' mass extinctions can account for only approximately 4% of all Phanerozoic extinctions, the entire period beginning with the Cambrian and comprising the Paleozoic, Mesozoic, and Cenozoic eras (Raup 1992).

At least in the public's mind, the 'sudden' extinction event that led to the demise of the dinosaurs is the best known. The Cretaceous-Tertiary (K-T) extinction 66 million years ago extinguished three-quarters of the plant and animal species on Earth and has been attributed to an asteroid impact causing the massive Chicxulub crater in the Yucatán peninsula of Mexico. In the popular imagination, this impact caused the near-instantaneous extinction of the dinosaurs. Notably, however, the impact can only be placed stratigraphically within 33,000 years of the actual end of the dinosaurs (Renne et al. 2013). Although the dinosaur extinction would be considered abrupt in geological terms, that extinction was not swift in terms of population dynamics. Further, the extinction of many other taxa causally linked to that impact was delayed for hundreds of thousands of years, possibly also relating to massive shifts in ecosystems due to volcanism (Brusatte et al. 2015). Consequently, some scientists believe that other causes for the extinction of the dinosaurs beyond catastrophic events should be considered, including epidemic diseases (Bakker 1986).

In their controversial 2007 book, *'What Bugged the Dinosaurs?'*, George and Roberta Poinar argued that the dinosaur extinction was due to disease-carrying insects, citing evidence of insects preserved in amber that overlap the disappearance of the dinosaurs (Poinar and Poinar 2007). They contend that dinosaurs declined over hundreds of thousands of years, perhaps even longer, due to insects carrying novel pathogens. Notably, the accompanying spread of flowering plants over this extended period, many of which depend on pollination by insects, also comports with the timeline of dinosaur extinctions. Additional suggestive evidence has been found in dinosaur eggshell samples. Electron microscopic scans have revealed unusual fossilized objects inside the pneumatic canals of the eggshells, representing the first evidence of fungal contamination of dinosaur eggs (Agnolin et al. 2012). Furthermore, an in-depth examination of full thickness erosions found in the mandible of a Tyrannosaurus Rex fossil is strongly suggestive of actinomycosis infection, known to afflict modern birds (Wolff et al. 2009).

Although there is no lack of potential causes that have been championed in the past to account for either catastrophic mass extinctions or the ongoing rate of extinction events that characterize our planet, devising a single, comprehensive extinction paradigm has been elusive. The recently introduced theory of 'multigenerational, attritional loss of reproductive fitness' (MALF) partially overcomes this deficiency (Wiens and Worsley 2016). Evidence from flowering plants and animals supports this new narrative, arguing that the commonplace attribution of extinctions to catastrophic events is untenable when their actual timelines are properly accounted.

MALF proposes a simple alternative. All populations must reproduce to survive as it is the "singular process responsible for the continuity of life and is a function common to all organisms that ever existed" (Wiens and Worsley 2016, p. 1097). If that reproductive vigor is absent, those populations are subject to extinction. This mandate has been traditionally overlooked since naturalists and biologists, beginning with Darwin and Wallace, have emphasized excess reproduction as a fundamental evolutionary tenet, viewing life as a planetary struggle. However, that narrative is unduly restrictive, given that excess is not always the case. For example, some flowering plants have extremely high embryonic abortion rates, ranging from 97–99%, which interestingly also coincides with elevated rates of genetic variation (Wiens et al. 2012). Further, some populations have naturally slow reproductive replacement rates and can still thrive. Consequently, there is a need to have a more comprehensive viewpoint about extinction, and MALF suggests a more pluralistic approach as the best solution, contending that extinctions are due to the accumulated weight of predation, disease, genetic factors, and climate changes. Accordingly, catastrophic extinctions associated with nearly instantaneous losses are a relatively minor factor in biohistory compared to MALF as a 'universal mechanism' of population decline and eventual extinction (Wiens et al. 2020).

MALF theory has also been applied to an ecological framework (Congreve et al. 2018). Within that theory, biological systems are ecological units that are 'nested and interconnected' into biological and evolutionary systems. In this type of hierarchical structure, MALF experienced at a lower level can spill over into larger patterns of differential species survival at the clade level dependent on the initiating mechanisms shaped by environmentally mediated patterns.

No theory of extinction or evolutionary development should be considered complete unless it can provide a discrete set of answers for extinctions as a phenomenon so commonplace that more than 99% of species that have existed are now extinct (Raup 1991). When extinctions and speciation are always ongoing, the answer to their causes must be sought within natural processes that overlap that same timeline. Van Valen (1973) observed that the probability of extinctions for organisms remains constant over evolutionary time scales. Further, no specific correlation can be found between extinction rates and the species' age. When these relevant factors are weighed, the most fundamental biological planetary dynamic that most determines reproductive success and the health of populations should be considered first. That operative agency is omnipresent infectious assault and defense, its protean biological manifestations, and its varied immunological consequences. As infectious disease dynamics underlie the fluid exchanges of genetic material that stimulate evolutionary development and speciation, it must apply equally to extinctions as one other aspect of the broad spectrum of potential outcomes of the replete panoply of biological manifestations of infectious interchanges (Miller 2013).

Cognition-Based Evolution permits the coherent integration of MALF with infectious interchanges, becoming a complete extinction paradigm based on several linked, supporting concepts. In our biological system, infectious disease dynamics is the primary process driving population health and evolution. Therefore, infectious disease dynamics must intrinsically relate to any general concept of reproductive fitness and its progressive decline. A generalized loss of reproductive fitness in populations could be stimulated by infectious diseases from bacterial infections, viruses, or parasites. Further, the impact of these would be exacerbated by malnutrition, harsh climate conditions, or other stressors. These types of infectious interchanges could incur through several patterns of incidence, with all of them having the demonstrable capacity to substantially reduce or eliminate select populations. Those various types of infectious insults can be cataloged in five dominant patterns.

A. Fluctuating cycles of current endemic pathogens

Microorganisms that are a historical part of any ecological landscape can undergo intermittent cycles of increased virulence. For example, in 2006, a particularly virulent strain of Ebola devastated an estimated 90% of all gorilla species over four years, threatening extinction (Bermejo et al. 2006). As another example, and again just recently, an unusual infectious event of Big Horn sheep in Oregon in early 1980 and 1981 led to a loss of 75–80% of local sheep populations. In that instance, co-infection with Pasteurella species (*Pasteurella multocida, Corynebacterium pyogenes,* and *Protostrongylus stilesi Dikmans, 1931*) led to widespread bronchopneumonia with 80% die-off of lambs in one summer and a 100% mortality the next, followed by 67% lamb mortality for two further summers (Spraker et al. 1984, Schommer and Woolever 2008).

As another like-kind example, episodes of colony collapse disorder in bats from fungal-induced White Nose Syndrome have led to devastating bat population declines, extirpating them over large parts of their ranges (Blehert et al. 2009, Thogmartin et al. 2013). Similarly, amphibian populations worldwide are threatened by chytrid fungus, identified in 1998. This background fungus is mysteriously causing massive amphibian population losses within many species, with an estimated 40% facing extinction despite relatively stable ecosystems (Wake 2012).

Such abrupt infection-induced population declines are not restricted to animals. Among plants, the American chestnut was the dominant hardwood species in the eastern United States, with vast stands from Maine to Florida. However, beginning in the 19th century, a series of fungal infections plagued them, leading to their near extinction in North America.

A 2002 experiment revealed some of the dynamics of these types of infections. Researchers infected 24 diverse species of Drosophila with five different nematode species as obligate parasites and studied disease spread (Perlman and Jaenike 2003). The number of species that could be purposely infected in the lab was much greater than seen in nature, and curiously, one nematode species, Parasitylenchus nearcticus, sterilized all *Quinaria* flies. Notably, this same nematode is associated with diminished fecundity in other species, inducing sterility in females.

B. The introduction of novel or zoonotic infections

Our recent experience with SARS-CoV-2 is an acute example of the potential lethality of novel pathogens. But, of course, this is merely a contemporary manifestation of the devastating plagues that have afflicted all populations throughout history. In the recorded history of human plagues, Sars-CoV-2 was fortunately relatively mild. Bubonic plague during the Middle Ages led to total population mortality that was an order of magnitude greater. Over one-third of Europe's entire population died in successive waves, and the cumulative toll may have been even worse in Asia.

With these types of pandemics, the typical pattern of annual infectious spread of common diseases can be altered. For example, annual flu is a significant societal burden, but mortality is generally experienced among the elderly, infirm, or very young with immature immune systems. However, in the 1918–1919 Great Influenza, a novel strain, over 500 million people were estimated to have been symptomatically infected, and 50–100 million people died globally. For still unknown reasons, that influenza outbreak preferentially targeted the reproductive young and healthy compared to the elderly or pediatric populations (Martini et al. 2019).

Hyperdisease represents a variant form of zoonotic infection, denoting the introduction of novel infectious pathogens into an unprepared ecosystem by new participants with devastating consequences for the resident populations. In the case of hyperdisease zoonosis, the carriers are not highly susceptible themselves since the diseases they carry have long been part of the infectious disease milieu in their location of origin. The most widely known of these events is the devastating population collapses from measles, smallpox, and typhus among the immunologically unprepared Aztec and Native American tribes brought by European colonization (Thornton 1987, Bianchine and Russo 1992, Diamond 1997).

Another instance of hyperdisease is the extinction of an "ark full" of Australian animal species 50,000 years ago, coinciding with the arrival of humans (Monastersky 1999). Similarly, when the black rat was inadvertently introduced to Christmas Island in 1899 CE, murid trypanosome infection spread from the newly arrived black rats, wiping out the endemic Christmas Island rat (*Rattus macleari*) (Wyatt et al. 2008). It all happened in fewer than ten years. This event was the first instance of a directly identified and confirmed correlation between a new competitor carrying a deadly pathogen and a localized extinction (Wyatt et al. 2008).

Of course, the disruptive effects of the introduction of a novel predator do not have to only include infectious pathogens. The predatory brown tree snake *Boiga irregularis* was introduced into Guam during the early 1950s. The direct and indirect extinction of many native vertebrate species has been linked to the introduction of that effective, novel predator (Fritts and Rodda 1998). Nonetheless, even in this instance, the pressure of constant and unanticipated predation is an immunostressor and would lead to a diminution in disease resistance.

The hyperdisease concept as a mediator of extinctions has gained considerable attention since some scientists have attributed the extinction of the North American mammalian megafauna 12,000 years ago to that phenomenon. Macphee and Marx (1997) insist that the megafauna extinctions coincide with the arrival of humans and their domesticated animals, any of which could have been carriers of novel diseases to an immunologically unprepared animal ecosystem .

Indeed, there is a modern example of this phenomenon. In 1859, the European rabbit was introduced to Australia as a source of game. Since it has no native predators, its population exploded, becoming a significant pest and devastating agricultural grassland. To quell this vexing problem, rabbits infected with a deadly myxoma virus were introduced into the native rabbit population. Three months later, 99.8% of the rabbits of southeast Australia were dead (Kerr and Best 1998). Consequently, the contention by Poinar and Poinar (2007) that the dinosaur extinction was due to epidemic disease from novel pathogens coming from insects gains strength as an example of an extended period of vulnerability to hyperdisease. However, the death of the dinosaurs was far from instantaneous as previously noted. Collected evidence argues that the Chicxulub impact might have predated the complete extinction of the dinosaurs by as many as 300,000 years (Keller et al. 2009).

C. New predators secondary to speciation

The same narrative of novel pathogens from new ecosystem participants applies equally to any new species that form within ecosystems, as well as those that travel. The birth of new species represents the origination of a unique, cohesive, complex holobionic set of coadapted communities that includes obligatory microbial constituents. Consequently, if novel microbes are part of this new constituency, the birth of a new species might equally equate with the death of another. Of course, the same is especially true if that new species is a novel predator that can upset the balance of a previously stable ecosystem, like the human-caused extinction of the dodo bird. Since speciation is considered just another feature of infectious disease dynamics, any new species becomes a form of invasive species both by dint of phenotypic capability and by potentially carrying new microbial competitors into an ecology.

D. Sexually transmitted pathogens

Sexually transmitted diseases are often overlooked as a cause of death or infertility. Syphilis causes both and is also associated with congenital malformations that would contribute to a loss of reproductive fitness. In some species, sexually transmitted diseases have led to observable threats to species survival. For example, sexually transmitted *Chlamydia pecorum* infections represent a major threat to the survival of the koala, causing blindness, infertility, pneumonia, and urinary tract infection affecting over 50% of the population. Koala retrovirus infection has had epidemic spread, can be transmitted at birth, and suppresses the immune system causing Koala Immune Deficiency Syndrome, linking to high mortality from immune deficits, leukemia, lymphoma, and solid tumors. In some areas of Queensland, Australia, the infection rate is 100% (Simmons et al. 2012). The virus has even endogenized into the Koala genome, an event that has been connected to a significant increase in cancer incidence (McEwen et al. 2021).

E. Loss of a critical symbiont and related causes of immune dysregulation

Among holobionts, the loss of a critical symbiont could affect reproductive fitness. Several insect species are critically dependent on *Wolbachia* parasites to reproduce (Werren 2003). Research has only recently begun to explore the critical balances between holobionic participants in maintaining sexual capacity or sexual attraction. For example, microbial byproducts facilitate reproduction in the *Trichuris muris* (whipworm) nematodes that thrive in the microbiota-rich environment of the cecum of mice. If a mutant Escherichia coli is substituted for the mice's natural E. coli symbionts, the T. muris fails to lay viable eggs (Venzon et al. 2022).

Further, the interaction between critical symbionts and holobionts is based on immunological tolerance. Just as immunology is crucial to speciation, it is equally vital to reproduction. Losing a critical symbiont, adding another, or conflict among existing symbionts can lead to significant immunological mismatches, impairing successful reproduction. Immunological compatibilities are related to loss of reproductive fitness and have been identified as one of the causes of infertility among human couples. Immunological dysregulation is also a significant factor in implantation failure and female infertility (Sarkesh et al. 2022).

A common denominator can be identified that link all of the just enumerated infectious mechanisms. Any immunological incompatibility equates with cell-cell recognition mismatches. Since all biological activity proceeds via the cell-cell communication of information, immunological incompatibility directly links to an organism's capacity for cellular-environmental engineering, which can lead to failed environmental assimilation. Therefore, even subtle infectious processes could introduce a relative immune deficiency with little apparent immediate effect but could cascade over time and generations. These might easily be influenced by other subsequent epigenetic impacts, eventually diminishing reproductive fitness along with a generalized failure to thrive.

Quite pertinently, phenotype is measured environmental prediction. It is an expression of natural viral-cellular engineering for the coordinated interrogation of information space-time in the continuous defense of individual cellular self-identity. Therefore, extinctions occur when species phenotypes prove to be bad predictions. What is the most likely cause of impaired predictions? Undoubtedly, not typical environmental patterns, even at their extremes. Nearly all creatures are prepared for those variations by virtue of their continuous successful adaptations, permitting them to be enduring species in the first place. Instead, an extinction event would require an unexpected criticality, and the most common of these, by far, is an infectious onslaught for which the species is wholly unprepared.

When immunological incompatibilities compound and increasingly impact cell-cell recognition systems, stress on cellular and organismal information management systems rises leading to dysfunctional co-engineering. Immune functions are the most sensitive barometer of that status, and any dysfunctions can, over time, lead to a gradual breakdown in species robustness and ability to cope with stresses. In effect, the bandwidth of recoverable environmental insults diminishes, leading to a species type of 'failure to thrive', mirroring a well-accepted biological concept within pediatric growth and development.

In medical terms, failure to thrive denotes a situation in which no single health variable is drastically damaged, but the cumulative weight of a constellation of more minor deficiencies constitutes a significant detrimental health condition. Although originally formulated to indicate the failure of proper age-related pediatric weight gain due to poor nutrition or disease, current clinical concepts recognize failure to thrive as a global system deficiency encompassing weight, nutrition, height, glucose dysregulation, diminished motor skills, and psychosocial cognitive dysfunction (Jaffe 2011). These same deficiencies that can afflict individual children have been observed across human populations during times of famine or war, leading to plunging fertility rates. Nevertheless, this entire constellation represents phenotypic and immunological instability in the face of environmental stresses. In both individuals and populations, its ultimate long-term consequences are conspicuously associated with reproductive decline and loss of reproductive vigor due to well-documented growth, cognitive, and behavioral complications (Larson-Nath and Biank 2016). In such circumstances, the circulation of an infectious pathogen not ordinarily associated with significant mortality might become a highly damaging incremental stress on an already weakened immune system. Indeed, this is a not uncommon phenomenon since many of our companion microbes represent pathobionts, as discussed in Chapter 10.

No matter the specific triggers, whether due to localized epidemics, pandemics, or hyperdisease, if the mortality rate from the initiating incidence is great enough, the size of the interbreeding population may fall below a critical threshold and be unable to recover, ultimately leading to extinction (MacPhee and Mark 1998). There is no doubt that population declines risk species extinction. Statistical modeling indicates that disease affects general population declines, risk of extinction, and biodiversity through the Allee effect (Friedman and Yakubu 2012). The Allee threshold is the point at which the survival of a population is challenged due to the loss of genetic diversity due to inbreeding, decreased mating opportunities, and loss of communal survival strategies. Current examples include the Florida panther and the African wild dog.

Therefore, it is unnecessary for an epidemic infection to have 100% virulence to cause an extinction event. Infirmity of the surviving members is sufficient to weaken a population to the point that a previously stable predator-prey relationship is upset, ultimately leading to gradual species extinction. Indeed, those same factors influencing animal extinctions that capture our attention must apply equally to plants. However, the analysis of mass extinctions typically has little to say about how these occur. Others have noted this research gap. Willis and Bennett (1995, p. 309) studied this and concluded," Major changes appear to have occurred in the plant record, but not at the same time as the animal record, [or] in the same dramatic way, but they have still occurred. Without evidence of mass extinction to explain these changes, it is suggested that there must be other driving forces behind plant extinctions". Accordingly, there is a need for a comprehensive extinction narrative that can account for all observations.

An approach through a conceptual frame of a loss of reproductive fitness (MALF) linked to infectious disease dynamics offers just such a productive clarification.

When unusual infectious events weaken populations, inadequate nutrition and other cumulative environmental stresses serve as further immunological triggers. In effect, the phenomenon of extinction is a final expression of a population-wide 'failure to thrive'. In this way, both rare mass extinctions and the continuous background extinction rate are reconciled through different timelines of a collective loss in reproductive dynamism.

The signature feature of infectious disease dynamics is its cyclical chaotic waveforms, matching the highly variable pattern of extinctions across the kingdoms over evolutionary space-time, linking biological realities to environmental hazards in a single continuum. When extinction is viewed within this coherent frame, it yields an identifiable final common framework. No matter its instigation, extinction always represents a mismatch between phenotype and current environmental realities as experienced at its requisite cellular level. This mismatch necessarily represents a failure of cell-cell recognition competencies that exerts unavoidable consequences on established patterns of viral-cellular co-engineering that grant planetary survival. In this regard, the dynamic declining trajectory of a previously robust species well suited to its environmental niche resembles senescence and aging experienced by all organisms.

Extinction theory cannot reside disembodied from evolutionary theory. Extinction is as much a part of evolution as variation and speciation. Satisfying the principle of Occam's Razor, they all must be the separable expression of the same biological process as different manifestations of the eternal dynamic of infectious interactions and their impact on cell-cell communications and immunological competencies among the four domains.

References

Agnolin, F.L., Powell, J.E., Novas, F.E. and Kundrát, M. 2012. New alvarezsaurid (Dinosauria, Theropoda) from uppermost Cretaceous of north-western Patagonia with associated eggs. Cretaceous Research 35: 33–56.

Bakker, R.T. 1986. The Dinosaur Heresies: New Theories Unlocking the Mystery of the Dinosaurs and Their Extinction. New York City, NY, USA: William Morrow.

Bermejo, M., Rodríguez-Teijeiro, J.D., Illera, G., Barroso, A., Vilà, C. and Walsh, P.D. 2006. Ebola Outbreak Killed 5000 Gorillas. Science 314(5805): 1564.

Bianchine, P.J. and Russo, T.A. 1992. The Role of Epidemic Infectious Diseases in the Discovery of America. Allergy and Asthma Proceedings 13(5): 225–232.

Blehert, D.S., Hicks, A.C., Behr, M., Meteyer, C.U., Berlowski-Zier, B.M., Buckles, E.L., … Stone, W.B. 2009. Bat White-Nose Syndrome: An Emerging Fungal Pathogen? Science 323(5911): 227.

Brusatte, S.L., Butler, R.J., Barrett, P.M., Carrano, M.T., Evans, D.C., Lloyd, G.T., … Williamson, T.E. 2015. The extinction of the dinosaurs. Biological Reviews 90(2): 628–642.

Congreve, C.R., Falk, A.R. and Lamsdell, J.C. 2018. Biological hierarchies and the nature of extinction. Biological Reviews 93(2): 811–826.

Cowie, R.H., Bouchet, P. and Fontaine, B. 2022. The Sixth Mass Extinction: fact, fiction or speculation? Biological Reviews 97(2): 640–663.

Coyne, J.A. and Orr, H.A. 2004. Speciation. Sunderland, MA: Sinauer Associates.

Diamond, J. 1997. Guns, Germs, and Steel: The Fates of Human Societies. New York: W.W. Norton & Company.

Erwin, D.H. 2006. How life on earth nearly ended 250 million years ago. Princeton, New Jersey: Princeton University Press.

Friedman, A. and Yakubu, A.-A. 2012. Fatal disease and demographic Allee effect: population persistence and extinction. Journal of Biological Dynamics 6(2): 495–508.

Fritts, T.H. and Rodda, G.H. 1998. The Role Of Introduced Species *In*: The Degradation Of Island Ecosystems: A Case History Of Guam. Annual Review of Ecology and Systematics 29(1): 113–140.

Jaffe, A.C. 2011. Failure to Thrive: Current Clinical Concepts. Pediatrics in Review 32(3): 100–108.

Keller, G., Abramovich, S., Berner, Z. and Adatte, T. 2009. Biotic effects of the Chicxulub impact, K–T catastrophe and sea level change in Texas. Palaeogeography, Palaeoclimatology, Palaeoecology 271(1–2): 52–68.

Kerr, P.J. and Best, S.M. 1998. Myxoma virus in rabbits. Revue Scientifique et Technique de l'OIE 17(1): 256–268.

Kirschner, M.W. and Gerhart, J.C. 2005. The Plausibility of Life: Resolving Darwin's Dilemma. New Haven: Yale University Press.

Lamkin, M. and Miller, A.I. 2016. On the Challenge of Comparing Contemporary and Deep-Time Biological-Extinction Rates. BioScience 66(9): 785–789.

Larson-Nath, C. and Biank, V.F. 2016. Clinical Review of Failure to Thrive in Pediatric Patients. Pediatric Annals, 45(2).

Lawton, J.H. and May, R.M. (eds.). (1995). Extinction rates (Vol. 11). Oxford: Oxford University Press.

MacPhee, R.D.E. and Marx, P.A. 1997. The 40,000-Year Plague: Humans, Hyperdisease, and First-Contact Extinctions. pp. 169–217. *In*: S.M. Goodman and B.D. Patterson (eds.). Natural Change and Human Impact in Madagascar. Washington D.C.: Smithsonian Institution Scholarly Press.

MacPhee, R.D.E. and Marxs, P.A. 1998. Lightning Strikes Twice: Blitzkrieg, Hyperdisease, and Global Explanations of the Late Quaternary Catastrophic Extinctions. American Museum of Natural History.

Martini, M., Gazzaniga, V., Bragazzi, N.L. and Barberis, I. 2019. The Spanish Influenza Pandemic: A lesson from history 100 years after 1918. Journal of Preventive Medicine and Hygiene 60(1): E64–E67.

McEwen, G.K., Alquezar-Planas, D.E., Dayaram, A., Gillett, A., Tarlinton, R., Mongan, N., ... Greenwood, A.D. 2021. Retroviral integrations contribute to elevated host cancer rates during germline invasion. Nature Communications 12(1): 1316.

Miller, W.B. 2013. The Microcosm Within: Evolution and Extinction in the Hologenome. Boca Raton, Florida, United States: Universal Publishers.

Monastersky, R. 1999. The Killing Fields. Science News, 156(23): 360–361.

Perlman, S.J. and Jaenike, J. 2003. Infection Success In Novel Hosts: An Experimental And Phylogenetic Study of Drosophila-Parasitic Nematodes. Evolution 57(3): 544–557.

Poinar Jr., G. and Poinar, R. 2007. What Bugged the Dinosaurs?: Insects, Disease, and Death in the Cretaceous (Illustrate). Princeton: Princeton University Press.

Racki, G. 2015. Catastrophism And Neocatastrophism Versus Cosmic Hazard: Ager Versus Alvarez; Cuvier Versus Laplace. PALAIOS 30(6): 443–445.

Raup, D.M. 1991. A kill curve for Phanerozoic marine species. Paleobiology 17(1): 37–48.

Raup, D.M. 1992. Extinction: Bad Genes or Bad Luck? New York City, NY, USA: WW Norton & Company.

Renne, P.R., Deino, A.L., Hilgen, F.J., Kuiper, K.F., Mark, D.F., Mitchell, W.S., ... Smit, J. 2013. Time Scales of Critical Events Around the Cretaceous-Paleogene Boundary. Science 339(6120): 684–687.

Sarkesh, A., Sorkhabi, A.D., Ahmadi, H., Abdolmohammadi-Vahid, S., Parhizkar, F., Yousefi, M. and Aghebati-Maleki, L. 2022. Allogeneic lymphocytes immunotherapy in female infertility: Lessons learned and the road ahead. Life Sciences 299: 120503.

Schommer, T.J. and Woolever, M.M. 2008. A review of disease related conflicts between domestic sheep and goats and bighorn sheep. Gen. Tech. Rep. RMRS-GTR-209 Fort Collins, CO: U.S. Department of Agriculture, Forest Service, Rocky Mountain Research Station.

Shapiro, J.A. 2011. Evolution: A View from the 21st Century. Upper Saddle River, NJ: FT Press.

Simmons, G., Young, P., Hanger, J., Jones, K., Clarke, D., McKee, J. and Meers, J. 2012. Prevalence of koala retrovirus in geographically diverse populations in Australia. Australian Veterinary Journal, 90(10): 404–409.

Spraker, T.R., Hibler, C.P., Schoonveld, G.G. and Adney, W.S. 1984. Pathologic Changes and Microorganisms Found In Bighorn Sheep During A Stress-Related Die-Off. Journal of Wildlife Diseases 20(4): 319–327.

Thogmartin, W.E., Sanders-Reed, C.A., Szymanski, J.A., McKann, P.C., Pruitt, L., King, R.A., ... Russell, R.E. 2013. White-nose syndrome is likely to extirpate the endangered Indiana bat over large parts of its range. Biological Conservation 160: 162–172.

Thornton, R. 1987. American Indian Holocaust and Survival: A Population History since 1492. Norman, Oklahoma: University of Oklahoma Press.

Van Valen, L. 1973. A new evolutionary law. Evolution Theory 1: 1–30.

Venzon, M., Das, R., Luciano, D.J., Burnett, J., Park, H.S., Devlin, J.C., ... Cadwell, K. 2022. Microbial byproducts determine reproductive fitness of free-living and parasitic nematodes. Cell Host & Microbe 30(6): 786–797.e8.

Wake, D.B. 2012. Facing Extinction in Real Time. Science 335(6072): 1052–1053.

Werren, J.H. 2003. Invasion of the Gender Benders: by manipulating sex and reproduction in their hosts, many parasites improve their own odds of survival and may shape the evolution of sex itself. Natural History 112(1): 58.

Wiens, D., Allphin, L., Wall, M., Slaton, M.R. and Davis, S.D. 2012. Population decline in Adenostoma sparsifolium (Rosaceae): an ecogenetic hypothesis for background extinction‡. Biological Journal of the Linnean Society 105(2): 269–292.

Wiens, D., Sweet, T. and Worsley, T. 2020. Validating the New Paradigm for Extinction: Overcoming 200 Years of Historical Neglect, Philosophical Misconception, and Inadequate Language. The Quarterly Review of Biology 95(2): 109–124.

Wiens, D. and Worsley, T. 2016. Reproductive failure: a new paradigm for extinction. Biological Journal of the Linnean Society 119(4): 1096–1102.

Willis, K.J. and Bennett, K.D. 1995. Mass extinction, punctuated equilibrium and the fossil plant record. Trends in Ecology & Evolution 10(8): 308–309.

Wolff, E.D.S., Salisbury, S.W., Horner, J.R. and Varricchio, D.J. 2009. Common Avian Infection Plagued the Tyrant Dinosaurs. PLoS ONE 4(9): e7288.

Wyatt, K.B., Campos, P.F., Gilbert, M.T.P., Kolokotronis, S.-O., Hynes, W.H., DeSalle, R., ... Greenwood, A.D. 2008. Historical Mammal Extinction on Christmas Island (Indian Ocean) Correlates with Introduced Infectious Disease. PLoS ONE 3(11): e3602.

CHAPTER 16

Old Controversies Revisited

Where is the Cusp of Creativity?

Ever since Darwin launched the disciplined study of evolution in the mid-19th century, many puzzling aspects have bedeviled generations of scientists. For example, how do forms gain complexity, functions, and phenotypic competence? If the first organisms were the simplest ones, and man is the apex of evolutionary development, what linkages might have enabled a primitive jellyfish to eventually become a man? Must that process be gradual? Could there have been consequential evolutionary leaps permitting large evolutionary advances? Many contemporary scientists no longer concur with the standard Neodarwinian dictums that evolution is both random and necessarily gradual. Fortunately, ready reconciliations become apparent when each of these contentious issues is considered within the framework of Cognition-Based Evolution.

A. Terminal addition revived

One consequential contributor to these long-standing debates was the German physician, zoologist, and philosopher Ernst Haeckel (1834–1919). Like Darwin, Haeckel was an evolutionist. However, he disagreed that natural selection is the guiding force of diversity. Instead, he advanced an alternate theory that the environment shaped organisms in reproducible patterns, postulating that their evolutionary course could be better comprehended through embryology. This insight led to his well-known principle that "ontogeny recapitulates phylogeny", known as the Bioenergetic Law, maintaining that the embryonic development of an organism from conception onward (ontogeny) advances through successive stages that resemble the adult forms of that organism's evolutionary ancestors (phylogeny) (Olsson et al. 2017). Consequently, the embryological development of more 'advanced' species passed through the stages of the adult organisms of earlier and more 'primitive' ones. Although superficially appealing, that theory has been effectively dismissed. Modern embryology studies indicate that although there are superficial resemblances in the early embryological development among many species akin to earlier progenitors, it is not a general phenomenon (Gould 1977).

Over time, the entire notion that embryology had any explanatory value was dismissed by biologists. However, the concept of Terminal Addition, an accessory component of the Bioenergetic Law, remained credible although highly controversial (Levit et al. 2022). According to the Law of Terminal Addition, new species evolve by successive additions to prior ones by effectively linking a next step (Olsson et al. 2017). Accordingly, in many respects, evolution is successively linear.

However, there are readily apparent problems with this approach if based on forms. Central among these is the lack of any direct correlation between genotype and phenotype. Furthermore, the fossil record does not offer specific support for the concept. Therefore, within a narrow perspective centered on the evolutionary succession of observable phenotypes, the theory seems deficient. Consequently, the original conceptual base of Terminal Addition has definite limitations. However, adopting a different contextual framework can prove helpful. Terminal Addition has validity when examined from the perspective of cell-cell signaling capabilities and examining how that incremental capacity links to the dynamics of viral-cellular co-engineering that enables holobionts. In particular, Terminal Addition helps to explain the continuous primacy of the unicellular form despite our evolutionary concentration on the varied multicellular forms that capture our imaginations.

Several linked concepts are crucial to this reconceptualization. First, unlike the typical Neodarwinian appraisal of phenotype as structural form or physiological capacities, it is also the cellular means of exploring the environment to return environmental cues to the unicellular zygote (Torday and Miller 2016a). Second, the unicellular zygotic stage undergoes mandated recapitulation to recenter the information management system of the cell and its co-linked N-space attachments, correcting damaging perturbations in cell-cell signaling patterns. Further, the crucial zygotic stage represents a reiterating platform for adjudicating the wave of acquired epigenetic marks to determine which must be suppressed and those that might be retained and become potentially heritable (Torday and Miller 2016a,b, Miller et al. 2020a). Crucially, both of these obviously anatomical endpoints are cellular solutions to environmental problems (Miller 2016, 2018). And moreover, all of these coordinated activities depend on cell-cell signaling. So, the epicenter of Terminal Addition is that vital activity, not phenotype per se.

Justifying this assertion requires answers to several issues in evolutionary biology that are not commonly considered. First, whose problems are being solved? The answer is that the purpose of the holobionic form is the perpetuation of the three cellular domains and their companion/competitor virome across evolutionary space-time (Miller and Torday 2018). So, all cell-cell signaling is a means of attaining this permanency. Thus, Terminal Additions clarify as representing cell-cell signaling plasticity that leads to direct changes in phenotypic development, aiding in the cellular exploration of the planetary environment. Therefore, Terminal Addition is at work, layer by layer, evinced through successive patterns of variant cell-cell signaling capabilities rather than in macroscopic appearances.

Consequently, the concept of the addition of new traits should not be measured by visible morphology but through the sequential development of cell-cell signaling pathways throughout evolutionary development (Torday and Miller 2018a). In this manner, evolutionary development is reframed from seeking to identify discrete morphological resemblances toward a deeper understanding that new traits arise through cellular-molecular modularization that can be additive over time, best understood on a cellular-molecular basis than through genetic phylogenies or phenotype (Torday and Miller 2018a). In other words, Terminal Additions are one of the cellular means of exploring information space, which necessarily must migrate over time, depending on aggregate environmental stresses. Old solutions to stresses remain part of the organism's developmental palette, and new ones are applied to meet current environmental exigencies.

There is research support for this alternative viewpoint. For example, bilateral animals develop through a form of Terminal Addition since their growth derives from a posterior generative zone (Jacobs et al. 2005). Similar sequential changes to cellular signaling molecules have been linked to the activities of developmental morphogens known to trigger the cascade of embryological events (Thisse and Thisse 2015). In further support, there is evidence that the strongest conservation of form and function is at the early stages of embryogenesis, acting as a funnel, constraining early development along robust pathways and thereby accounting for any initial superficial similarities among species (Irie and Kuratani 2011).

Furthermore, eukaryotic plasma membrane compliance has traveled through a crucial cellular-molecular evolutionary continuum based on the successive deployment of cholesterol and its bioactive derivatives (Torday and Miller 2018b). For example, that pathway links unicellular eukaryotic endocytosis/exocytosis to the efficient functioning of the swim bladder of fish which is dependent on the secretion of lubricating cholesterol (Daniels and Orgeig 2003). From this base, supplemental shifts in cholesterol deployment propelled the development of animal physiology, such as the use of cholesterol-based lung surfactant, which is critically necessary to enable mammalian respiration (Avery and Mead 1959). In that transition, the development and phylogeny of the mammalian lung alveolus also depended on a separate series of terminal addition events via additional epithelial-mesenchymal interactions (Torday and Rehan 2007).

By this means, microevolution has direct links to macroevolution through the accretion of cell-cell signaling mechanisms to support changes in physiology and morphology, apart from supporting genetic shifts, and perhaps of equal or even greater importance (Torday and Miller 2018a). Indeed, it can be assumed that plastic cell-cell signaling mechanisms arise first and are then encoded in heritable, retrievable genomic memory thereafter. At each juncture, relevant terminal additions support cellular problem-solving to achieve continuous complementarity with environmental conditions or to explore new environments such as the water-land transition. At critical evolutionary moments, multiple body systems can be simultaneously affected. For example, it is believed that a 'greenhouse effect' of rising atmospheric carbon dioxide stimulated the water-land transition over 400 million years ago (Romer 1949). In response, critical amplifications of the Parathyroid Hormone-related Protein (PTHrP) signaling pathway supported novel lung and bone phenotypes (Torday and Rehan 2002) and assisted other required land-adaptive functions involving the skeletal system, kidneys, and skin (Torday and Miller 2018c). In this manner, at critical intervals, evolution can proceed with a cascade of connected changes en suite.

If evolution derives some developmental purchase from relevant terminal additions, then clearly, it can no longer be regarded as an entirely random process. Indeed, reverse evolution would not be possible if evolutionary development did not progress through non-random pathways involving some terminal additions. Until recently, reverse evolution was deemed impossible, codified as Dollo's Law of Irreversibility (Guex et al. 2020). In 1893, Louis Dollo, an eminent paleontologist, emphatically stated that "... an organism cannot return, even partially, to a previous state already realized in its ancestral series" (Dollo 1893). However, modern research confirms that reverse evolution is rather commonplace since specific terminal additions have occurred (Guex 2016). Instances of reverse evolution are well-documented, particularly during extinction crises and periods of sub-lethal environmental stress, with many examples among planktonic foraminifera, radiolarians, nautiloids, conodonts, corals, and ammonoids (Guex 2016). Less prominent retrograde progressions during ecologically stable periods have also been uncovered.

Even beyond morphological patterns, complex traits can be exempted from Dollo's law in certain circumstances. For instance, it is believed that viviparity (live-bearing) has repeatedly evolved, but oviparity (egg-laying) evolved only once. Nonetheless, common lizards can reverse from viviparity to oviparity in a form of reverse evolution in real-time (Recknagel et al. 2018). Recent research indicates that even major phenotypic features can undergo reversal. It has been shown that the wing evolution of Phasmids (part of the polyneopteran order of insects and a major winged insect lineage), is a reversible and dynamic process (Forni et al. 2021). Moreover, multiple reversals of avian bill length of Hawaiian honeycreeper birds over a 1.7 million-year period have also been documented (Freed et al. 2016). Further, the same phenomenon of a reversal of evolutionary status can be exhibited at the cellular level. The capacity for cancer cells to undergo reverse evolution to access an ancient unicellular toolbox is one of those features that permits cancer cells to out-compete normal cells (Chen et al. 2015, Miller and Torday 2018).

Fully comprehending the dynamics of terminal addition centers on realizing that phenotype and its supporting metabolism are not ends but the active means of environmental exploration to sustain organismal-environmental complementarity. All this activity is ultimately dedicated to perpetuating the unicellular form (Miller and Torday 2018). An important biological logic underlying the successive patterns of terminal addition can be identified. Adding supplemental and occasionally novel features at the end of an established developmental pathway is unlikely to risk intricate, highly coordinated, and integrated evolutionary and cell-cell signaling networks (Guex et al. 2020). Consequently, terminal addition is less likely to undercut cellular problem-solving or lead it down blind evolutionary paths.

Hence, evolutionary terminal additions represent environmentally-induced, periodic adjustments to cell-cell signaling, leading to variant developmental forms. Each of these biological expressions is a manifestation of mutualistic/competitive cellular niche construction as outward phenotype (Torday and Rehan 2011, Torday 2016, Torday and Miller 2018a, Torday and Miller 2020a). Each of these addition/adjustments mitigates through a stream of epigenetic environmental cues. Accordingly, cell-cell signaling pathways open, close, and then, if stressed or injured, can sometimes reopen as reverse evolution. Terminal Addition can now be reappraised from its previous morphological context to the successive incorporation of accreted cell-cell signaling paths directed towards cellular solutions to epigenetic stresses. In this process, recapitulation with terminal addition indeed does occur, not as historically challenged as a succession of embryological resemblances to ancestral forms, but embodied within the perpetuating unicellular zygote and its essential anchoring roots adjudicating pliant cell-cell signaling pathways.

B. Evolutionary gaps

Just as cellular-molecular pathways offer some reconciliation between the concept of Terminal Addition and evolutionary development, a similar channel exists to resolve the persistent debate between the Neodarwinian insistence on gradualism in evolution and those scientists that entertain the possibility of substantial sudden evolutionary gaps. In that regard, the concept of Dollo's point pertains. Dollo's point refers to that atavistic plateau at which retrograde progression beyond that point would yield lethal events (Guex et al. 2020). The presence of Dollo's point helps to explain the limits of retrograde evolutionary processes. However, its further implication is that there has to be intermittent evolutionary ratcheting which can be considered a significant gap in an otherwise gradual, continuous series of adaptations. These types of ratcheting instances may explain the cause of the gaps in the fossil record that remain unexplained. For example, bats have no direct antecedent fossils and appear abruptly in the Eocene fossil record approximately 50 million years ago (Gunnell and Simmons 2005). Although this puzzle is typically presumed to be due to the delicacy of bat bones, it is curious that the most ancient specimens are among the best preserved. Although rarely emphasized, the fossil record is not supportive of gradualism but is "episodic and characterized by geologically abrupt changes in the nature and distribution of organisms." (Shapiro 2011, p. 143).

A protracted experiment with bacteria offered surprising observations. When twelve separate populations of *Escherichia coli* were propagated for 10,000 generations in identical environments, a varying tempo of evolutionary divergences was observed, fitting a model of punctuated equilibrium (Lenski and Travisano 1994). The fastest changes occurred in the first 2,000 generations, then began to slow with virtual stasis in the last 5,000 generations. This same conclusion was amplified by two recent studies with *Saccharomyces cerevisiae*, whose diploid genome was shown to undergo separate bursts of rapid genomic evolution generated by systemic genomic instability compatible with the punctuated equilibrium model proposed by Gould and Eldredge in 1972 (Heasley et al. 2021).

Recent research indicates that there have been intermittent instances of large-scale genomic rearrangements in some organisms from chromosomal inversions, polyploidy, and a variety of horizontal transfers, including transposons and retrotransposons. These large-scale contemporaneous genetic shifts can yield substantial molecular domain shuffling sufficient to result in discontinuous changes in cell-cell signaling pathways (Shapiro 2011). Genomic analysis confirms that TEs can contribute to novel rewiring of genomic networks, significantly contributing to major evolutionary transitions (Shapiro 2022). Crucially, these TEs are subject to rapid activation that can trigger widespread genomic changes.

Drosophila research confirms that substantial transposase-induced chromosomal rearrangements mediate genomic deletions, duplications, inversions, and translocations with observable, large evolutionary consequences (Shapiro 2011). Notably, all eukaryotic genomes have evidence of chromosomal rearrangements, such as duplications, compatible with discontinuous abrupt evolutionary events (Shapiro 2011). None of this would have surprised some earlier biologists who had long championed the presence of abrupt consequential evolutionary transitions alongside more gradual genomic shifts (Stebbins 1951, Margulis 1971, McClintock 1984). Accordingly, this once heretical stance is now well-accepted. This change in perspective also has quite practical consequences. Phenotypic plasticity among cancer cells is now attributed to a combination of gradual and punctuated genome evolution (Graham and Sottoriva 2017).

Episodes of punctuated phenotypic evolution have been identified in holobionts that extend beyond shifts in a central genome. Jiggins and Hurst (2011) have reported the rapid spread of a *Rickettsia* bacterial symbiont over a 6-year period. In a population of the whitefly *Bemisia tabaci*, the presence of that symbiont was associated with a dramatic increase in survival and fecundity as an altered phenotype. When all multicellular eukaryotes are holobionic eukaryotic-microbial partnerships, a significant shift in the constituency of that partnership can yield a substantial change in cell-cell signaling pathways for that holobiont.

The resolution of these prior controversies can only be suitably reached when evolution is placed in a correct framework acknowledging its primarily non-random nature. Evolution is partially reversible because it is a non-random process of cellular co-engineering. As with any coordinated engineering process, it generally proceeds through more minor changes through terminal additions, which in the case of biology, are typically from epigenetic impacts (Guex 2016, Torday and Miller 2020a). However, that relatively tranquil process undergoes periodic, abrupt transitions of varying amplitude, ranging from the genomic interpolation of a transposable element or a retrovirus insertion to more substantial intermittently triggered events, including genetic deletions, inversions, and even large-scale genomic duplications. It is reasonable to suppose that these latter larger events account Dollo's point as a regression obstacle.

Naturally, any genomic changes impact the cellular information management system of cells, and perforce has a reciprocating influence on cellular information fields and an organism's N-space Episenome (Miller et al. 2020b). An explicit example of an abrupt, momentous shift in an informational operating system is the endosymbiotic origin of Eukaryota (Baluška and Lyons 2018a,b). An entirely new, permanent cellular domain resulted. Conversely, smaller genomic variation might only yield less appreciable shifts in an otherwise stable informational architecture. Undoubtedly, the specifics of the genetic variation count. If it affects a robust regulatory pathway or network, its impact on cell-cell signaling and cellular co-engineering could have substantial ramifications beyond the specific amount of code. In this way, the lack of apparent correspondence between genotype and phenotype fully reconciles. Phenoytpe is a manifestation of an aspect of the larger cellular information management system. Depending on the genomic target, even a slight genetic shift might initiate an entire cascade of viral-cellular and cell-cell co-engineering interactions yielding a consequential phenotypic variation if it involves a crucial regulatory pathway. Accordingly, there may be no exact correspondence between the amount of discrete genomic variation and its overt phenotypic expression.

Although mutations are predominantly non-random, typically yielding morphologic changes that are gradual in scope, there can be isolated genetic mutations that produce surprising evolutionary changes. For example, recent research on a polymorphic population of *Aquilegia coerulea*, a species of flowering plant in the buttercup family, has found that an isolated mutation in a single gene (APETALA3-3) can result in large-scale abrupt, discontinuous morphological heritable change. In this instance, an entire sepal (part of flowering angiosperms that protect the flower in bud and support petals when in bloom) has replaced a nectar spur necessary for pollination (Cabin et al. 2022).

C. Creativity = plasticity, constraints, and criticalities

Of all topics in evolution, none has been as contentious as the origin of evolutionary novelty. Yet undoubtedly, any incidence of terminal addition is an act of evolutionary creativity. Any evolutionary adjustments to an existing stable organismal form, of any scope, must actuate across multiple co-linked and co-dependent levels as an intricately coordinated variation. Some clarity about biological novelty emerges from that perspective. The emergence of novelty as an act of biological creativity demands both license and countervailing constraints to avoid potentially deleterious cellular disruption. Indeed, this is obvious since the epigenome, genome, and proteome are all impacted and must seamlessly coordinate.

Since an exquisite level of coordination is necessary and must be channeled to avoid disruptions, natural viral-cellular engineering cannot be a random process. As the result of concurring cellular action, evolutionary novelty can now be seen as a product of non-random viral-cellular co-engineering. In this manner, inevitable random inputs can be leveraged by a 'harnessing of stochasticity' (Noble 2021). Evolution is non-random cellular co-engineering and, just as with human engineering, occasional serendipitous random (stochastic) inputs can be channeled into constructive engineering outputs. To accomplish that productive variation along coordinated co-engineering pathways, several simultaneous processes must constructively counterweight each other as checks and balances to maintain compatibility at successive holobionic levels. These concurring and balanced forces are biological plasticity and constraints.

Adaptive diversification is commonly attributed to phenotypic plasticity, most typically framed with respect to genes (Schneider and Meyer 2017). The 'Baldwin effect' asserts that organisms reciprocate with their environment, gradually expanding their epigenetic and developmental repertoire, thereby increasing the possibility of an emergence of advantageous variation. Consequently, plasticity should confer increasing fitness to the next generation (Hendry 2016). Since genes are tools of discriminating cells, it becomes self-evident that the central context of biological plasticity is the variety of means by which intelligent cells can collaborate and negotiate their specific individual status to meet common ends in the multicellular collective form. Therefore, the true feedstock of plasticity is the continuous interchanges among intelligent cells in reaction to environmental cues. Thus, in holobionts, plasticity is multilevel flexibility permitting a channeled evolutionary course as reliable variations linked to changes in environmental quality (Collins et al. 2020). Evolution can thereby proceed along paths of 'differential canalization' (Kriegman et al. 2018).

Accordingly, the frequently discussed issues of epistasis (a gene at one particular locus modifies the phenotypic expression of an independently inherited gene at another) and pleiotropy (a single gene controls or influences multiple phenotypic traits) can now be re-appraised as aspects of the cellular toolkit in dealing with variable and often uncertain environmental conditions (Miller et al. 2020b). Since the cellular deployment of its tools, including its genes, depends on the measuring assessment of information, a shared reference system across an organizational structure for systematic concordant measurement is mandated in multicellular holobionts (Miller et al. 2020b). Therefore, any deployment of genomic variation as epistasis, pleiotropy, or genetic rearrangements must have its counterpart in N-space architecture. Since an N-space Episenome is specifically modeled as a developmental architecture, it can be expected that it would represent a vital constraining force on variation, forcing all forms of variational plasticity from any source to conform to its environmentally-tested and heritable developmental architecture and its enmeshed chronological thresholds. Thus, the N-space Episenome can license cellular flexibility but also serves as a direct form of constraint (canalization).

It is well-acknowledged that evolution largely proceeds as iterative preadaptations and exaptations of unicellular faculties (Torday and Miller 2020b). The cellular changes during evolution are derived from preexisting genetic mechanisms reallocated to adapt to new conditions (Torday and Miller 2018a). For example, the development of the sensory receptors of hair cells in human hearing is based on an exaptation of the same pou-iv gene used for mechanosensory reception in sea anemone (*Nematostella vectensis*) tentacles (Ozment et al. 2021). For any such exaptation, coherent evolutionary variation as productive phenotypic adjustment must transmit along cell-cell communication pathways in response to environmental stresses (Miller 2016, 2018, Torday and Miller 2018a, Miller et al. 2020a). However, some environmental impacts can represent a highly significant exogenous stress, including high-amplitude physical environmental or epigenetic impacts, which can encompass various genomic incursions ranging from individual TEs to retroviral incursions. Any of these might yield large-scale genomic rearrangements. (Miller 2018, Miller and Torday 2018, Baluška et al. 2021). Yet, all must still be accommodated to enable coherent problem-solving engineering.

Although these events represent cellular stresses, they also create opportunities as imposed criticalities that trigger novel cellular expressions. Larger-scale genomic arrangements represent one example of that type of criticality that might serve as a threshold toward creative evolutionary novelty. Intermittent substantial genomic shifts are stressors that can stimulate a cascading shift in crucial cell-cell signaling patterns, releasing pathways for productive and potentially novel viral-cellular co-engineering.

There is no doubt that criticalities are an essential aspect of physical phenomena. The Danish physicist Per Bak and colleagues (1987) developed the concept of self-organizing criticalities, explaining that the natural order resulted from a critical phase transition as a precise tipping point, and meticulously applied that idea to other physical phenomena like earthquakes, avalanches, financial markets, and biological evolution. Moreover, recent research in rat brains and other species confirms that brain activity occurs in neuronal spikes that are avalanches of activity, best modeled as critical or near-criticality states hovering around a threshold between synchronized and unsynchronized brain activity (Fontenele et al. 2019, Fosque et al. 2021).

Other types of criticalities have been observed in macroorganic biologic systems (Miller 2013). In 1874, an unprecedented Rocky Mountain locust swarm plagued mid-western farmers. The sky was blackened, and the earth was blanketed with an unfathomable number of insects, much beyond anything previously seen in the Americas. This unparalleled locust swarm stretched 100 miles north-south and 1800 miles across the continent with an estimated insect population of 12.5 trillion (Wagner 2008). By 1902, all these Rocky Mountain locusts were extinguished, and the cause has never been determined (Ryckman 1999, Lockwood 2005). Curiously, no further episodes have been witnessed.

Even information can be considered a type of criticality. Hankey (2018) proposes that any understanding of biology requires an exploration of a 'new' information paradigm, neither quantum nor digital, but representing a form of critical instability capable of optimizing regulatory responses to stressful environments. Pertinently, this posited, uniquely biological form of information can be considered as constituting self-referential awareness, permitting homeorhetic balance in the face of environmental fluxes. Thus, this new type of information with its embedded criticalities differentiates information in non-biological systems from biological data with its ambiguities that require choices and predictions (Miller et al. 2020a). Therefore, the process of settling those ambiguities can be considered a critical state of phase transition. Thus, biology is conditioned on 'inevitable incompatibilities' embedded within an unavoidable gap between the initial detection of information and its linked deployment (Matsuno 2017). Breaching that gap represents a criticality.

The issue of speciation lends itself to a framework of criticalities. The average species' longevity is estimated to be between 5 and 10 million years. Yet, some species have existed uninterruptedly for hundreds of millions of years. Speciation requires a specific major trigger or a series of accumulating minor triggers within the genome, proteome, and immune sub-systems. Eventually, these latent triggers activate together as a criticality, and like an avalanche, a cascade initiates across linked cell-cell signaling mechanisms sufficient to effect a speciation event (Miller 2013). Thus, speciation reflects a triggered co-engineering shift that stimulates a reproductive barrier either due to a single powerful genomic rearrangement or the concatenated accretion of smaller shifts in cell-cell signaling and information management, ultimately reaching a criticality. Thus, its expected tempo would be chaotic, which is what is observed.

Given myriad influences, including criticalities, the robust plasticity that supports cellular creativity must be properly channeled. Consequently, all aspects of information management must have embedded constraints. Indeed, every aspect of the cellular information cycle has such constraints. First, self-referential information is ambiguous. Further, the measured cellular assessment of information is energy-intensive, and cellular energy must be conserved. Moreover, any energy expenditure by a cell, even just to measure information, becomes an information signature to other self-referential cells. That action initiates an obligatory work channel which also requires energy, which is not unlimited. Thus, constraints form an integral aspect of any engineering cycle, and suitable constraints are an essential feature of information management at all levels.

Variations must represent constrained sets of individual cellular and collective responses to environmental cues as a vital aspect of evolution (Torday and Miller 2020a). Accordingly, the necessity of constraints is now well-recognized as a foundational aspect of evolution (Shakhnovich and Koonin 2006, Futuyma 2010, Murren et al. 2015). The obligatory recapitulation of the unicellular zygote for multicellular eukaryotic reproduction is an explicit form of constraint with its recentering of cellular life and adjudication of epigenetic impacts to avoid damaging drift (Miller 2018). Further, all cell-cell activities require energetic coherences and mutualizing reciprocations. Meiosis, with its obliged halving of genomes, imposes species averaging in the context of populations, representing a prominent form of constraint. Certainly, too, immunological protections are another potent form of explicit multilevel constraint (Miller 2016, Miller and Torday 2018). All genetic interchanges between organisms reciprocate and are constrained by immunological rules. These delimiting guideposts energize the continuing compact between cells and microbes that sustain holobionts.

One further potent force constrains biology, although not typically considered as such in a traditional evolutionary framework. As previously indicated, selection is not a propulsive force in evolution since it is always post-facto to variation by definition. Selection is an environmental filter, demanding organismal-environmental complementarity, which is an evident constraint (Miller 2016, Miller et al. 2020a). Planetary conditions impose boundaries and limits, and through those constraints, selection acts as a consistent counter force against significant variation. Selection rigorously culls the extremes of variation that emerge within any given fitness landscape (Miller and Torday 2018). Any extreme variation represents 'overfitting' to immediate environmental stresses. However, these are nearly always temporary, and when the cycle inevitably returns to its mean, those over-adapted organisms are substantially out of synchrony with the reestablished average environment. Selection is 'reversion to the mean' writ to biology and thereby constitutes a significant instance of a biological 'attractional' effect (Agnati et al. 2009). Therefore, the multiplicity of constraints within our biological system consistently reinforces limitations on the extent of variations, continuously attracting each species back toward its mean.

Considering selection at the level of prokaryotes illustrates its dominant constraining effects. Conventionally, the frequent recombinations at the prokaryotic scale have been considered a source of continuous variation. It certainly is when applied to its first rounds of interactions. However, just as in eukaryotic meiosis with an obliged continual halving over successive generations, when prokaryotic recombinations are appraised at the level of extremely large populations over time, abundant recombinations through horizontal transfers counteract any significant variations through their own form of consistent prokaryotic genetic averaging. For this reason, many microbiologists do not regard individual bacterial strains as actual species (Torday and Miller 2020a). Reiterating recombinations ensures the collective transfer of pertinent environmental information with few significant divergences, ultimately maintaining the prokaryotic population genetic average (Dixit et al. 2017). Thus selection, at both unicellular and multicellular levels, exerts a consistent counter force against any significant variations, acting as a potent constraint.

The cusp of biological creativity is centered in the range of extraordinary cellular-molecular solutions to problems that self-referential cognition energizes (Miller 2016, Miller et al. 2020a). Within that overarching narrative, terminal additions are successive waves of ratification of cellular solutions to cellular problems, as tinkering with the system (Jacob 1977). Selection assures that these solutions continue to meet the current amplitude of environmental fluxes through an unyielding imposition of a single standard: cellular measurements must be correct. All cellular measurements must co-align along cellular-molecular signaling and co-engineering pathways that are coherent enough to meet environmental exigencies. Thus, both plasticity and constraints are built into living systems. It is their juxtaposition that enables creativity. Absent a system of cellular checks and balances that extends to their co-partnering virome, reproducible cellular life, as we can readily observe, would be impossible.

The material expression of any biological creativity is, by definition, a product of contingent, variant natural cellular engineering and niche construction. In this regard, cells have an advantage over humans; they have no fears or human egos. Their cognition-centered creativity can directly link to their biological expression as novelty. All cellular engineering must be intricately coordinated to achieve useful novelty. Therefore, cellular engineering must be the product of non-random intelligent cellular solutions to cellular problems. Nonetheless, all productive biological expressions must exist within stipulated boundaries. Among these, there is one overarching, explicitly imposed circumscription. All cells must deal with informational ambiguity. However, this injunction is far from an impediment. On the contrary, it is an absolute requirement of creativity. Hodgkin (1998) has argued that pleiotropic alterations can be viewed as a major constraint on mutation since pleiotropic interactions require complex genetic and physiological reciprocations and interconnections. Hodgkin supported that non-traditional perception by referencing William Empson (1906–1984), an English literary critic. As a literary authority, Empson had analyzed the nuanced meaning of ambiguity in literature, concluding that although ambiguity can confound at one level by raising uncertainties, it also opened unexpected avenues of fresh thought and insight with the potential of discovering new connections through unpredictable mental resonances.

Consequently, it can be safely advanced that the constraints of informational ambiguity are the wellspring of cellular creativity. Evolutionary novelty and creativity emerge from cellular-molecular-genomic plasticity and periodic criticalities, serving as tools of intelligent cells for improved cellular problem-solving. These solutions are then disciplined through a categorical framework of reciprocal, demanding multilevel constraints. To reach that balance and sustain it requires intelligence. Thus, the origin of creativity and biological novelty crystallizes. Creativity flows from intelligent, measuring cells.

"Genius means little more than the faculty of perceiving in an unhabitual way".

—William James, author

References

Agnati, L.F., Baluška, F., Barlow, P.W. and Guidolin, D. 2009. Mosaic, self-similarity logic and biological attraction principles. Communicative & Integrative Biology 2(6): 552–563.

Avery, M.E. and Mead, J. 1959. Surface Properties in Relation to Atelectasis and Hyaline Membrane Disease. Archives of Pediatrics & Adolescent Medicine 97(5_PART_I): 517–523.

Bak, P., Tang, C. and Wiesenfeld, K. 1987. Self-organized criticality: An explanation of the 1/ f noise. Physical Review Letters 59(4): 381–384.

Baluška, F. and Lyons, S. 2018a. Energide–cell body as smallest unit of eukaryotic life. Annals of Botany 122(5): 741–745.

Baluška, F. and Lyons, S. 2018b. Symbiotic Origin of Eukaryotic Nucleus: From Cell Body to Neo-Energide. pp. 39–66. *In*: V.P. Sahi and F. Baluška (eds.). Concepts in Cell Biology – History and Evolution. Cham, Switzerland: Springer.

Baluška, F., Miller, W.B. and Reber, A.S. 2021. Biomolecular Basis of Cellular Consciousness via Subcellular Nanobrains. International Journal of Molecular Sciences 22(5): 2545.

Cabin, Z., Derieg, N.J., Garton, A., Ngo, T., Quezada, A., Gasseholm, C., ... Hodges, S.A. 2022. Non-pollinator selection for a floral homeotic mutant conferring loss of nectar reward in Aquilegia coerulea. Current Biology 32(6): 1332–1341.e5.

Chen, H., Lin, F., Xing, K. and He, X. 2015. The reverse evolution from multicellularity to unicellularity during carcinogenesis. Nature Communications 6(1): 1–10.

Collins, S., Boyd, P.W. and Doblin, M.A. 2020. Evolution, Microbes, and Changing Ocean Conditions. Annual Review of Marine Science 12: 181–208.

Daniels, C.B. and Orgeig, S. 2003. Pulmonary Surfactant: The Key to the Evolution of Air Breathing. Physiology 18(4): 151–157.

Dixit, P.D., Pang, T.Y. and Maslov, S. 2017. Recombination-Driven Genome Evolution and Stability of Bacterial Species. Genetics 207(1): 281–295.

Dollo, L. 1893. The Laws of Evolution. Extract from the Bulletin de La Société Belge de Géologie de Paléontologie & D'Hydrologie 7: 164–166.

Fontenele, A.J., de Vasconcelos, N.A.P., Feliciano, T., Aguiar, L.A.A., Soares-Cunha, C., Coimbra, B., ... Copelli, M. 2019. Criticality between Cortical States. Physical Review Letters 122(20): 208101.

Forni, G., Martelossi, J., Valero, P., Hennemann, F.H., Conle, O., Luchetti, A. and Mantovani, B. 2021. Macroevolutionary Analyses Provide New Evidence of Phasmid Wings Evolution as a Reversible Process. Systematic Biology.

Fosque, L.J., Williams-García, R.V., Beggs, J.M. and Ortiz, G. 2021. Evidence for Quasicritical Brain Dynamics. Physical Review Letters 126(9): 098101.

Freed, L.A., Medeiros, M.C.I. and Cann, R.L. 2016. Multiple Reversals of Bill Length over 1.7 Million Years in a Hawaiian Bird Lineage. The American Naturalist 187(3): 363–371.

Futuyma, D.J. 2010. Evolutionary Constraint And Ecological Consequences. Evolution: International Journal of Organic Evolution 64(7): 1865–1884.

Gould, S.J. 1977. The Return of Hopeful Monsters. Natural History 86(June/July): 22–30.

Gould, S.J. and Eldredge, N. 1972. Punctuated equilibria: an alternative to phyletic gradualism. Models in paleobiology 1972: 82–115.

Graham, T.A. and Sottoriva, A. 2017. Measuring cancer evolution from the genome. The Journal of Pathology 241(2): 183–191.

Guex, J. 2016. Retrograde Evolution During Major Extinction Crises. Heidelberg: Springer.

Guex, J., S. Torday, J. and Miller, W.B. (eds.). 2020. Morphogenesis, Environmental Stress and Reverse Evolution. Cham: Springer.

Gunnell, G.F. and Simmons, N.B. 2005. Fossil Evidence and the Origin of Bats. Journal of Mammalian Evolution 12(1–2): 209–246.

Hankey, A. 2018. A Mathematical Model of Free Will Based on Experience Information in a Quantum Universe. pp. 549–556. *In*: Unified Field Mechanics II: Formulations and Empirical Tests. World Scientific.

Heasley, L.R., Sampaio, N.M.V. and Argueso, J.L. 2021. Systemic and rapid restructuring of the genome: a new perspective on punctuated equilibrium. Current Genetics 67(1): 57–63.

Hendry, A.P. 2016. Key Questions on the Role of Phenotypic Plasticity in Eco-Evolutionary Dynamics. Journal of Heredity 107(1): 25–41.

Hodgkin, J. 1998. Seven types of pleiotropy. The International Journal of Developmental Biology 42(3): 501–505.

Irie, N. and Kuratani, S. 2011. Comparative transcriptome analysis reveals vertebrate phylotypic period during organogenesis. Nature Communications 2(248).

Jacobs, D.K., Hughes, N.C., Fitz-Gibbon, S.T. and Winchell, C.J. 2005. Terminal addition, the Cambrian radiation and the Phanerozoic evolution of bilaterian form. Evolution & Development 7(6): 498–514.

Jacob, F. 1977. Evolution and Tinkering. Science 196(4295): 1161–1166.

Jiggins, F.M. and Hurst, G.D.D. 2011. Rapid Insect Evolution by Symbiont Transfer. Science 332(6026): 185–186.

Kriegman, S., Cheney, N. and Bongard, J. 2018. How morphological development can guide evolution. Scientific Reports 8: 13934.

Lenski, R.E. and Travisano, M. 1994. Dynamics of adaptation and diversification: a 10,000-generation experiment with bacterial populations. Proceedings of the National Academy of Sciences 91(15): 6808–6814.

Levit, G.S., Hoßfeld, U., Naumann, B., Lukas, P. and Olsson, L. 2022. The biogenetic law and the Gastraea theory: From Ernst Haeckel's discoveries to contemporary views. Journal of Experimental Zoology Part B: Molecular and Developmental Evolution 338(1–2): 13–27.

Lockwood, J.A. 2005. Locust: The Devastating Rise and Mysterious Disappearance of the Insect that Shaped the American Frontier. New York: Basic Books.

Margulis, L. 1971. Symbiosis and Evolution. Scientific American 225(2): 48–57.

Matsuno, K. 2017. From quantum measurement to biology via retrocausality. Progress in Biophysics and Molecular Biology 131: 131–140.

McClintock, B. 1984. The Significance of Responses of the Genome to Challenge. Science 226(4676): 792–801.

Miller, W.B. 2013. The Microcosm Within: Evolution and Extinction in the Hologenome. Boca Raton, Florida, United States: Universal Publishers.

Miller, W.B. 2016. Cognition, Information Fields and Hologenomic Entanglement: Evolution in Light and Shadow. Biology 5(2): 21.

Miller, W.B. 2018. Biological information systems: Evolution as cognition-based information management. Progress in Biophysics and Molecular Biology 134: 1–26.

Miller, W.B., Baluška, F. and Torday, J.S. 2020a. Cellular senomic measurements in Cognition-Based Evolution. Progress in Biophysics and Molecular Biology 156: 20–33.

Miller, W.B. and Torday, J.S. 2018. Four domains: The fundamental unicell and Post-Darwinian Cognition-Based Evolution. Progress in Biophysics and Molecular Biology 140: 49–73.

Miller, W.B., Torday, J.S. and Baluška, F. 2020b. The N-space Episenome unifies cellular information space-time within cognition-based evolution. Progress in Biophysics and Molecular Biology 150: 112–139.

Murren, C.J., Auld, J.R., Callahan, H., Ghalambor, C.K., Handelsman, C.A., Heskel, M.A., ... Schlichting, C.D. 2015. Constraints on the evolution of phenotypic plasticity: limits and costs of phenotype and plasticity. Heredity 115(4): 293–301.

Noble, D. 2021. Cellular Darwinism: Regulatory networks, stochasticity, and selection in cancer development. Progress in Biophysics and Molecular Biology 165: 66–71.

Olsson, L., Levit, G.S. and Hoßfeld, U. 2017. The "Biogenetic Law" in zoology: from Ernst Haeckel's formulation to current approaches. Theory in Biosciences 136(1–2): 19–29.

Ozment, E., Tamvacakis, A.N., Zhou, J., Rosiles-Loeza, P.Y., Escobar-Hernandez, E.E., Fernandez-Valverde, S.L. and Nakanishi, N. 2021. Cnidarian hair cell development illuminates an ancient role for the class IV POU transcription factor in defining mechanoreceptor identity. ELife, 10.

Recknagel, H., Kamenos, N.A. and Elmer, K.R. 2018. Common lizards break Dollo's law of irreversibility: Genome-wide phylogenomics support a single origin of viviparity and re-evolution of oviparity. Molecular Phylogenetics and Evolution 127: 579–588.

Romer, A.S. 1949. The Vertebrate Story. Chicago: University of Chicago Press.

Ryckman, L.L. 1999. The Great Locust Mystery. Rocky Mountain News. Archived from the Original on February 28, 2007.

Schneider, R.F. and Meyer, A. 2017. How plasticity, genetic assimilation and cryptic genetic variation may contribute to adaptive radiations. Molecular Ecology 26(1): 330–350.

Shakhnovich, B.E. and Koonin, E.V. 2006. Origins and impact of constraints in evolution of gene families. Genome Research 16(12): 1529–1536.

Shapiro, J.A. 2022. What we have learned about evolutionary genome change in the past 7 decades. Biosystems 215–216: 104669.

Shapiro, J.A. 2011. Evolution: A View from the 21st Century. Upper Saddle River, NJ: FT Press.

Stebbins Jr., G.L. 1951. Cataclysmic Evolution. Scientific American 184(4): 54–59.

Thisse, B. and Thisse, C. 2015. Formation of the vertebrate embryo: Moving beyond the Spemann organizer. Seminars in Cell & Developmental Biology 42: 94–102.

Torday, J. 2016. The Cell as the First Niche Construction. Biology 5(2): 19.

Torday, J. and Miller, W. 2016a. Phenotype as Agent for Epigenetic Inheritance. Biology 5(3): 30.

Torday, J.S. and Miller, W.B. 2016b. The Unicellular State as a Point Source in a Quantum Biological System. Biology 5(2): 25.

Torday, J.S. and Miller, W.B. 2018a. Terminal addition in a cellular world. Progress in Biophysics and Molecular Biology 135: 1–10.

Torday, J.S. and Miller, W.B. 2018b. A systems approach to physiologic evolution: From micelles to consciousness. Journal of Cellular Physiology 233(1): 162–167.

Torday, J.S. and Miller, Jr, W.B. 2018c. Unitary Physiology. Comprehensive Physiology 8(2): 761–771.

Torday, J. and Miller, W. 2020a. Cellular-Molecular Mechanisms in Epigenetic Evolutionary Biology. Cham, Switzerland: Springer.

Torday, J.S. and Miller Jr., W.B. 2020b. The Singularity of Nature: A Convergence of Biology, Chemistry and Physics. United Kingdom: Royal Society of Chemistry.

Torday, J.S. and Rehan, V. K. 2002. Mechanotransduction Determines the Structure and Function of Lung and Bone : A Theoretical Model for the Pathophysiology of Chronic Disease. Cell Biochemistry and Biophysics 37(3): 235–246.

Torday, J.S. and Rehan, V.K. 2007. The evolutionary continuum from lung development to homeostasis and repair. American Journal of Physiology-Lung Cellular and Molecular Physiology 292(3): L608–L611.

Torday, J.S. and Rehan, V.K. 2011. A Cell-Molecular Approach Predicts Vertebrate Evolution. Molecular Biology and Evolution 28(11): 2973–2981.

Wagner, A.M. 2008. Grasshoppered: America's Response to the 1874 Rocky Mountain Locust Invasion. Nebraska History 89: 154–167.

CHAPTER 17

The Primacy of Cellular Consciousness

In his 1879 'The Descent of Man, and Selection in Relation to Sex', Darwin offered his perspective on intelligence: " The difference in mind between man and the higher animals, great as it is, certainly is one of degree and not of kind" (Darwin, 2008, p. 105). Now, nearly 150 years later, this dictum must be revised, illuminating as it was in its time. This principle applies not only between man and 'higher animals' but between man and cells, from the first instantiation of conscious self-awareness forward. For many biologists, it is now well-established that cognitive self-awareness as consciousness is present in the cellular form with its discriminating plasma membrane, existing in unbroken continuity since the beginning of life (Reber 2019, Baluška et al. 2022). Indeed, life is defined by that self-referential cognition (Shapiro 2011, Miller 2016, 2018).

Although neuroscientists had initially been resistant to accepting this perspective, many now agree that any consistent philosophy of mind requires acknowledging that cognitive self-awareness as an elemental form of consciousness exists in bacteria and other basal cellular systems (Reber 2019, Fields et al. 2021). When direct observation is combined with integrated information theory, quantum theory, and computational thermodynamics, it is clear that selfhood is scale-free and innate among all living things. Our human autonoetic cognitive/conscious awareness derives from that basal aliquot as a continuum (Miller 2016, 2018, Reber 2019, Fields et al. 2021, Baluška et al. 2021, 2022).

Surprisingly, there is no consensus across the scientific disciplines regarding what consciousness precisely represents or how to define it explicitly. Naturally, then, there are many conflicting theories of consciousness. Throughout the 20th century, all had been afflicted with the same limitation. None of those theories were specifically centered at the level of cognitive cells. Within the last two decades, Integrated Information Theory (ITT) has offered a productive alternative pathway for considering consciousness within a cellular frame since it regards experiential qualia as integrated information based on "informational relationships generated by a complex of elements" (Tononi 2004). Integrated information is defined as the amount of information from a 'complex of elements', including their relevant connections, existing over and above the information content of its separate elements

(Tononi 2008). Accordingly, consciousness can be conceived as dimensional cellular awareness of information invested with both quantitative (number) and qualitative (qualia as interrelationships) aspects as causal interactions among its parts.

Several advantages accrue from this schema. Consciousness can be regarded as a unique intrinsic state of integrated information that remains axiomatically related to universal fundamentals such as mass, charge, or energy as the essentials of a universal relational informational matrix. This fundamental informational matrix is conceptualized as N-space, representing the interactions between matter and energy whose meaning among self-referential organisms can only be derived from their interactions. Matter and energy in and of themselves do not represent information. Perceived information exists only through their interaction, making them available to be sensed at the atomic-molecular level. N-space represents that entire architecture of those potential interactions. A living organism can only perceive a fraction of them, according to its specific senomic apparatus. As discussed in Chapter 4, every cell has its individual Pervasive Information Field (PIF) that represents its effective compartmentalization of N-space. These PIFs aggregate in multicellularity to form a holobionic N-space Episenome as a collective attachment to N-space.

Consequently, consciousness can now be seen as intrinsic to all living forms as integrated information. The self-referential integration of information defines the living state since it is based on an attachment to a universal information matrix (N-space) through self-awareness, separating it from computers or robots.

Notably, the ITT frame supports several basic consonant assumptions within CBE. Integrated information at all cellular scales is CBE's absolute necessity, enabling all unicellular and multicellular life. Further, Tononi (2008) emphasizes that integrated information requires a "highly structured set of mechanisms' that permit many nested discriminations (choices)". Clearly, any discrimination of sensory data entails its measurement. However, its measured validity is limited since biological information is ambiguous, which propels its collective appraisal. Indeed, the collective assessment of information is exactly representative of integrated information. However, and significantly, that type of collective integrated and measured information among individual self-referential cells implies an information space-time architecture that could permit conjoining collective measurements through a shared reference platform. The biological integration of the individual cellular measurement of ambiguous information entails N-space as a fundamental universal information fabric. In the living state, N-space is compartmentalized as PIFs or an N-space Episenome as the basis of any consistent cellular measurement. These informational platforms are the essence of integrated information among self-referential cells (Miller 2016, Miller et al. 2020a,b). Therefore, cellular consciousness is an attachment of a competent cell to its relevant N-space compartment that permits its self-referential assessment of ambiguous informational cues and energizes its collaborative, integrated assessment.

The issue of the informational quantity and quality as potentially separate characteristics of information has been particularly vexing. Oceans of words have been written about the so-called 'hard problem' of consciousness, proposed by David Chalmers in 1995, and now widely accepted as valid in consciousness studies (Chalmers 1995). As proposed, consciousness can be divided into two elements that must be analyzed separately. The 'easy' problem of wiring connections can be cataloged through physical properties and processes. It can be conceived as largely a quantity issue. However, the 'hard' problem of information quality as experiences that connect to subjective sensations and emotions defy any facile examination. Chalmers has maintained that only some organisms have experiences. However, meticulous research contradicts this viewpoint by substantiating that all living organisms have their idiosyncratic types of experiences and attendant ambiguities according to scope (Mashour and Alkire 2013, Boly et al. 2013, Klein and Barron 2016, Miller 2016, 2018, Baluška and Miller 2018, Baluška et al. 2022).

However, once self-referential cognitive awareness is correctly understood as instilled within the cellular form, there is no need for this artificial categorization. Both types are embodied as basal sentient dualism (Reber 2019, Baluška et al. 2021, 2022a). Cellular cognitive/sentient capacities must naturally encompass both types of conscious awarenesses that are continuously operative in cellular life, existing across all scales. An organism receiving and generating complex integrated information is a conscious self-referential agent, acquiring and deploying integrated information. As an internal measurement by definition, this self-generated information is an experience that is both irreducible and exclusive. Accordingly, an experience cannot be reduced to independent parts since it exists only through integrated information through measurement within the cellular interior. And further, each such experience excludes all other's experiences as each is a product of its own idiosyncratic internal analysis (Oizumi et al. 2014).

There is a further justification why integrated information occasions experiences. There is no doubt that cells of all types are sentient and can measure informational input (Miller et al. 2020a,b, Balušk and Reber 2021, Baluška et al. 2021, 2022). But most importantly, all information that any cell has is ambiguous, which is precisely why it must be measured. After all, perfect information would not require the cellular expense of its limited energy to conduct its measuring assessment. However, and quite crucially, that process of measuring information is entirely internal, preceding any communication of its value to any other cell. Hence, all the information that any cell has at its disposal is the product of its self-produced analysis as previously noted (Cárdenas-García 2020, Miller et al. 2020a). Necessarily, the self-produced analysis of an uncertain environmental cue sufficient to lead to a measurement certainly represents an experience at scale.

The ability of any cell to conduct such an internal analysis of information depends on an entire suite of capabilities, but two are paramount. First, each cell must have an 'intelligent', competent outer membrane; second, this membrane must link to retrievable and deployable memory. To this end, the cell membrane is the linchpin of the cellular senome and the gateway to its entire senomic apparatus, serving as the essential cognitive conduit between the external and internal cellular environments (Baluška and Miller 2018, Miller et al. 2020a,b, Baluška et al. 2021, 2022). Accordingly, all cellular information that underpins conscious self-reference inextricably links to the internal cellular self-production of information. Axiomatically then, this self-appraisal represents the meaning of its information in context. By definition, each cell produces its version of reality based on its exclusive internal measurements. That measurement defines its relevant reality in terms of both its conditional ambiguity and being a direct internal assay of internal molecular and energetic fluxes. A remarkable entailment follows from this living essential—all cells create (self-produce) their own reality, and since we are cellular beings, so do we.

When any cognitive self-aware cell conducts its measured assessment of ambiguous environmental cues, thereby determining its value as meaning toward its contingent deployment of scarce cellular resources as a productive reaction, that activity represents a prediction through inference. Inference is a prediction representing a form of anticipation, and that anticipation channels the deployment of cellular resources based on environmental cues. Undoubtedly though, within self-referential cells, anticipation is an assignment of meaning through measurement. And necessarily, that activity is an experience. That assignment of meaning has no exact measuring value until its qualitative contingent aspects are accounted for within its context. Thus, cellular cognition/consciousness/self-reference embodies experience because it is inseparable from its measuring process, that encompasses both quantitative and qualitative features. Such measurements immediately become a prediction that results in the contingent deployment of resources, denoting qualia at scale as well as a quantitative assessment. We, humans, experience qualia at our scale with our idiosyncratic manifestations as a specific species aggregate of the trillions of cells that comprise a human holobiont, including both our eukaryotic body cells and associated microbiome.

Furthermore, the biomolecular and energetic particulars of the sensory information apparatus of the cell necessarily link to the cellular 'knowing' of the conditional ambiguities of biological information. If it were otherwise, cells would not expend energy to measure, accepting its presenting value. Cells have myriad sensory capacities and cellular signaling pathways, many of which can be readily evaluated. Until very recently, there was no means of interrogating the specific cellular sensing of informational ambiguity However, research now confirms that the individual neurons of the brain often fail to respond to inputs that should provoke responses and instead, demonstrate unexpected frequent periods of silence (Hodassman et al. 2022). Unlike transistors or other electronic devices in which inputs and outputs are always the same, these neurons react to identical stimuli differently based on their contextual status. Additionally, this research found that this individualized pattern of responses to similar stimuli does not undermine brain function but is critical to its operations, which is an important finding for achieving higher levels of artificial intelligence.

Accordingly, these experimentally observed differences between how cells and computers respond to information are the cusp of an uncertainty relationship. Further, this evidence of responses to informational cues based on individual status as the gap between precise and ambiguous information should be regarded as experiential. This difference can be roughly likened to the distinction between digital and analog information. Graphically, digital information has orthogonal up and down spikes defining bits of information that are precise and completely reproducible. Analog information has sinusoidal sloping curves, indicating the time-dependent acquisition limitations of analog processes. The observable difference between these two types of informational inputs helps to conceptualize informational ambiguity, framing the differences that stimulate experiences in the self-referential frame.

Therefore, experiences are the summated cellular measurements of imprecise information at all scales. For example, music can be electronically interrogated and accurately measured as a set of vibrational sonic patterns representing the tones that impact our sensory organs. Yet, the embedded ambiguities in those notes provide our experiential relationship with that music, including such subjective feelings like delight. Consequently, the sharp focus of experientiality is informational ambiguity. If the impact of musical information had the precision of computer data, it would be an analytical set but not a subjective experience any more than any endless clatter. That informational input would be considered background noise filtered from our conscious self-referential state and would not represent an overt experience. Importantly, cells clearly understand that their receptive, sensory information is ambiguous. Hence, cells have experiences since both are bindingly linked. Fortunately, this perspective effectively reduces. Any stimulus that crosses a cell membrane beyond background noise is an experience.

Consequently, all aspects of consciousness arose simultaneously. Accordingly, minds do not create consciousness. Minds are one manifestation of consciousness. This congruity yields an obliged, productive integration. A theory of mind is an extension of cell theory. From the preceding, there is no 'hard problem' of consciousness. All aspects of the consciousness we enjoy are instantiated in our cells. Our life form embellishes it as our specific aggregation of cellular consciousness into our own transcendental experiences. The 'hard problem' was a brilliantly seductive idea. In its facile appeal, a generation of scholars little interested in cellular biology was traduced. Yet, for all the acclaim, it was merely an incorrect surmise based on an insufficient understanding of cellular cognition and its anchoring roots.

In the complicated milieu of cellular life with its astounding complex internals, it can be expected that the relative aspects of quantity and quality of information, the latter as qualia, might be approached by different cellular dynamics. For example, the quantity aspect might travel within Cartesian dynamics along familiar pathways of classical realism through the discrete transfers of energy and bioactive molecules. The additive component of feelings, preferences or experiential sentience might alternately dwell within the realm of various quantum phenomena such as non-locality, entanglement, coherences, quantum biological electron tunneling, antipodal drift, or anticipated information not received. (Miller et al. 2019). The justification for this perspective is that the latter is more embedded within ambiguities and doubt than mere quantity, which might be expected to be measured with greater fidelity.

Any separation between informational quality and quantity is naturally an artificial device for categorization. Consciousness is a 'whole' as the continuous self-referential collapse of superimposed implicates into explicates. Certainly, though, any self-referential observer might actively sense only a part of any full range of implicates prior to choosing to collapse those superpositions into explicates. In such circumstances, these remaining uncollapsed implicates might represent our subliminal self-consciousness, representing two sides of a single coin of an inherent total ensemble of self-referential awareness that defines cellular life (Reber 2019, Baluška and Reber 2019, Miller et al. 2019).

Within this framework, a better understanding of cellular measurement as a metric of consciousness clarifies. Hameroff and Penrose have previously argued that consciousness can be linked to the settling of quantum superpositions through quantum coherences centered within cellular microtubules in neurons (Hameroff 2007). Consequently, quantum coherences can be envisioned as areas of 'coherent' information density, i.e., where informational ambiguities are settled through cellular measurement. In this case, quantum coherences are zones of higher measuring validity. Cellular measurement can now be better understood as primarily based on extracting coherences from background noise. This framework clearly distinguishes cellular measurement from computer data analysis or thermostatic function, explaining why cellular measurement is not like a measurement in chemistry, and helping to explain why brain cells do not react like electronic devices. All measurements in a cell are a coherent set of instantaneous resolutions of ambiguities sufficient to justify the deployment of cellular resources. The difference is that in computer/transistor terms, the threshold-action criteria are fixed with the specific aim of the maximum reproducibility of the same precise presenting value or defined within a small range. Therein lies the crucial difference between a computer and the living state. A cell need never use the same boundaries of presenting values as its thresholds to settle ambiguities. It is always context-dependent, which is the heart of the concept of the 'knowing' cell, requiring the boundary condition of a competent plasma membrane. From this background, cognitive self-awareness can be appreciated as a quantum state arising from and dependent upon coherences induced within a boundary condition in which resonant informational energies are entrained (Miller 2016). Undoubtedly, however, this capability is memory dependent. Consequently, self-referential consciousness requires an intelligent membrane and retrievable and deployable memory.

The basic cell is an embodiment of the conscious 'self' and a repository of information space-time memory. This elusive quality of selfhood encompasses three intertwined elements (Miller et al. 2019). First, a self-referential organism must 'know that it knows'. Self-awareness entails grasping an essential separateness from the environment requiring individual contingent reactions to it. Secondly, since multicellular life depends on shared measurements, self-aware cells must also 'know that others know'. Thirdly, and very importantly, cellular self-awareness must enfold a sense that others like themselves 'know in self-similar patterns'.

All three must integrate as essential elements of self-referential consciousness. All those aspects of self-reference are required for multicellularity, forming the ground state for the mutualistic assessment of integrated information, either within biofilms or among holobionts. No matter our emphasis on our human conscious abstractions, the focal point of consciousness is ever and always problem-solving. Abstractions are only an alternative means of problem-solving. Necessarily too, all aspects of self-reference are information-dependent. Consequently, self-referential problem-solving is always a matter of resolving choice contingencies. Thus, self-referential consciousness for cells can be compactly defined as the self-directed ability to uphold selfhood by using ambiguous information to solve problems (Miller et al. 2019). No matter how cleverly designed, the distinction between computers and any artificial intelligence becomes stark. Only living organisms have a sense of self as a conditional state of 'knowing' that their information is uncertain. No machine has those qualities, no matter how seductively it mimics human thought.

One aspect of cognition receives surprisingly little attention, although it occupies the central aspect of our conscious choices. The role of preference is little considered. One central aspect of preference is prediction. For many scientists, cognition is centered on inferences from which problem-solving predictions can be made (Friston et al. 2011, Miller 2018, Marshall 2021). Indeed, Friston (2017) stated that: "inference is actually quite close to a theory of everything—including evolution, consciousness, and life itself."

Cooke (2021) has proposed that consciousness is a form of inference performed by all living systems at all scales, including its qualitative aspects as subjective states. In this regard, the concept of preference, when applied to cells, helps clarify the contentious issue of whether or not cells are sentient and capable of subjective experiences. Any cell's cognitive actions are directed to sustain itself through problem-solving. Doing this successfully depends on inferences gleaned from the measured assessment of ambiguous information that can become cellular predictions. The purpose of that cellular problem-solving is to maintain a state of internal homeorhetic equipoise. However, that sustained state of homeorhetic flux in a cognitive cell is, by definition, its state of preference since it is being continuously defended by its limited resources. Thus, cellular cognition/consciousness is 'knowing' a preferred state compared to another. Nonetheless, such a knowing, preferred state is undoubtedly an experience. Therefore, cells have experiences that are sufficient to sustain their preferred states. This conscious appreciation of status emanates from individual cells and aggregates across levels to become holobionts such as ourselves that exert our own contingent choices to reach our own total holobionic states of preference by one means or another.

Accordingly, homeorhetic equipoise equates with the flow of information and energy utilization upon which any state of preference must depend. As a necessary derivative manifestation of that ongoing physical dynamic, preference can be effectively connected to the minimization of variational free energy and the suppression of surprisal (unpredictable outcomes) (Friston et al. 2006). Based on this free energy principle, cellular preference can be well understood as correlating with states of 'least uncertainties' that directly correspond with maximized *EI (non-random effective information that permits a cell to act in its own behalf), as previously discussed in Chapter 4 (Miller et al. 2019). Cellular preference as prediction, therefore, represents a crucial entanglement between variational free energy derivative of thermodynamics and cellular self-measurement of information that eventually resolves into biological form.

Notably, this perspective offers clarity for our understanding of biological processes. Cellular preference is that condition of homeorhetic flux that is perceived by the 'knowing' cell within its conditional limits as being in a state of comportment with environmental stresses. Thus, cellular preference is exactly coincident with continuous organismal-environmental complementarity and its successful assimilation. From this, natural selection comes into sharper focus. Natural selection assures that the 'knowing' cell centers its measurements to meet environmental exigencies, post-facto to living choices and its contingent variations. Thus, selection channels cellular preferences enacted through the measurement of information and thereby consistently reinforces consciousness. Therefore, the purpose of living consciousness is to remain in harmony with its environmental context at all times.

Naturally, each organismal type resolves uncertainties among its constituent cells in its specific idiosyncratic manner. When preference is placed into a construct of Integrated Information Theory, it can be productively linked to phenotype within CBE. In this frame, aggregated consciousness, as we ourselves experience and which is necessarily specific to us as humans, can be considered a form of conscious phenotype (Miller et al. 2019). Recall from earlier chapters that the purpose of phenotype is the cellular means of exploring the environment and bringing back environmental experiences and informational cues to the entire cellular constituency and, ultimately, the recapitulating zygote (Torday and Miller 2016) Conscious phenotype can be represented as an ensemble of cellular self-referential awarenesses based on their senomic framework that is species-specific and encompasses its experiential repertoire. Furthermore, this connects directly to preferential minimization of variational free-energy to maintain homeorhetic equipoise, which is coincident with cellular thermodynamic limitations.

In that framework, unanticipated external environmental perturbations beyond expectations would place that organism outside of its upper boundary of surprisal (a low probability of prediction). It would deviate beyond normative cellular homeorhetic patterns as non-minimization of variational free energy that would skew away from a preferred state of least uncertainties. When considered this way, conscious phenotype becomes another means of an organism's exploration of the environment, just like all other types of phenotype. Further, the process of forming the conscious phenotype of that species becomes another instance of cellular niche construction through cellular co-engineering, centering on the deployment of integrated information (Miller et al. 2019). Thus, the cellular preferential drive toward attaining a state of suppression of surprisal and co-linked minimization of variational free energy is a form of niche construction of a conscious phenotype. Notably, this same framework has been applied to all other phenotypes (Constant et al. 2018).

This background helps to illuminate the source of biological creativity. The cusp of creativity can now be identified as a burst of activity within that conscious phenotype as a novel exploration of the implicates of information space and its veiled non-localities (Miller et al. 2019). Consequently, that environmental-informational exploration and its ultimate deployment as biological expression is a form of niche construction of conscious phenotype that is species-specific as it is directly based on a summation of the particular participants in that specific holobiont. Inevitably, every macro consciousness, such as our own, is an idiosyncratic amalgam of the aggregated selves of its body cells and microbial constituents and their entangled connections within its species boundaries (Miller et al. 2019). This viewpoint yields an important derivative. When consciousness is an endowed property of all cells, and self-referential awareness depends on integrated information, human conscious phenotype and the human mind does not entirely relate to a brain and nervous tissue, but can only be understood as an integrated whole body phenomenon (Fuchs 2012). We cannot escape our holobionic selves. Every one of our cells is an embodiment of 'self' and a repository of continuous information space-time memory. Necessarily, we are their aggregate. Accordingly, our form of human intelligence, with its unique capacity for cognitive richness, abstraction, and creative problem-solving, is a combinatorial product of the individual 'state' function of consciousness of each of our constituent cells, including the contribution from our holobionic microbial partners. Together, they yield an integrated system consciousness (Miller et al. 2019). Naturally, then, every holobiont would manifest its own unique integrated summation of the consciousness of each of its cells.

When this perspective is acknowledged, and it is recalled that all the information that any cell has about its reality is internally self-produced, there is an entailing conclusion that some might find unsettling. Just as every cell creates its version of reality through self-produced information, then we, as obligatory cellular aggregates, must be doing the same. We do not know an objective outward reality. Our reality is self-produced and unique to each of us. As humans, our perceptions of what we consider reality is a function of our personal conscious phenotype and unique to each of us. That conscious phenotype can never be identical between any two individuals. When properly considered, we discover a pathway to human creativity. We each explore our realities and derive different informational cues from them. A potential linkage between our enriched human capacity for abstraction and creativity emerges through this framework. Both of these conscious manifestations represent the activities of human combinations of cells to both predict and still postpone the settlement of sets of internally generated superimposed implicates as an alternative mode of problem-solving (Miller et al. 2019).

The concepts of extended consciousness and extended mind have been the subject of significant investigation (Tollefsen 2006, Theiner et al. 2010). Given our cellular framework, we can expand our understanding of both concepts to encompass a concept of 'extended cellular self'. For example, single-celled diatoms are considered opportunistic organisms that maximally reproduce in favorable environmental conditions. However, recent research indicates that diatoms will make altruistic decisions benefiting the larger ecology even when not immediately stressed (Annunziata et al. 2022). Some diatoms will voluntarily suppress sexual reproduction and block their nutrient acquisition and growth even under nutrient-rich conditions. This apparent paradoxical behavior can be resolved by entertaining that free-living diatoms experience their own form of group cognition as a form of collective unicellular consciousness. This collective action can be regarded as an augmented form of selfhood as a basal expression of extended mind.

Many are disposed to enthusiastically validate collective consciousness and extended mind for humans. Yet, it is not overreaching to regard multicellularity at any scale as its basal form. On the other hand, is there then an irreducible minimum of consciousness? Yes, assuredly, there must be, yet the scale is certainly not clear. Data suggests high levels of competencies even at the molecular level, including r-protein networks in ribosomes (Timsit and Grégoire 2021). Further, the crucial protein complex mTOR (mammalian target of rapamycin) is a major regulator of growth and metabolic processes, exhibiting surprising contingent reactions and a remarkable range of functions in holobionts (Lieff 2020). Undoubtedly, how these critical molecules engage with the cellular senome must be a key determinant in any conscious activity.

It has been proposed that senomic connections integrate the internal cellular apparatus (vesicles, cytoskelton, signaling molecules) to form subcellular nanobrains that participate in quantum interactions to enable aggregate conscious representation (Baluška et al. 2021). In such circumstances, our intentionality as our form of conscious phenotype is a concatenated process of the aggregations of nano-intentionality, expressed first at the molecular level (Fitch 2008). This means that some signature molecules would have to be critical components of the measurement of senomic information at the cellular level.

Sub-categorizing along these lines links quite satisfactorily with the senome concept as a whole. Just as the genome is composed of genes which are a form of informational compaction, it has been indicated in Chapter 5 that the senome is comprised of 'senes' as another type of information storage and transmission (Baluška and Miller 2018). At least theoretically, then, subcellular nanobrains and nano-intentionality could achieve physical expression through senes that connect to the genome, epigenome, and proteome. Consequently, consciousness could be modeled as the buildup of senes as one manifestation of integrated information.

When viewed in this manner, the debate over the potential for consciousness in recently derived human brain organoids (three-dimensional cortical neural tissues created from human pluripotent stem cells) can be quickly resolved (Niikawa et al. 2022). Each participant cell is conscious as it is experienced at its scale and has a competent senomic apparatus that permits its self-organization. Therefore, their aggregate form has its own conscious phenotype, as limited as it may be.

Placed in this framework, consciousness cannot be disconnected from quantum processes since the senome relies on them. Hence, our experiences are quantum-related and, therefore, inherently based on ambiguities as superimpositions of possibilities. This cascading process is what we 'experience'. Definitionally, no two individuals react to the same environmental stimulus identically since all cellular information obliges internal self-referential measurement. Instead, each will view the environment subjectively and necessarily differently.

Synesthesia is an extreme example of that natural variability. Certain individuals are synesthetics, responding to triggered stimuli with unusual variant secondary sensory experiences, such as 'seeing' musical notes as colors (Tilot et al. 2019). Therefore, synesthetic experiences are self-produced sensory patterns that follow different pathways from the norm. Within CBE, the complex topic of consciousness can be productively simplified. Of the greatest importance is understanding that the instantiation of conscious self-reference preceded biological self-organization sufficient to enable a fully competent cell. The cellular form does not merely embody consciousness but relies on it as an integrated whole. Consciousness is a state of self-awareness permitting contingent problem-solving in the context of informational ambiguity. Cellular life depends on this informational flow as its ineluctable requirement. From this, a definition of life ensues. Life is self-referential cognition (Miller 2016, Reber 2019, Baluška and Reber 2021). Self-reference is the inception of biological observer status as the birth of the living implicate realm that permits living choice contingencies (Miller et al. 2020b) Insofar as life dwells within that implicate realm, there is only one certainty. What is real for the living is doubt. Consciousness is doubt. Conscious organisms 'know' that, make the best of their insecurities (least uncertainties), and deliberately get on with their lives. Accordingly, N-space can now be seen as 'the realm of doubt'.

The concept of N-space is meaningless for computers since computers are neither alive nor conscious. Computers have neither doubts nor preferences. There is no such thing as ambiguous information, only digital data. Our insistence on dividing consciousness into separate participles of qualia, abstract thought, or capacity for language are academic exercises. All aspects of consciousness are based on cell-centered self-produced information in which the quality of that information is explicitly flavored through that act of self-production. Self-produced information has the inherent dimensions of quantity and quality, constituting its total measuring value based on ambiguous cues. All that information is directed toward contingent problem-solving to maintain preferential homeorhetic flux. Hence, cognitive, measuring cells sustain life as preferential (self-measured) homeorhetic flux. Hence, axiomatically, preference derived from informational ambiguity has embedded qualia which are experiences. The exact self-produced interpretation of that qualia depends on cellular context because all cells are parts of collaborative living informational networks as integrated informational interactome (Miller 2018). For cells, quantity has a quality all its own. Accordingly, the need for cohesive information space attachment (N-space) to permit conjoining measurements emerges as a default necessity for consciousness.

There are only two assured realities. Life's conditional ambiguities assure that life exists within uncertainties (Miller et al. 2019). The essence of self-referential consciousness is choice contingency in its perpetual context of ambiguity Thus, self-referential subjectivity equals knowing 'doubt' as all varieties of informational uncertainty which manifest idiosyncratically as its essential qualia. Self-reference can now be well-defined. It is the enigmatic property of 'knowing that information need not be deliberate'. From that, selfhood can be represented as the base system repertoire of any living organism's organized attachment to biological information space-time with its constitutive superpostiions of possibilities as uncertainties (Miller et al. 2019). Consequently, self-reference is scale-free across all living forms, encompassing all cells, plants, and animals.

Therefore, biological expression is never an exactly reproducible one-to-one correspondence between any stimulus and a deployment of cellular resources, as has already been demonstrated with neurons. There is always an obligatory gap as ambiguities and their attendant superimposition of possibilities (Miller 2018). The great philosophers may not have had our scientific background, yet they intuited this same principle. The philosopher Henri Bortoft acknowledged this by emphasizing a comment on morphogenesis by the polymath/ poet/philosopher Goethe: "The whole emerges simultaneously with the accumulation of the parts, not because it is a sum of the parts, but because it is immanent within them" (Simeone and Ehresmann 2017). That immanence can now be identified as embodied within the measuring cell as self-referential consciousness.

Basal cognition is scale-free. The essence of cognition is knowing one state versus another, enabling contingent problem-solving that permits settling informational ambiguities to achieve one state over another. All forms of cognition harmonize through an entanglement of cellular quantum coherences, resonances, criticalities, and instabilities in consistent confrontation with a stressful environment. Therefore, the proposition of 'what is life?' can be reduced. It is fundamental epitomic cellular self-referential cognition and the means through which that consciousness links in self-iterative patterns (Miller et al. 2019).

"Simplicity is the highest goal, achievable when you have overcome all difficulties. After one has played a vast quantity of notes and more notes, it is simplicity that emerges as the crowning reward of art."

Frederic Chopin (1810–1849), pianist and composer

References

Annunziata, R., Mele, B.H., Marotta, P., Volpe, M., Entrambasaguas, L., Mager, S., … Ferrante, M.I. 2022. Trade-off between sex and growth in diatoms: Molecular mechanisms and demographic implications. Science Advances 8(3).

Baluška, F. and Miller, Jr, W.B. 2018. Senomic view of the cell: Senome versus Genome. Communicative & Integrative Biology 11(3): 1–9.

Baluška, F., Miller, W.B. and Reber, A.S. 2021. Biomolecular Basis of Cellular Consciousness via Subcellular Nanobrains. International Journal of Molecular Sciences 22(5): 2545.

Baluška, F., Miller, W.B. and Reber, A.S. 2022. Cellular and evolutionary perspectives on organismal cognition: from unicellular to multicellular organisms. Biological Journal of the Linnean Society (blac005).

Baluška, F. and Reber, A. 2019. Sentience and Consciousness in Single Cells: How the First Minds Emerged in Unicellular Species. BioEssays 41(3): 1800229.

Baluška, F. and Reber, A.S. 2021. The Biomolecular Basis for Plant and Animal Sentience: Senomic and Ephaptic Principles of Cellular Consciousness. Journal of Consciousness Studies 28(1–2): 31–49.

Baluška, F., Reber, A.S. and Miller, W.B. 2022. Cellular sentience as the primary source of biological order and evolution. Biosystems 218: 104694.

Boly, M., Seth, A.K., Wilke, M., Ingmundson, P., Baars, B., Laureys, S., … Tsuchiya, N. 2013. Consciousness in humans and non-human animals: recent advances and future directions. Frontiers in Psychology 4(625).

Cárdenas-García, J.F. 2020. The Process of Info-Autopoiesis – the Source of all Information. Biosemiotics 13(2): 199–221.

Chalmers, D.D. 1995. Facing Up to the Problem of Consciousness. Journal of Consciousness Studies 2(3): 200–219.

Constant, A., Ramstead, M.J.D., Veissière, S.P.L., Campbell, J.O. and Friston, K.J. 2018. A variational approach to niche construction. Journal of The Royal Society Interface 15(141): 20170685.

Cooke, J.E. 2021. What Is Consciousness? Integrated Information vs. Inference. Entropy 23(8): 1032.

Darwin, C. 2008. The Descent of Man, and Selection in Relation to Sex. Princeton, New Jersey, USA: Princeton University Press.

Fields, C., Glazebrook, J.F. and Levin, M. 2021. Minimal physicalism as a scale-free substrate for cognition and consciousness. Neuroscience of Consciousness 2021(2): niab013.

Fitch, W.T. 2008. Nano-intentionality: a defense of intrinsic intentionality. Biology & Philosophy 23(2): 157–177.

Friston, K. 2017. The mathematics of mind-time. Aeon magazine.

Friston, K., Kilner, J. and Harrison, L. 2006. A free energy principle for the brain. Journal of Physiology-Paris 100(1–3): 70–87.

Friston, K., Mattout, J. and Kilner, J. 2011. Action understanding and active inference. Biological Cybernetics 104(1–2): 137–160.

Fuchs, T. 2012. Feelings of Being Alive. pp. 149–165. In: J. Fingerhut and S. Marienberg (eds.). Feelings of being Alive. Berlin, Germany: De Gruyter.

Hodassman, S., Meir, Y., Kisos, K., Ben-Noam, I., Tugendhaft, Y., Goldental, A., ... and Kanter, I. 2022. Brain inspired neuronal silencing mechanism to enable reliable sequence identification. Scientific Reports 12: 16003.

Hameroff, S. 2007. Orchestrated Reduction of Quantum Coherence in Brain Microtubules: A Model for Consciousness. NeuroQuantology 5(1).

Klein, C. and Barron, A.B. 2016. Insects have the capacity for subjective experience. Animal Sentience 1(9).

Lieff, J. 2020. The Secret Language of Cells: What Biological Conversations Tell Us About the Brain-Body Connection, the Future of Medicine, and Life Itself. Texas: Ben Bella Books.

Marshall, P. 2021. Biology transcends the limits of computation. Progress in Biophysics and Molecular Biology 165: 88–101.

Mashour, G.A. and Alkire, M.T. 2013. Evolution of consciousness: Phylogeny, ontogeny, and emergence from general anesthesia. Proceedings of the National Academy of Sciences 110(supplement_2): 10357–10364.

Miller, W.B. 2016. Cognition, Information Fields and Hologenomic Entanglement: Evolution in Light and Shadow. Biology 5(2): 21.

Miller, W.B. 2018. Biological information systems: Evolution as cognition-based information management. Progress in Biophysics and Molecular Biology 134: 1–26.

Miller, W.B., Baluška, F. and Torday, J.S. 2020b. Cellular Senomic Measurements in Cognition-Based Evolution. Progress in Biophysics and Molecular Biology 156: 20–33.

Miller, W.B., Torday, J.S. and Baluška, F. 2019. Biological evolution as defense of "self". Progress in Biophysics and Molecular Biology 142: 54–74.

Miller, W.B., Torday, J.S. and Baluška, F. 2020a. The N-space Episenome unifies cellular information space-time within cognition-based evolution. Progress in Biophysics and Molecular Biology 150: 112–139.

Niikawa, T., Hayashi, Y., Shepherd, J. and Sawai, T. 2022. Human Brain Organoids and Consciousness. Neuroethics 15(1): 5.

Oizumi, M., Albantakis, L. and Tononi, G. 2014. From the Phenomenology to the Mechanisms of Consciousness: Integrated Information Theory 3.0. PLoS Computational Biology 10(5): e1003588.

Reber, A.S. 2019. The First Minds: Caterpillars, Karyotes, and Consciousness (Illustrate). New York, NY, USA: Oxford University Press.

Shapiro, J.A. 2011. Evolution: A View from the 21st Century. Upper Saddle River, NJ: FT Press.

Simeonov, P.L. and Ehresmann, A.C. 2017. Some resonances between Eastern thought and Integral Biomathics in the framework of the WLIMES formalism for modeling living systems. Progress in Biophysics and Molecular Biology 131: 193–212.

Theiner, G., Allen, C. and Goldstone, R.L. 2010. Recognizing group cognition. Cognitive Systems Research 11(4): 378–395.

Tilot, A.K., Vino, A., Kucera, K.S., Carmichael, D.A., van den Heuvel, L., den Hoed, J., … Fisher, S. E. (2019). Investigating genetic links between grapheme–colour synaesthesia and neuropsychiatric traits. Philosophical Transactions of the Royal Society B: Biological Sciences 374(1787): 20190026.

Timsit, Y. and Grégoire, S.-P. 2021. Towards the Idea of Molecular Brains. International Journal of Molecular Sciences 22(21): 11868.

Tollefsen, D.P. 2006. From extended mind to collective mind. Cognitive Systems Research 7(2–3): 140–150.

Tononi, G. 2004. An Information Integration Theory of Consciousness. BMC Neuroscience 5(42): 1–22.

Tononi, G. 2008. Consciousness as Integrated Information: a Provisional Manifesto. The Biological Bulletin 215(3): 216–242.

Torday, J. and Miller, W. 2016. Phenotype as Agent for Epigenetic Inheritance. Biology 5(3): 30.

Chapter 18

How Might Biology Instruct Physics?

Undoubtedly, physics has informed biology. For example, a better understanding of cellular enzymatic function was gained by incorporating quantum biological electron tunneling into the modeling of protein configurations, overcoming limitations based on classical chemical dynamics (Xin et al. 2019). Conversely, biology can reciprocally inform physics. Indeed, this proposal may be anathema to some scientists. After all, it was the peerless New Zealand physicist and Nobel Prize winner Ernest Rutherford, the father of nuclear physics, who acidly remarked, "All science is either physics or stamp collecting". In apposition, there is a productive avenue of reverse exploration of the abiotic state from the insights about the requisites of biotic systems; structural order depends on the measurement of information within a referencing system that necessitates boundary mechanisms. Accordingly, the flexible palette of the competent cellular measurement of stimuli and the informational architecture required to support this intelligent activity can be examined, yielding a deeper understanding of abiotic interactions.

As this book exemplifies, it has only recently become clear that the essence of biology is the cellular measurement of information as continuous cellular problem-solving in perpetual confrontation with environmental pressures. The context of that living measurement is informational ambiguity due to the inherent inaccuracies of observer/participant subjective evaluation enumerated in Chapter 4 relating to classical Newtonian and quantum mechanisms. Accordingly, physics and biology would productively reconcile if forms of measurement take place across abiotic states, having shared features with cellular measurements.

As previously explained, in the living realm, a robust definition of information is a 'difference that makes a difference'. Crucially, any such difference is necessarily a measurement. A measurement is required to make that distinction. However, a referencing system is pivotal to any valid measurement given informational ambiguities. In the living state, a discrete inside vs. outside boundary mechanism as an effective plasma membrane helps secure that referencing system. The cellular plasma membrane permits the cellular compartmentalization of universal N-space as a localized Pervasive Information Field essential to living structural order. Consequently, for abiotic states to share any equivalency with biotic ones, a form of measurement should be present in abiotic states, thus entailing some corresponding boundary mechanism. Accordingly, it is proposed that N-space compartmentalizations also occur in the abiotic realm and contribute to structural order, just as a Pervasive Information Field and N-space Episenome are functional partitioning of universal N-space within living systems.

For this to be operative, there should be three concurrent features across our universal system. First, a holographic universal relational information matrix (N-space) must be present as a universal informational referencing system upon which all others depend. Secondly, an updated concept of what information represents must be presented that is equally applicable to both abiotic and biotic states. Third, the concept of the partitioning of N-space must be systematized so that it applies to abiotic states.

The concept of N-space as a universal holographic relational information matrix was introduced in Chapter 6. Integral to this concept is that information represents a basic property of the universe, constituting its universal fabric. Stonier (1996) maintained that energy and information are interconvertible. Based on that equivalency, it is possible to conceive of N-space as a scale-free universal relational information matrix as the universal interconvertible connections between matter and energy (Miller et al. 2020a). Hence, N-space, as an informational fabric, bears some resemblance to the type of universal connectedness that metaphysicians have considered as constituting universal consciousness (Dodig-Crnkovic 2013, Chopra and Kafatos 2017). However, to emphasize, postulating some epistemic agency is not necessary; even atoms can accumulate information to become physical composites that engage in their own form of selective interactions (Akama 2014).

Universal consciousness is defended as the means of supplying the necessary observer to settle quantum relationships in abiotic states. However, it is not necessary to impute a conscious observer if abiotic interactions can be appropriately asserted as capable of satisfying the requisite observer status to settle entanglements in abiotic states. Still, for this to be the case, several requirements must be met. At the outset, information must be conceived as a universal, fundamental property representing the inherent interconnection between matter and energy. Those connections are both classical and quantum. Crucially, once those relationships become the proper focal point of existence, neither matter nor energy has any explicit meaning in and of themselves. Instead, the only reality is in their interrelationships since measurement of differences is perforce based on those interrelationships (Miller et al. 2022). Matter without energetic interactions would equate with a completely inert universe. However, no place in the universe has been identified at zero degrees Kelvin, and even if it existed, it could not be detected without perturbing it through an energetic relational context. Consequently, reality is solely within its interconnections as entanglements between matter and energy. Furthermore, those interconnections can be productively considered a form of universal connective tissue forming a fabric of universal information.

Proceeding within that framework, the features of the N-space holographic relational information matrix that undergirds biologic information assessments, bearing a resemblance to *k*-space architecture, helps satisfy the necessary abiotic observer requirement (Miller et al. 2020a, Baluška et al. 2022). In Chapter 6, N-space was described in terms of the *k*-space architecture of MRI scanning in which each of the data partitions contains contributions from every point in the data field. Using that conceptual model, universal N-space is conceived as a scale-free relational information matrix with holographic features similar to *k*-space. Every universal location has some element of a contribution from all other universal locations.

When it is accepted that information is everywhere as interactions between matter and energy, it follows that their informational meaning derives from them as superimpositions of implicates (potentials) to an observer. If information is everywhere as interrelated connections, its meaning must reside within any differences that can be detected. Notably, Bateson's useful definition of information as a 'difference that makes a difference' implies that time is relevant to that distinction (Schroeder 2017). From this definition, two further derivatives issue. Differences over time correspond to detector/observer context (Miller et al. 2022). Further, and deserving re-emphasis, any 'difference that makes a difference' is necessarily a measurement as the only means of permitting a discrimination. Given these essentials, it can be proposed that the same definition of information can be appropriately applied to interactions in both biotic and abiotic states.

To fully satisfy that measurement exists in abiotic states, a further requisite requires elaboration. N-space partitions of universal relational information space must also exist in the abiotic sphere and contribute to structural order. When information is deemed the fundamental fabric of the universe, the concentrations of matter and energy as identifiable cosmic physical structures, such as galaxies with their millions of stars can now be considered areas of relatively concentrated information in universal information space-time. As such, within the abiotic sphere, these can be regarded as compartmentalizations of N-space akin to cellular PIFs and multicellular N-space Episenomes.

To justify this latter stance, the concept of Markov blankets and boundaries applied to living states as an intrinsic part of the Free Energy Principle and active inference can be productively applied to abiotic states. A Markov blanket defines the boundaries of a Bayesian network as a statistical set, interacting with other Markov blankets in a nesting configuration at different scales (Palacios et al. 2020). Although each nesting Markov blanket is interrelated, it still maintains conditional independence from others. Markov blankets help explain how living systems self-organize through statistical compartmentalization into internal and external states. Markov blankets represent the states that separate them and enjoin their conditional independence, inducing active inference as noted in Chapter 5 (Kirchhoff et al. 2018). This architecture mirrors the cell, which depends on its Markov blanket to distinguish itself from everything else and yet, permits internal and external states to influence one another. Notably, this contingent separation between an inside and an outside supports the autopoietic self-generation of information (Kirchhoff et al. 2018). Further, a living system of nested Markov blankets is scale-free, extending uninterruptedly from DNA to organelles, cells, holobionts, and out into the external environment as collective holobionic activities. Consequently, Markov blankets and boundaries are applicable to all types of boundary systems, from the inside organization of cells with their highly discrete, intelligent cellular plasma membranes to any other type of system that can become a self-generating organized whole.

Analogously, when placed in an informational frame, abiotic structures can be regarded as having forms of boundary systems akin to living systems. For example, solar systems and galaxies have effective gravitational boundaries and are organized in periodic orbits and trajectories. Arms of galaxies have myriad constituent stars and planets in definable relationships. Similarly, entire galaxies and galaxy clusters have their own patterns and self-contained influences as they are widely separated from one another. Accordingly, each constituent universal organized structure can be regarded as having its own Markov blanket and boundary representing its type of self-ordering information structure as compactions of universal information density. In this way, galactic borders represent general organizational boundaries that separate galaxies from one another and contribute to structural order. Nonetheless, each galaxy remains related to other galaxies. Each is its individual Markov blanket and boundary system within a universal nesting system that ultimately resides in universal informational N-space connected by myriad quantum interrelationships. In this manner, abiotic structure and biotic systems have firm resemblances.

Thus, when information is regarded as fundamental to the universe along with matter and energy in a scale-free relational information matrix (N-space), all organized cosmic physical structures, such as galaxies, can be considered their own N-space partition. Through entanglements and coherences, this structural ordering based on information serves as a cohering relational platform linking partitioned abiotic structures which both receive and send information. Moreover, in that specific regard, each such compartment, such as a galaxy, represents a separate, conditionally independent inertial quantum reference frame linked to other separate galactic compartments.

This partitioning of N-space into different quantum reference frames provides a useful conceptual avenue for reconciliation between physics and biology. It provides the necessary requirements for information in abiotic states to represent a 'difference that makes a difference' that can apply as much to an electron as a human. In a quantum rotating system, the state of any entanglement with respect to an observer may vary between reference frames. Quantum information becomes the superposition of two or even more states, and crucially, these entanglements would be observed differently between separated observers (Giacomini et al. 2019). Consequently, when information is placed in a rotating frame, independent, separated observers will receive 'off-set' information compared to one another, creating inherent informational asymmetries (Giacomini et al. 2019). Notably, in a holographic universal informational matrix, there is no shortage of observers in the abiotic sphere. Potential observers within separated reference frames are universally distributed due to the principle of quantum non-locality. Therefore, it can be asserted that information in abiotic states has quantitative and qualitative aspects, the latter relating to the differential 'experience' of the information value of the entanglement correlations between different observers in separate Markov blanket N-space compartment quantum reference frames. No two observers settle quantum asymmetries in identical ways if they are in separate distant non-inertial quantum frames, unlike the artificial observational structures that characterize our quantum investigations by laboratory means.

Thus, it can be construed that information has differential meaning among separate observers within abiotic contexts as 'differences that make a difference' among observers in different inertial quantum reference frames. Naturally, these differential quantum observer-dependent assessments permit resolutions of quantum implicates as superpositions and asymmetries that settle into coherences and explicates. In this manner, the universe exists through abiotic observer participation irrespective of living observation, as it must since the universe existed for billions of years before life began on our small planet.

With that background, the probabilistic settlement of entanglements within separated compartmentalized abiotic quantum reference frames becomes conditional. Each is an individual non-inertial quantum frame of reference. Any observer that might settle an informational non-local entanglement in its quantum frame would not necessarily experience that interaction in the same way as a different observer in its separate quantum reference frame. Consequently, abiotic quantum observers can be construed as participating in an abiotic form of measurement in their differential settling of non-local and local quantum entanglements since they are participants in 'differences that make a difference'. Since this is a contextual discrimination, it is a measurement. Further, that settlement of quantum superpositions has meaning. Each resolution impacts and assimilates within the universal N-space holographic information matrix, thereby contributing to universal order through both local Cartesian relationships and non-locality. In this way, information represents the connective tissue between matter and energy across the universe, whose interactions directly contribute to universal physical and structural order, which is meaningful.

There are several principles from biology that sanction the possibility that measurement should be active in abiotic systems as well as the living frame. It is now well-accepted that multicellular evolution is a series of exaptations of basic unicellular processes (Torday and Miller 2018, 2020). Accordingly, it can be further extrapolated that those elemental unicellular cellular faculties that are being exapted over evolutionary space-time to permit multicellular life must relate to other exaptations of earlier prevenient abiotic features.

Following that line of reasoning, since multicellular biological structure depends on N-space compartmentalization, N-space partitions should also be a component of the abiotic realm, permitting their forward exaptation in living systems. Abiotic N-space partitions with their relative Markov blankets and boundaries contribute to universal order as both are parts of conjoining N-space as a universal relational information fabric. Accordingly, biological N-space partitions would be an exaptation of a preexisting abiotic physical information architecture that contributes structural order within the abiotic realm just as it does in biology. In that circumstance, structural order in both abiotic states and biology would relate to the resolution of quantum entanglement asymmetries and superpositions that cohere across universal space-time. Based on information, this quantum ordering might be considered a universal interrelational glue that merges the abiotic and biotic spheres. Further, if that were the case, then quantum non-locality transforms from merely a remarkable quantum phenomenon to an essential component of structural order as a significant part of universal information within an overarching universal holographic N-space informational, relational matrix.

Further, the concept of separate quantum reference frames is a specific feature of the contemporary modeling of living systems. Fields et al. (2022) consider neural actions as hierarchies of quantum reference frames. If this is accepted, then every cell with its own Pervasive Information Field as its individualized N-space informational attachment should be considered its own specific quantum frame of reference. That frame references both to outward N-space and its relevant N-space Episenome. The latter is a local collective partition of N-space that serves as an aggregate cellular referencing matrix to center the individual constituent cellular internal measuring assessment of information cues, enabling concordant multicellular information assessment. Necessarily too, that N-space Episenome must maintain its own attachment to total universal N-space. Each biological partition of N-space, either as a PIF or N-space Episenome is also a Markov blanket and boundary (Miller et al. 2020a, Baluška et al. 2022). Importantly though, there is no reason to suppose that these compartmentalizations of N-space as either PIFs or N-space Episenomes are either an originating or endpoint iteration. Instead, it should be presumed that they are part of a universal system of nested reiterated architecture as an entailed feature of the concept of nested Markov blankets.

In this manner, a seamless continuum between abiotic and biotic states can be envisioned. Both abiotic and biotic states have a shared quotient of assessing fundamental information as quantum correlations between matter and energy, each in its own way. Both can be productively placed within a framework of the partitioning of universal holographic N-space. What separates these states? When information is a fundamental property of the universe, then necessarily, the meaning of that information is directly related to the characteristics and limitations of the observer/detector. Those attributes determine how that observer can receive the inputs of any informational quantum space-time interrelationships as potential entanglements, asymmetries, correlations, resonances, or coherences and, in turn, how those can be settled.

Fortunately, those differences are readily appraised. There is a demonstrable and measurable difference between abiotic states and living ones. Abiotic states can be characterized as circumscribed by a partitioning of N-space in which the boundaries are inexact, just as galactic boundaries are, reflecting how physical phenomena such as gravity gradually decrease over distances. In contrast, the boundary separations in the living frame are much more discrete as the requisite cellular plasma membrane. Further, that intelligent membrane is linked to significant retrievable and deployable memory, again unlike abiotic states.

Consequently, abiotic particles detect minimal qualitative differences in information content within their systems' architectural limits. Absent useful memory, any settling of quantum fundamental uncertainty relationships is bounded as a series of at-the-moment interactions. These interactions contribute to local and universal order and are thereby meaningful. However, there is no entailing effective memory apart from some hysteretic properties. Hysteresis is the dependence of the state of a system on its history, such as seen in magnetic induction. However, abiotic states have no internal self-reflection or capacity for adaptation. Consequently, physical particles represent contextual observers of quantum interactions sufficient to satisfy the quantum resolution of entanglements and coherences and undergo classical

Newtonian interactions as a valid observer, and no more. However, living systems are demonstrably different, possessing intelligent, discrete membranous boundaries whose resonances permit internal self-referencing associated with memory utilizing an array of cellular tools. Significantly, the ultimate sources of information are the same within abiotic and biotic states. The difference is entirely within the qualities of the detectors (Miller et al. 2022).

Many proposals have been advanced to better explain how the quantum realm integrates into biological expression. For example, states of gauge symmetry can be used to model biological information considering it as a vectoral bundle that satisfies particular symmetries in linear space (Longo and Montévil 2011, Miller et al. 2019). For particles in those states, a shift in space-time direction between points correlates with their gauge potentials. Furthermore, some experimental evidence has concluded that brain activity may derive from the stability or collapse of these gauge symmetries (Tozzi and Peters 2018).

Others have proposed that symmetry breaking, as modeled by experimental particle physics, has a direct relationship to experiential biological states through the application of three universally applicable principles: complementarity, recursion, and sentience (Kafatos 2014). In quantum mechanics, complementarity is the unifying factor between apparent opposites, embodied in the concept of antipodal information, with its en face and opposite sides, as previously discussed. Separate self-referential observers may only be able to derive information from the same phenomenon from opposite vantage points. Recursion presumes 'as here, so elsewhere' asserting a principle of self-similarity within quantum terms, indicating the presence of reiterative, cohesive principles that are scale-free. This principle ensures that self-referential organisms react to informational stimuli in self-similar patterns enabling collective measurement. The principle of sentience represents the crucial correspondence between the 'inside' and 'outside' that enables cellular life.

Naturally, these underlying principles all support cellular homeorhetic preference just as much as chemiosmosis and other life-sustaining processes. In the living frame, all processes are aspects of self-referential awareness. When it is acknowledged that information is always imprecise in a biological context, then all information has embedded implicates as contingent anticipation and predictions, considered as deviations from sets of random variables within a probabilistic quantum system (Miller et al. 2019).

In this frame, uncertainty can be alternatively assessed as incomplete information about an operative system (Khrennikov 2007). Therefore, any measurement of ambiguities is a cluster of quantum and classical thermodynamic inferences that yields a set of predictions that represent anticipatory apprehension to any self-aware entity. Notably, these linkages entirely reduce to cellular problem-solving in which the upper and lower set of expectational boundaries of an informational matrix resolves a set of implicates into explicate cellular action (Gunji et al. 2016). Thus, biology exists through both freedoms of action and constraints, always attempting to achieve maximal predictability in space-time (Igamberdiev and Shkolvskiy-Kordi 2017). To attain this, self-referential cells measure imprecise information as predictions through classical and quantum inferences.

Conventional quantum theory presumes that a sufficient exploration of quantum probabilities should culminate in discovering objective features of nature as an ultimate reality. The physicist Christopher Fuchs (2011) disagrees. No objective reality can be defined through quantum wave functions. Instead, even these convincing mathematical wave functions exist as Bayesian probabilities, described as a Quantum Bayesian system or QBism. In this frame, rather than reflecting any actual reality, every derived wave function is subject to "degrees of belief about the system" (Gefter 2015). In QBism, quantum wave functions can never lead to a specific reproducible physical reality. Instead, they are an inherent reflection of the descriptive capacities of an observer updating its 'beliefs' after making its self-referential measurements (Fuchs 2011). In this framework, the universe is not an objective physical object but is instead an 'at the moment' subjective encounter with it as a succession of constantly updated successive measurements. Accordingly, all those experienced quantum wave functions are not any ultimate objective reality, but only an observer's immediate 'subjective' representation of that knowledge since it is steeped within obligatory self-reference. This stance conforms with the central biological factor of cellular life. All the information that any cell has is self-produced (Cárdenas-García 2020, Miller et al. 2020a).

Recent studies have added to the complexities of the issue of measurement in biological systems. Experiments indicate that humans can directly perceive quantum phenomena, including individual light photons, based on their properties of entanglement and coherences (Castelvecchi 2016). This surprising revelation carries a curious derivative. Quantum perception may require the observer to share characteristics with what is being observed in an as-yet unexplained self-reflexive posture (Rosen 2021). Consequently, the possibility looms that something exceptional is going on beyond our current construct of quantum mechanics.

The eminent philosopher and mathematician Alfred North Whitehead and the brilliant theoretic physicist David Bohm believed that all entities in the universe have an embedded experiential sense-awareness, even within the inanimate realm (Whitehead, 1967, 1985, Bohm and Hiley 1975). For them, all reality is experiential. In their terms, the universe is defined through a relational universal 'sense-awareness' (Rusu 2011). Many other scientists were highly influenced by this perspective and Whitehead and Bohm's search for a universal 'underlying wholeness' (Griffin 1985).

Bohm further explored this thought experiment through a novel concept of a holomovement (Lohrey and Boreham 2021). In Bohm's view, wholeness was universal, unified, and interconnected through an indivisible implicate order. In these terms, 'implicate' means 'to fold inwards', indicating that each separate space-time location contains the total universal structure 'enfolded' within it (Lohrey and Boreham 2021). Whatever physical objects we might observe within the continuum of space-time are representations of the explicate order that has unfolded from the implicate order. Of course, these concepts are not new at a metaphysical level. The pre-Socratic Greek philosopher Anaxagoras postulated that "everything is in everything" in the 5th century B.C.E, arguing that while matter is infinitely divisible, every fragment of it necessarily carries mixtures of all possible qualities within it, just in different proportions (Seager 2018).

The structure of a hologram resembles this, although only inexactly. In a hologram, every local region of the projected three-dimensional image contains the entire image mapped at lower resolution, i.e., "each fragment of the image is not only a part of the whole but also an instance of the whole" (Lohrey and Boreham, 2021). This is precisely the conceptual matrix that has been used to envision N-space architecture which includes that theoretical base of the enfolding of implicates but does not suggest universal consciousness (Miller et al. 2020a,b). Everything is connected, and it is experienced according to the capacities and limitations of the relevant observers. Aboitic states have measured connections but insufficient structural ordering or memory to enable self-reference.

The concept of 'meaning' also has special implications for Bohm. Meaning is more fundamental than matter or energy (Lohrey and Boreham 2021). Meaning has agency and 'enfolds' matter and energy within it. Since we are conscious, certainly our life has self-referential meaning. If consciousness is universal, then the universe, according to Bohm, has meaning as a particular enfolding of matter and energy in unique combinations. Hence, instead of considering cell-cell communication as an exchange of information, it is actually an exchange of meaning. For example, genetic code is information, but it is also a physical embodiment of meaning as an enfolded representation of a set of solutions to problems for conscious organisms. A period on a page is information as a dot of ink on a white background but has an enfolded meaning as the conclusion of a thought process. That meaning has three inherent aspects: context, unity, and interconnection. Further yet, and crucially, meaning has 'implications'. These can be seen as the superposition of possibilities carried forward from meaningful communication. Hence biology is where the enfolding of consciousness as meaning expresses as physicality, becoming a unity of the implicit and explicate realms. In such circumstances, as Lohrey and Boreham (2021, p. 226) assert, "meaning represents the content of consciousness", which " in the context of the holomovement is universal consciousness".

However, as has been detailed, there is no requirement for meaning to be exclusively invested within consciousness. The novel conception of construing abiotic meaning by virtue of compartmentalized separate inertial rotational quantum reference frames as abiotic N-space partitions offer that alternative. In that frame, quantum asymmetries and superimpositions settle into quantum coherences and resonances with qualitative, conditional, and contingent features contributing to universal structure. Within a universal holographic informational reference system, the settling of quantum non-localities. even within abiotic states contributes to universal order by enfolding within universal N-space. In this circumstance, quantum non-locality becomes part of a universal informational matrix as part of its essential fabric (Fig. 1)

When this framework is embraced, a clear narrative of universal connectedness emerges that need not demand universal consciousness. Certainly, informational connections are abundantly evident to us as they comprise our living experiences. Yet, there is no justification for assuming that the living frame initiates these as a localized phenomenon. Instead, our experientiality represents a by-product of a precedent and encompassing process of universal connections and, significantly, has further connections (Torday and Miller 2018). Among the living, that connectedness

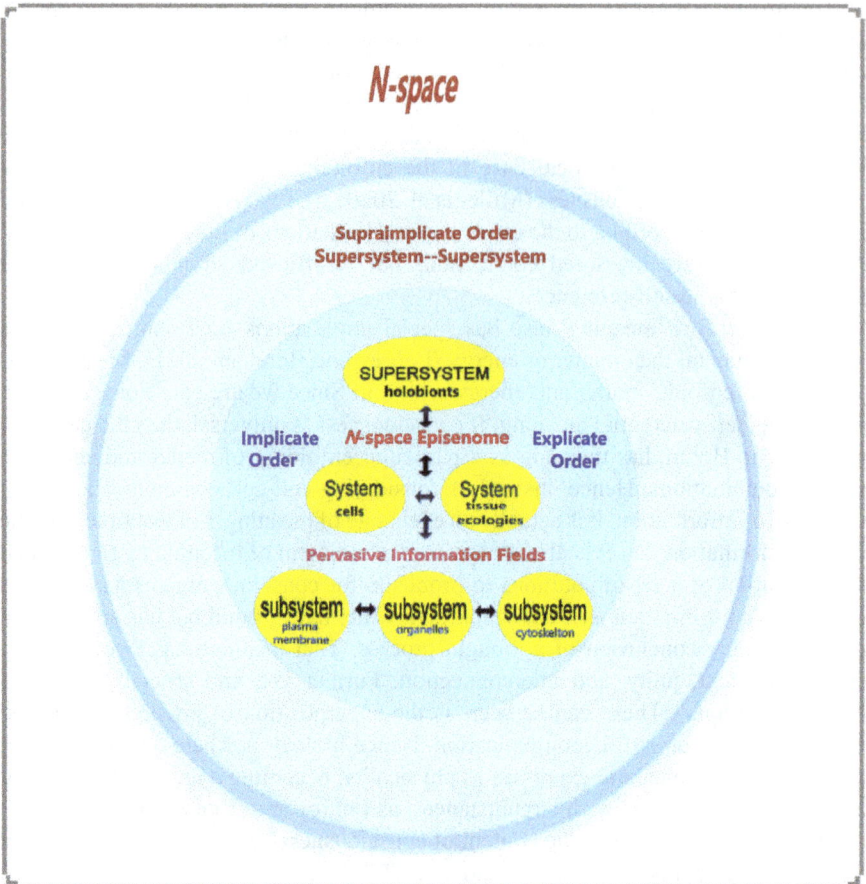

Fig. 1. Legend – The enfolding of systems within N-space.

N-space represents a universal relational information matrix to which all systems attach. Intracellular structures as subsystems interconnect through a cellular Pervasive Information Field as their relevant information space. These cells aggregate in multicellularity into cellular ecological systems that connect through their concordant N-space Episenomes as supersystem holobionts. Supersystem holobionts function together in groups as collections of supersystems as part of a supraimplicate order that can be considered a form of extended consciousness. All levels explore information space, settling implicates into explicate biological activity.

can now be identified as the singularly effective partitioning of universal N-space, embodied in cellular Pervasive Information Fields, and collectively aggregated in multicellularity and transferring from generation to generation as its heritable informational N-space Episenome. Nonetheless, that living set of connections can be assumed as an exaptation of earlier informational connections established between separate non-inertial quantum reference frames as their form of N-space partitioning in the abiotic realm.

Whitehead (1967, 1985) insisted that the entire universe represents the timeless enfolding of implicates in quantum space-time that connects every object to all other universal objects, even within the inanimate realm. The universe thereby summates as the energetic interactions and correlations within its constitutive universal informational matrix. Therefore, the universe must be characterized as a verb rather than a noun. The universe is not defined by matter or energy but by the actions of its observer/detectors, living or not.

Yet, this need not equate with consciousness in abiotic states. Universal connections are real, but overt consciousness is not required to effect the settlement of quantum implicates. Superpositions can settle through different forms of contextual measurement. However, both abiotic and biotic states share a universal connectedness by contributing to scale-free universal relational informational N-space as part of a holographic conceptualiztion of the universe. Lohrey and Boreham (2021) emphasize that this different viewpoint has practical significance to evolutionary biology. The distinction between the Darwinian framework of competition and the model of endosymbiotic cooperation articulated by Lynn Margulis represents the difference between explicate and implicate realms. Traditional Neodarwinism tends toward the Cartesian explicate frame, insisting on a "world of separation and division" by emphasizing competition. Conversely, Margulis opened a path towards framing evolution as a series of superimposed implicates that stresses the underlying unity of all systems, empowered by symbiosis, symbiogenesis, and mutualizing cooperation. Clearly, the latter better fits the nested architecture of interrelated Markov blankets that can now be understood as essential to both the physical and living realms.

Biology teaches that all the information that a cell knows is self-produced. Each cell constructs its individual reality. Necessarily too, physicists are cellular beings. Physics, too, is a product of self-produced information and is potentially and perhaps inseparably chained to its own self-referential interpretation of date and experimental design. It opens the possibility that correctly understanding biology, as much as encumbered self-referential beings might be licensed, requires a new physics. In 1939, Arthur Eddington had presciently written, "Physical science consists of purely structural knowledge, so that we know only the structure of the universe it describes" (Clement 1953, p. 266). Hence, the meaning of physics may need to percolate from the study of biology and its quantum tools, accepting a holistic frame that describes a unique form of information that upholds an essential unity between abiotic and biotic states (Seager 2018).

Others have pointed out that the currently inexplicable interrelationship between mind and matter should be re-explored through fresh paths that do not conform with current mainstream methods and perspectives (Lohrey and Boreham 2021). What is certain about biology is that life is problem-solving and knowledge accumulating (De Loof 2017, Miller 2018, Torday and Miller 2020, Baluška et al. 2021). Recognizing this, Nagel (2012) weathered substantial criticism, forthrightly stating that a new physics is required to understand the living frame. Despite all the daunting difficulties that search will impose, the pathway of discovery is not clouded. It resides within the most profound discernment of the elusive characteristics of information as fundamental universal connections that structure abiotic states and impel life. Assuredly, the specific direction of that exploration crystallizes. We must seek to fully comprehend the ravishing, profound, ineluctable mystery of self-referential consciousness.

"When self-reference is properly regarded as the fundamental essence of biology, then biology and physics are directly reconciled. They are both differing modes of measurement".

Miller et al. 2019

References

Akama, K. 2014. A Universal Origin of Information Accumulation in Nature arXiv preprint. ArXiv 1408.6201.

Baluška, F., Miller, W.B. and Reber, A.S. 2021. Biomolecular Basis of Cellular Consciousness via Subcellular Nanobrains. International Journal of Molecular Sciences 22(5): 2545.

Baluška, F., Reber, A.S. and Miller, W.B. 2022. Cellular sentience as the primary source of biological order and evolution. Biosystems 218: 104694.

Bohm, D.J. and Hiley, B.J. 1975. On the intuitive understanding of nonlocality as implied by quantum theory. Foundations of Physics 5(1): 93–109.

Cárdenas-García, J.F. 2020. The Process of Info-Autopoiesis—the Source of all Information. Biosemiotics 13(2): 199–221.

Castelvecchi, D. 2016. People can sense single photons. Nature.

Chopra, D. and Kafatos, M.C. 2017. You Are the Universe: Discovering Your Cosmic Self and Why It Matters (Reprint). New York: Harmony.

Clement, W.C. 1953. Russell's Structuralist Thesis. The Philosophical Review 62(2): 266.

De Loof, A. 2017. The evolution of "Life": A Metadarwinian integrative approach. Communicative & Integrative Biology 10(3): e1301335.

Dodig-Crnkovic, G. 2013. Alan Turing's Legacy: Info-computational Philosophy of Nature. pp. 115–123. In: G. Dodig-Crnkovic and R. Giovagnoli (eds.). Computing Nature. Berlin, Heidelberg: Springer.

Eddington, S.A. 1939. The Philosophy of Physical Science: Tarner Lectures (1938). CUP Archive. Cambridge , UK: Cambridge University Press.

Fields, C., Glazebrook, J.F. and Levin, M. 2022. Neurons as hierarchies of quantum reference frames. Biosystems 219: 104714.

Fuchs, C.A. 2011. Coming of Age With Quantum Information: Notes on a Paulian Idea. Cambridge , UK: Cambridge University Press.

Gefter, A. 2015. A Private View of Quantum Reality. Retrieved from https://www.quantamagazine.org/quantum-bayesianism-explained-by-its-founder-20150604/

Giacomini, F., Castro-Ruiz, E. and Brukner, Č. 2019. Quantum mechanics and the covariance of physical laws in quantum reference frames. Nature Communications 10: 494.

Griffin, D.R. 1985. Bohm And Whitehead On Wholeness, Freedom, Causality, And Time. Zygon 20(2): 165–191.

Gunji, Y.-P., Sonoda, K. and Basios, V. 2016. Quantum cognition based on an ambiguous representation derived from a rough set approximation. Biosystems 141: 55–66.

Igamberdiev, A.U. and Shklovskiy-Kordi, N.E. 2017. The quantum basis of spatiotemporality in perception and consciousness. Progress in Biophysics and Molecular Biology 130: 15–25.

Kafatos, M.C. 2014. Physics and consciousness: quantum measurement, observation and experience. Frontiers in Consciousness Research 4: 1–14.

Khrennikov, A. 2007. Can Quantum Information be Processed by Macroscopic Systems? Quantum Information Processing 6(6): 401–429.

Kirchhoff, M., Parr, T., Palacios, E., Friston, K. and Kiverstein, J. 2018. The Markov blankets of life: autonomy, active inference and the free energy principle. Journal of The royal society interface 15(138): 20170792.

Lohrey, A. and Boreham, B. 2021. Lifting the veil on Bohm's holomovement. Communicative & Integrative Biology 14(1): 221–229.

Longo, G. and Montévil, M. 2011. From physics to biology by extending criticality and symmetry breakings. Progress in Biophysics and Molecular Biology 106(2): 340–347.

Miller, W.B. 2018. Biological information systems: Evolution as cognition-based information management. Progress in Biophysics and Molecular Biology 134: 1–26.

Miller, W.B., Baluška, F. and Torday, J.S. 2020b. Cellular Senomic Measurements in Cognition-Based Evolution. Progress in Biophysics and Molecular Biology 156: 20–33.

Miller, W.B., Baluška, F. and Reber, A.S. 2022. A scale-free universal relational information matrix (N-space) reconciles the information problem: N-space as the fabric of reality. Submitted.

Miller, W.B., Torday, J.S. and Baluška, F. 2019. Biological evolution as defense of "self". Progress in Biophysics and Molecular Biology 142: 54–74.

Miller, W.B., Torday, J.S. and Baluška, F. 2020a. The N-space Episenome unifies cellular information space-time within cognition-based evolution. Progress in Biophysics and Molecular Biology 150: 112–139.

Nagel, T. 2012. Mind & Cosmos: Why the Materialist Neo-Darwinian Conception of Nature is Almost Certainly False. Oxford, UK: Oxford University Press.

Palacios, E.R., Razi, A., Parr, T., Kirchhoff, M. and Friston, K. 2020. On Markov blankets and hierarchical self-organisation. Journal of Theoretical Biology 486: 110089.

Rosen, S.M. 2021. The Strange Nature of Quantum Perception: To See a Photon, One Must Be a Photon. The Institute of Mind and Behavior, Inc. The Journal of Mind and Behavior 42(3 & 4): 229–270.

Rusu, B. (2011). Whitehead and Green: The Metaphysics of Universal Relatedness. Chromatikon: Annales de La Philosophie En Procès/Yearbook of Philosophy in Process 7: 169–201.

Schroeder, M.J. 2017. The difference that makes a difference for the conceptualization of information. Multidisciplinary Digital Publishing Institute Proceedings 1(3): 221.

Seager, W. 2018. The Philosophical and Scientific Metaphysics of David Bohm. Entropy 20(7): 493.

Stonier, T. 1996. Information as a basic property of the universe. Biosystems 38(2–3): 135–140.

Torday, J.S. and Miller, W.B. 2018. The Cosmologic continuum from physics to consciousness. Progress in Biophysics and Molecular Biology 140: 41–48.

Torday, J. and Miller, W. 2020. Cellular-Molecular Mechanisms in Epigenetic Evolutionary Biology. Cham, Switzerland: Springer.

Tozzi, A. and Peters, J.F. 2018. What is it like to be "the same"? Progress in Biophysics and Molecular Biology 133: 30–35.

Whitehead, A.N. 1967. Science and the Modern World. New York: Free Press.

Whitehead, A.N. 1985. Symbolism: Its Meaning and Effect. New York: Fordham University Press.

Xin, H., Sim, W.J., Namgung, B., Choi, Y., Li, B. and Lee, L.P. 2019. Quantum biological tunnel junction for electron transfer imaging in live cells. Nature Communications 10(1): 3245.

CHAPTER 19

A Separation from the Past

"We must therefore carefully distinguish between two quite different doctrines which Darwin popularised, the doctrine of evolution, and that of natural selection. It is quite possible to hold the first and not the second".

J.B.S. Haldane, 1892–1964
The Causes of Evolution, 1932

There is a persistent belief that natural selection creates adaptations (Price 2017). Of course, this could never be the case since variations must occur first to be presented for filtering selection. Consequently, evolutionary biology must be based on the scientific study of variations, not selection. Nature concurs insofar as there many biological mechanisms to produce variations through natural cellular engineering, epigenetic impacts, genomic rearrangements, and horizontal genetic transfers. It was evident to Darwin that variation came before selection, even though he had no idea what created those variant 'sports' that he discussed. Nor was Darwin a specific proponent of what became the Neodarwinist canon that mutations are blind and phenotypic variation is unbiased. However, as direct observational evidence of highly integrated functional biology accumulated that seemed to contradict random mechanisms, a concept of developmental bias emerged to account for the essential reciprocations between the environment and phenotype necessary to sustain evolutionary development (Uller et al. 2018, Corning 2020). Within this construct, certain ontogenetic trajectories influence the direction and outcome of evolutionary change towards one set of evolutionary traits over another as a developmental bias that somehow influences the stream of random genetic mutations (Arthur 2004).

Despite minor adjustments, mainstream biology has been very slow to accept that the current concepts of the Modern Synthesis are deeply flawed. The resistance to any fresh narrative has been strong. The noted evolutionary biologist and geneticist Richard Lewontin (1970) opined about that deficiency, noting that natural selection does not explain the forms of life that can be observed. Consequently, there has to be an immense amount of evolutionary biological detail beyond Neodarwinism. Many other scientists have argued similarly (Jablonka and Lamb 1999, 2008, Pigliucci 2007, Baluška 2009, Witzany 2010, Shapiro 2011, Torday 2015, Miller 2016, 2018, Miller and Torday 2018, Miller et al. 2019, Baluška and Reber 2019, Corning 2020).

Indeed, it has been strenuously argued that the "Modern Synthesis obscures and sometimes seriously misrepresents the underlying causal dynamics in living systems, and in evolution" (Corning 2020, p. 2).

A. Re-envisioning genetic inheritance

Over the last several decades, there has been substantial progress in reshaping premature assumptions about genetic inheritance and initial characterization of the nuclear genome as largely static and changing slowly due to replication errors. Because biology has remained frustrated by the lack of congruence between phenotype and genotype, a willingness slowly developed to explore alternative narratives such as symbiogenesis and natural genetic engineering (Noble et al. 2014). In addition, long-held beliefs, such as Crick's Central Dogma insisting on the unilateral flow of information from DNA to RNA to proteome, are now acknowledged as untenable (Shapiro 2009, 2011). This unduly limiting narrative has been replaced by a more nuanced understanding of genomic flexibility, opportunities for epigenetic adjustments, and the contributions of a wide range of other genetic participants, including myriad small RNAs.

Although once regarded as a rarely mutable source of hereditary stability, the genome is now understood to be a reciprocating, read-write flexible palette that is fully responsive to environmental stresses supporting pan-cellular interactions (Dreze et al. 2009, Witzany 2009, Shapiro 2011, 2017, Miller et al. 2020a). Accordingly, the genome can be considered akin to a developmental organ that continuously adjusts through serial context-dependent accretions, exaptations, and deletions that directly relate to internally generated cellular processes and external pressures (Lamm 2014).

Moreover, the virome is a crucial co-respondent with the genome as a source of fluid transkingdom communication of information that is more likely to be a symbiotic partner in viral-cellular co-adaptations than an overt pathogen (Goldenfeld and Woese 2007, Ryan 2009, Koonin 2009, Miller 2016). In this way, genes can be reappraised as cellular tools that can be contingently deployed in response to both non-stochastic and stochastic inputs yielding productive outputs (Miller 2016, 2018, Miller et al. 2020a, Torday and Miller 2020). When viewed within that revised role, a genome can be considered a lineal information repository with 'common interests' that supports the same or closely aligned phenotypes over time as a cooperative program (Stencel and Crespi 2013). In that circumstance, a genome is another iteration of niche construction whose cellular purpose is meaningful, evolution-tested information supporting cellular problem-solving in the face of environmental heterogeneity (Hoffmeyer 2002). Undoubtedly then, the essence of cellular problem-solving through the creative use of cellular tools is the development of new ecological partnerships that enhance cellular survival characteristics. A flexible genome, epigenome, and any of their proteomic outputs are essential to these creative biological outcomes.

An advantage genes have over human tools like hammers is that genes are subject to independent rearrangement. It would be as if our hammers could self-modify to enable new shapes and cellular connections. However, even as our views of genes have become much more expansive, it is essential to recall that genes do not 'do' anything on their own (Oyama 2000). As Lewontin insisted, "DNA becomes 'information' about the organism only in the actual process of cell function" (Oyama 2000, p. xiii). Further, a growing consensus argues that cellular expression and behaviors are subordinate to a variety of inputs that are 'unmediated' by genes (Walsh 2014). As these consequential sources of heritable input multiply, the need for a coordinating mechanism becomes apparent. Life's complexity and causal structure require coherent 'mediation' across levels (Walker and Davies 2016). Accordingly, it has been asserted that a 'control kernel' is needed with the "requisite property of connecting informational architecture to the causal mechanisms of the network" (Walker et al. 2016, p. 13). That requisite property is a phase shift in causal structure where information gains efficacy over matter (Walker et al. 2016). The identity of that 'control kernel' is now apparent. Self-referential consciousness, whose instantiation is coincident with the origin of life, is that epicenter of causality and control in biological development (Miller 2016, 2018, Reber 2019, Baluska and Reber 2019, Miller et al. 2020a).

B. Cognition takes center-stage

As the known consequential sources of heritable input have multiplied based on emerging research techniques, the need for a non-random coordinating mechanism has become readily apparent. Consequently, there is a growing acceptance that the dominant Neodarwinian narrative requires a substantial rewrite distancing itself from evolution based on selection and genes to another more robust explanatory framework based on the intelligent use of information across all biological scales. A new coherent narrative is apparent in which self-referential cognition is central. When placed in that crucial framework, the correct biological order focuses. Genes are not evolutionary drivers but can be readily seen to be the tools of intelligent, problem-solving cells.

The cognitive faculties of cells have already been documented. An appreciation of the range of faculties and aptitudes across scales is now rapidly emerging. Fruit flies have highly sophisticated cognitive systems displaying faculties that were believed to be confined to mammals, including sensitive conscious awareness, variable attention, conditioning, and working memory (Grover et al. 2022). With cognition validated across all living scales, the separating cusp between physical systems and biological ones clarifies: " Self-reference is a specific feature of biological systems and not physical systems....." (Goldenfeld and Woese 2011, p. 9). When cells are affirmed as self-aware, the exquisite narrative of biology and its evolution resolves. Cells are cognitive, self-organizing, and problem-solving. Biology's entire narrative is a story of their continuous perpetuation over billions of years, melding into self-generated advantaged forms to do so. That survival is predicated on the reception, assessment, and deployment of information by cells in their continuous confrontation with their external environment, alone or collaboratively. The perpetuation of self-reference requires information management. Genes serve cells as crucial participants within that requisite information management system.

When cognition is fundamental and is inherently dependent on information, then the quality of that information becomes paramount. Consequently, the measurement of information to assess its validity is key. It is now undoubted that cells measure. However, some particular implications of measurement are unobvious. In a self-referential system, communication is a form of work, and "work is energy exchanged in the least randomly distributed way possible" (Berry 2019). Since communication adds order to living systems, and order has measuring value, then communication is itself a form of measurement. Any communication has its particular measuring value, which confers meaning. However, in a cognitive frame, meaning is purposed toward predictions sufficient to stimulate the contingent deployment of scant cellular resources. Therefore, for a self-referential entity, communication is an exchange of information as a measurement directed toward prediction (Miller et al. 2020a).

Undoubtedly, then, meaningful information is the common currency of this new biology. The molecular biologist and geneticist Christoph Adami (2016) offers a crucial definition of biological information as: "that thing that allows a living thing to make predictions with accuracy better than chance alone" (Adami 2016, p. 1). The difference between biological information and computer information thereby crystallizes. Prediction and anticipation are implicit within self-reference as its living quality, separating it from purely physical systems. Importantly, these qualifying self-referential capacities have been proved at all scales, including the unicellular level (Hellingworf 2005, Ben-Jacob and Levine 2006, Ben-Jacob 2009, Ford 2004, 2009, 2017, Miller and Torday 2018). Bacteria, amoebae, and scavenging T-cells all demonstrate the capacity to anticipate and predict. Necessarily, they can only accomplish that through measurement (Xavier et al. 2011, Miller 2018, Miller et al. 2019, 2020a). Indeed, bacteria demonstrate sophisticated predictive, associative learning at levels equal to that of metazoan nervous systems (Tagkopoulos et al. 2008, Goo et al. 2012).

It is important to ask, what is it that cells are measuring? Obviously, they are measuring the quantity and quality of bioactive molecular and energetic inputs. However, that measuring assessment as prediction is directed towards the consistent maintenance of cellular homeorhetic equipoise. However, and not immediately apparent, in a self-referential frame, this sustaining state of flux equates with a state of preference (Miller 2013). Significantly then, cells are monitoring their state of preference and how environmental cues affect that through cellular measurement. Accordingly, the impetus for natural cellular engineering and niche constructions, and the cooperation, collaboration, co-dependencies, and competitions that energize these living processes can now be further understood as a direct reflection of a drive to attain and sustain preferential cellular states. And notably, that is what we humans also iteratively do. Preference drives engineering. Consequently, when cognition is at the center of biology, all macroorganisms represent the physical expression of a functioning, harmonious collective state of cellular (and even perhaps viral) preferences in which mutualized preferences meet environmental proscriptions. Accordingly, reproduction is an epiphenomenon of cellular preference. Selection assures that cellular preferences conform to environmental constraints. This is by no means theoretical at any scale and is essential to living success. Microbial communities populated by multiple species will co-exist even when they share metabolic pathways dependent on the same nutrients. To compensate, they exhibit patterns of diauxic growth, consuming the available nutrient resources one at a time, each according to its preferences in patterns of spontaneous self-organization (Wang et al. 2021).

From this, it is evident the context of life as a verb equates with the active use of information to sustain change towards preferential conditions at all scales. Further, yet, when states of cellular preference are acknowledged as the driving self-referential evolutionary force, the endpoint of evolutionary development can no longer be placed into a simple narrative of climbing a fitness landscape. To a cell, preference might outweigh how we judge fitness. Phenotypic exploration is not merely for fitness but to gain information towards improved niche constructions to further develop, satisfy, and maintain cellular states of preference. When cognition is central, preference takes precedence.

C. A different path to complexity—Redefining multicellular life as co-engineering and niche construction

1. Shared measurements

All cells must measure to survive and co-engineer. Yet, multicellular eukaryotic engineering exclusively manifests as obligatory holobionts comprised of populations of individual self-referential cells of mixed species with differing genomes. Two critical ramifications ensue. Since all these cellular constituents participate in concordant growth and development, and all cellular information is self-produced, there must be a platform where measured environmental cues can be productively referenced among all participants. Further, any engineering requires a general architectural patterning accessible to all who might participate and is modifiable by them. The success of any engineering depends on the relevancy of the available information. In living systems, local control has priority. That reason is simple. There is higher information density and quality at the local cellular level. Thus, a shared informational matrix (N-space Episenome) that is adjustable over time, heritable, and transferable must be an embedded component of the information management system of cells to assure the compatibility of the mixed cellular types and species that comprise local tissue ecologies in holobionts (Miller et al. 2019). Simply put, N-space is where cellular measurements meet (Miller et al. 2020b).

2. Shared rules

Just as a consonant measuring platform is necessary, successful co-engineering requires conforming rules. Indeed, essential cellular processes must be continuously maintained over evolutionary space-time to sustain cellular life, including chemiosmosis, an ordered state, and homeorhetic balance (Torday and Rehan 2012). The obligatory recapitulation of the unicellular zygote is the active biological manifestation of the need for cellular life to consistently recenter according to basic principles and governing rules. That recentering is necessary to avoid any life-threatening skew through the flexible adjustment and adjudication of acquired epigenetic impacts (Torday and Miller 2016).

This reiterating impulse to sustain originating rules is an example of the 'principle of optimality', indicating that the best path for decision-making at successive scales proceeds by linking to smaller sub-problems solved at earlier levels (Igamberdiev and Shklovskiy-Kordi 2017). Attendant within that principle is a particular imperative. That optimal path must conform with an initial advantageous decision. The first decisions are the most important for humans, computer programs, and cells. However, remaining in concert with those initiating principles is just as paramount. Importantly too, for successful navigation of an external environment that can be chaotic, following that optimal path after its first initiation requires valid information along the way. Among cells, this is best achieved through collective assessment, as emphasized in previous chapters.

It can be reflected that the collective activities upon which multicellular harmony rests might easily be in conflict with individual needs. Nonetheless, the first rule of cellular life is the protection of self-integrity (Miller et al. 2019). The cell membrane is explicitly purposed to that vital drive and does so through boundary conditions that promote the maximal predictability of information for that cell. However, that necessity must also be balanced within the further imperative of cooperation that permits co-ordinate co-engineering and niche constructions and those symbioses that drive biology and evolution. Evidence indicates that both of these requirements, which might seem at odds, accord with the Principle of Optimality. Cells learn that their protection of self-integrity is best served by being cooperative and in service to others to a sufficient extent to receive mutualizing reciprocations. That balanced duality is the optimal path. Notably, that path is preconditioned on self-referential awareness of individual status and that of others.

There can be no doubt that these are elemental rules. Detailed examinations of bacterial biofilms surprisingly reveal that the development of bacterial biofilms is comparable to animal embryogenesis (Futo et al. 2021). In effect, biofilms are a true form of multicellular life and were so from their first emergence billions of years ago.

3. All rules requires consistent reinforcement—the role of immunology

In any ordered system, rules must be consistently followed and outliers disciplined. This highly coordinated activity is the role of immunology in its continuous defense of self-integrity. No multicellularity can be sustained if competitive interlopers gain access and disrupt the consistent application of local rules. Furthermore, immunology is the crucial linchpin of reproduction, whose success depends on the integrity of self-similar recognition through immunological compatibility. Indeed, immunological discriminations define the operating characteristics of our biological system since they are pre-variational compared to post-facto fitness. Therefore, selection can be understood as a derivative function of the immunological enforcement of self-referential identity, sustained through adherence to cellular life-sustaining principles over evolutionary space-time. This is the essence of cellular strategic information management. Consequently, immunology can now be judged as another form of information and communication that guide cellular decision-making within any information matrix (Tauber 2000). Thus, immunology represents another tool of cognition.

D. An animating endpoint—Biology and evolution recast

Placing information at the center of biology is not novel (Walker and Davies 2013). However, placing biology within an encompassing context of self-referential cognition and then further exploiting that cohering crucible to a full extent is assuredly so. The result is a nearly alien frame when contrasted with the constricting lens of conventional Neodarwinism. In so doing, biology is projected beyond materiality and placed in its proper context as a 'verb' propelled through cellular communication directed towards local problem-solving (De Loof 2015).

Previous chapters have emphasized the wide-ranging differences between Neodarwinism and CBE. Among the many discriminating elements that have been enumerated, one critical differential feature separates the two worldviews. When cellular cognition becomes biology's base, the cellular measurement of information and the adaptive variations that ensue become the propulsive thrust of biological and evolutionary development. All of biology follows. Cells measure because all biological information is ambiguous. In all its myriad forms, our observable biological world is based on this collective assessment of environmental information to maximize its measuring value as inference and prediction. Its direction is always towards sustaining states of individual self-referential homeorhetic preference. Notably, this reduces to an equality with cellular problem-solving and also, explicitly, the defense of cellular self-integrity.

Consequently, multicellularity is strategic cellular information management to assure continuous organismal-environmental assimilation as constant cellular problem-solving representing sustaining states of preferential flux. Within this coherent framework, our evolutionary narrative clarifies: it is a succession of those various multicellular problem-solving strategies over space-time. Crucially, all these measured actions are directed toward the continuous defense of self-referential identity (Miller 2018). There is no explicit destination. Cellular problem-solving and all viral-cellular co-engineering are the cellular means of assuring continuous cellular-environmental complementary for the never-ceasing requirement of all cells to continuously assimilate the environment for the perpetual survival of the four domains (Miller and Torday 2018). Further, and non-trivially, any material expression of biological variation through measured information assessment must have its concurrent adjustment in N-space. Consequently, material multicellular biological variations must be mirrored by a concomitant reciprocal shift of its relevant N-space Episenome (Miller et al. 2020b).

There is a further entailing derivative from these principles of CBE, separating it from all prior evolutionary narratives. CBE insists on the primacy of cellular cognition and the measurement of information by intelligent cells. Consequently, its premises can be subjected to validation or refutation. The theory itself should be robustly predictive of what researchers will ultimately find. If correct, then CBE should be a triggering stimulus for the necessary transformation of biology from a descriptive exercise to a predictive discipline. Further, that research should not merely confirm the enumerated principles of CBE but should also provide the correct theoretical framework for exploiting eukaryotic cells and microbes as capable co-engineering partners for our betterment.

CBE emphasizes the self-organizing capacities of cells, their measuring capacities and measuring tools, their predictive abilities, the meaning of doubt in cellular terms and nuanced preferences in context, the co-existent requirement of templating in N-space, and the shared rules of behavior among our cellular constituents. All of these should serve as biology's predictive protocols to meet contemporary stresses. Fortunately, this is not merely conjectural. Substantial progress in that direction is already underway. Biological robots from frog (*Xenopus laevis*) cells are currently being constructed that exploit cellular self-organizing behaviors to self-assemble capable nano-robots as xenobot swarms that exhibit coordinated locomotion with their cilia (Blackiston et al. 2021). Crucially, these capable xenobots can competently navigate in aqueous environments, heal after damage, and demonstrate a variety of emergent group behaviors. Further, a computational model is being developed to predict the range of functional collective behaviors that can be elicited from these xenobot swarms. This type of research is the specific reward of our separation from the past. The Modern Synthesis may be elegant but offers little or nothing towards that productive narrative.

What might be considered the essential distance between CBE and Neodarwinism? Cells are not minuscule automatons. Instead, all cells dwell within an unbending realm. Being is doubt. The living capacity to resolve doubt ever and always lies at the level of cells. Thus, life's story on the planet from its first tentative beginnings can be summed. Evolution is irresolution.

"Once we consider multicellular living organisms as communities of coordinated-but inherently autonomous entities, the nature of life is more meaningfully revealed".

(Ford 2009, p. 351)

References

Adami, C. 2016. What is information? Philosophical Transactions of the Royal Society A: Mathematical, Physical and Engineering Sciences 374(2063): 20150230.

Arthur, W. 2004. Biased Embryos and Evolution. Cambridge , UK: Cambridge University Press.

Baluška, F. 2009. Cell-Cell Channels, Viruses, and Evolution. Annals of the New York Academy of Sciences 1178(1): 106–119.

Baluška, F. and Reber, A. 2019. Sentience and Consciousness in Single Cells: How the First Minds Emerged in Unicellular Species. BioEssays 41(3): 1800229.

Ben-Jacob, E. 2009. Learning from Bacteria about Natural Information Processing. Annals of the New York Academy of Sciences 1178(1): 78–90.

Ben-Jacob, E. and Levine, H. 2006. Self-engineering capabilities of bacteria. Journal of The Royal Society Interface 3(6): 197–214.

Berry, R.S. 2019. Three Laws of Nature: A Little Book on Thermodynamics. New Haven: Yale University Press.

Blackiston, D., Lederer, E., Kriegman, S., Garnier, S., Bongard, J. and Levin, M. 2021. A cellular platform for the development of synthetic living machines. Science Robotics 6(52).

Corning, P.A. 2020. Beyond the modern synthesis: A framework for a more inclusive biological synthesis. Progress in Biophysics and Molecular Biology 153: 5–12.

De Loof, A. (2015). Organic and Cultural Evolution can be Seamlessly Integrated Using the Principles of Communication and Problem-Solving: The Foundations for an Extended Evolutionary Synthesis (EES) as Outlined in the Mega-Evolution Concept. Life: The Excitement of Biology 2(4): 247–269.

Dreze, M., Charloteaux, B., Milstein, S., Vidalain, P.-O., Yildirim, M. A., Zhong, Q., ... Vidal, M. 2009. "Edgetic" perturbation of a C. elegans BCL2 ortholog. Nature Methods 6(11): 843–849.

Ford., B.J. 2004. Are Cells Ingenious? Microscope 52(3/4): 135–144.

Ford, B.J. 2009. On Intelligence in Cells: The Case for Whole Cell Biology. Interdisciplinary Science Reviews 34(4): 350–365.

Ford, B.J. 2017. Cellular intelligence: Microphenomenology and the realities of being. Progress in Biophysics and Molecular Biology 131: 273–287.

Futo, M., Opašić, L., Koska, S., Čorak, N., Široki, T., Ravikumar, V., ... Domazet-Lošo, T. 2021. Embryo-Like Features in Developing Bacillus subtilis Biofilms. Molecular Biology and Evolution 38(1): 31–47.

Goldenfeld, N. and Woese, C. 2007. Biology's next revolution. Nature 445(7126): 369–369.

Goldenfeld, N. and Woese, C. 2011. Life is Physics: Evolution as a Collective Phenomenon Far From Equilibrium. Annual Review of Condensed Matter Physics 2(1): 375–399.

Goo, E., Majerczyk, C.D., An, J.H., Chandler, J.R., Seo, Y.-S., Ham, H., ... Hwang, I. 2012. Bacterial quorum sensing, cooperativity, and anticipation of stationary-phase stress. Proceedings of the National Academy of Sciences 109(48): 19775–19780.

Grover, D., Chen, J.-Y., Xie, J., Li, J., Changeux, J.-P. and Greenspan, R.J. 2022. Differential mechanisms underlie trace and delay conditioning in Drosophila. Nature 603(7900): 302–308.

Hellingwerf, K.J. 2005. Bacterial observations: a rudimentary form of intelligence? Trends in Microbiology 13(4): 152–158.

Hoffmeyer, J. 2002. The central dogma: A joke that became real. Semiotica 2002(138): 1–13.

Igamberdiev, A.U. and Shklovskiy-Kordi, N.E. 2017. The quantum basis of spatiotemporality in perception and consciousness. Progress in Biophysics and Molecular Biology 130: 15–25.

Jablonka, E. and Lamb, M.J. 1999. Epigenetic Inheritance and Evolution: The Lamarckian Dimension. Oxford, UK: Oxford University Press.

Jablonka, Eva and Lamb, M.J. 2008. The Epigenome In Evolution: Beyond The Modern Synthesis. Vosgis Herald 12.

Koonin, E.V. 2009. Darwinian evolution in the light of genomics. Nucleic Acids Research 37(4): 1011–1034.

Lamm, E. 2014. The genome as a developmental organ. The Journal of Physiology 592(11): 2283–2293.

Lewontin, R.C. 1970. The Units of Selection. Annual Review of Ecology and Systematics 1(1): 1–18.

Miller, W.B. 2013. The Microcosm Within: Evolution and Extinction in the Hologenome. Boca Raton, Florida, United States: Universal Publishers.

Miller, W.B. 2016. Cognition, Information Fields and Hologenomic Entanglement: Evolution in Light and Shadow. Biology 5(2): 21.

Miller, W.B. 2018. Biological information systems: Evolution as cognition-based information management. Progress in Biophysics and Molecular Biology 134: 1–26.

Miller, W.B., Baluška, F. and Torday, J.S. 2020a. Cellular Senomic Measurements in Cognition-Based Evolution. Progress in Biophysics and Molecular Biology 156: 20–33.

Miller, W.B. and Torday, J.S. 2018. The Origins of Biological Deception: Ambiguous Information and Human Behavior. Minding Nature 11(3).

Miller, W.B., Torday, J.S. and Baluška, F. 2019. Biological Evolution as Defense of "Self". Progress in Biophysics and Molecular Biology 142: 54–74.

Miller, W.B., Torday, J.S. and Baluška, F. 2020b. The N-space Episenome unifies cellular information space-time within cognition-based evolution. Progress in Biophysics and Molecular Biology 150: 112–139.

Noble, D., Jablonka, E., Joyner, M.J., Müller, G.B. and Omholt, S.W. 2014. Evolution evolves: physiology returns to centre stage. The Journal of Physiology 592(11): 2237–2244.

Oyama, S. 2000. The Ontogeny of Information: Developmental Systems and Evolution. Durham, North Carolina: Duke University Press.

Pigliucci, M. 2007. Do We Need An Extended Evolutionary Synthesis? Evolution 61(12): 2743–2749.

Price, M.E. 2017. Entropy and Selection: Life as an Adaptation for Universe Replication. Complexity 2017: 1–4.

Reber, A.S. 2019. The First Minds: Caterpillars, Karyotes, and Consciousness (Illustrate). New York City, NY, USA: Oxford University Press.

Ryan, F. 2009. Virolution. New York, NY, USA: HarperCollins Publishers.

Shapiro, J.A. 2009. Revisiting the Central Dogma in the 21st Century. Annals of the New York Academy of Sciences 1178(1): 6–28.

Shapiro, J.A. 2011. Evolution: A View from the 21st Century. Upper Saddle River, NJ: FT Press.

Shapiro, J. 2017. Living Organisms Author Their Read-Write Genomes in Evolution. Biology 6(4): 42.

Stencel, A. and Crespi, B. 2013. What is a genome? Molecular Ecology 22(13): 3437–3443.

Tagkopoulos, I., Liu, Y.-C. and Tavazoie, S. 2008. Predictive Behavior Within Microbial Genetic Networks. Science 320(5881): 1313–1317.

Tauber, A.I. 2000. Moving beyond the immune self? Seminars in Immunology 12(3): 241–248.

Torday, J.S. 2015. A Central Theory of Biology. Medical Hypotheses 85(1): 49–57.

Torday, J. and Miller, W. 2016. The Unicellular State as a Point Source in a Quantum Biological System. Biology 5(2): 25.

Torday, J. and Miller, W. 2020. Cellular-Molecular Mechanisms in Epigenetic Evolutionary Biology. Cham, Switzerland: Springer.

Torday, J.S. and Rehan, V.K. 2012. Evolutionary Biology, Cell-Cell Communication and Complex Disease. Hoboken, New Jersey: Wiley-Blackwell.

Uller, T., Moczek, A.P., Watson, R.A., Brakefield, P.M. and Laland, K.N. 2018. Developmental Bias and Evolution: A Regulatory Network Perspective. Genetics 209(4): 949–966.

Walker, S.I. and Davies, P.C.W. 2013. The algorithmic origins of life. Journal of The Royal Society Interface 10(79): 20120869.

Walker, S.I. and Davies, P.C.W. 2016. The "Hard Problem" of Life. ArXiv Preprint ArXiv:1606.07184.

Walker, S.I., Kim, H. and Davies, P.C.W. 2016. The informational architecture of the cell. Philosophical Transactions of the Royal Society A: Mathematical, Physical and Engineering Sciences 374(2063): 20150057.

Walsh, D.M. 2014. The negotiated organism: inheritance, development, and the method of difference. Biological Journal of the Linnean Society 112(2): 295–305.

Wang, Z., Goyal, A., Dubinkina, V., George, A.B., Wang, T., Fridman, Y. and Maslov, S. 2021. Complementary resource preferences spontaneously emerge in diauxic microbial communities. Nature Communications 12(1): 6661.

Witzany, G. 2010. Biocommunication and natural genome editing. World Journal of Biological Chemistry 1(11): 348.

Witzany, G. (ed.). 2009. Natural Genetic Engineering and Natural Genome Editing. Hoboken, New Jersey: Wiley-Blackwell.

Xavier, J.B. 2011. Social interaction in synthetic and natural microbial communities. Molecular Systems Biology 7(1): 483.

Chapter 20

Conclusion—The Intelligent Measuring Cell

"I often say that when you can measure what you are speaking about, and express it in numbers, you know something about it; but when you cannot measure it, when you cannot express it in numbers, your knowledge is of a meagre and unsatisfactory kind; it may be the beginning of knowledge, but you have scarcely in your thoughts advanced to the state of Science, whatever the matter may be".

Lord Kelvin, 1883

Measure seven times, cut once

Russian Proverb

When introducing any complex and unfamiliar concept, a common expectation is that it should be amenable to reduction to a manageable, compact bullet point. In venture capital, the equivalent challenge is the formidable 30-second 'elevator pitch'. Commensurately, the intricacies of Darwinian theory have been often framed by the catchphrase, 'survival of the fittest' as a convenient, rough quintessence of Darwinism. Unarguably, any catchphrase is an immoderate over-simplification. Yet, in the case of Darwinian evolutionary biology, 'survival of the fittest' insinuated into the public's imagination with uncommon facility and resolutely imprinted on generations of academics. If this gripping tautology is a reasonable metric of potential success, what similar handy signature could be offered for CBE? Fortunately, a single sentence suffices to adequately summarize CBE's entire scope—"Intelligent cells measure information and communicate effectively to engineer collective biological outcomes to uphold themselves". Admittedly, it's not as arresting as "survival of the fittest". Nor is it as poetic as Tennyson's "Nature red in tooth and claw", co-opted from a poem written ten years before Darwin's 'Origin' but swiftly appropriated by those favoring natural selection. Happily, in the case of CBE, a further reductive distillation can be offered that still manages to encompass CBE with suitable rigor: "Intelligent cells measure". For both Darwinism and CBE, these respective taglines serve well enough to encapsulate the essence of their respective stances with the proviso that each has many entailing implications.

As limited as these shibboleths may be, both serve well by symbolically emphasizing the ample gulf between their respective scientific worldviews. The Modern Synthesis remains tethered to a central conviction that biological variation emerges from endless successions of predominantly random genetic mutations, principally due to replication errors disciplined by selective pressures. In effect, it is biology through blind, serial approximations. CBE insists otherwise. Dual requisites propel evolutionary variation. First, self-referential intelligent cells must measure ambiguous environmental cues. Second, these internal assessments must be communicated to others for collective judgment to improve their chances of sustaining their defining self-identity. Consequently, CBE's foundational emphasis is on purposeful, non-random collaborative action with genes serving as tools.

Even though both syntheses agree that natural selection has a significant role, they analyze it contrarily. Conventional Neodarwinism regards selection as an implicit evolutionary driver. CBE stipulates that selection is post-facto to variation, having no propulsive evolutionary force, but influencing biological outcomes as a constraining environmental filter. Selection assures that cellular measurements match planetary proscriptions. This relationship explains why the concept of natural selection as a function of fitness is a tautology. Cells must meet unyielding but cyclical planetary requirements. That is merely a default of our intelligent, living processes. Evolution is the narrative of how biological variations arise to be subject to selection. When evolutionary development is firmly placed within the context of viral-cellular engineering and niche construction, there is no requirement for any measure of absolute fitness. It is not 'survival of the fittest;, but 'survival of the fit enough to survive' through conjoint cellular problem-solving. Imperfect variations emerge as collective cellular co-engineering solutions to environmental problems and can still do well enough to grant survival. Selection assures that those collaborative outputs are sufficient to meet the environment with all its vagaries.

Unassailable contemporary scientific evidence corroborates that biology and evolutionary development proceed primarily through non-random cellular activities. Intelligent measuring cells deliberately collaborate to improve the validity of their measurements of environmental cues. Every creature that we can see with our eyes is its product. Therein lies the overriding gap between Neodarwinism and Cognition-Based Evolution. Each is a mind-set rooted in difference scales. CBE is a cell theory focusing on the intelligent cellular use of information. Neodarwinism remains grounded within the 20th century's romance with genes and selection and how these affect the gorgeous macroorganisms that roam beside us.

Despite these substantial differences, CBE has no intention of diminishing Darwin's visionary insights or momentous contributions to biology. Still, all science requires unbiased re-examination when sufficient data justifies a divergent benchmark. After all, supplanting long-held dogmas with fresh evidence is the defining characteristic of scientific progress. At the turn of the 21st century, biology's éminence grise Carl Woese (2002, p. 8742) contended this exact point: "We cannot expect to explain cellular evolution if we stay locked into the classical Darwinian mode of thinking".

That arresting proclamation has now been validated. Evolution cannot be correctly delineated in the absence of a coherent cellular theory of evolution.

Throughout the 20th century, there has been an entrenched belief that the complexities of living processes could be better grasped within a machine metaphor. However, in the 21st century, that machine metaphor is recognized as inapt (Bongard and Levin 2021). Goldenfeld and Woese (2011, p. 386) highlighted this over a decade ago: "Self-reference should be an integral part of a proper understanding of evolution, but it is rarely considered explicitly".

Cognition-Based Evolution corrects that deficiency and thereby represents a unique and comprehensive reconsideration of evolution. This extensive amendment is necessary. The problems with the still-dominant Modern Synthesis have become entirely self-evident. Noble (2021) insists that the Modern Synthesis, formulated in 1942 and before modern genetics, remains steeped within four governing illusions: the absolute primacy of natural selection, the inviolable separation between somatic and germ cells, Crick's Central Dogma of a unidirectional flow of genetic information, and a rejection of Darwin's gemmules, hypothesized as discrete particles permitting communication between the body and germline. All these initial concepts have been proved incorrect. Even Darwin's hypothesis of gemmules has renewed validity, reinvigorated in the form of exosomes (Noble 2021). Crkvenjakov and Heng (2022) concur about these deficiencies, adding that the faith-based assumptions that microevolution through the gradual accumulation of gene mutations could plausibly lead to macroevolution must be entirely re-examined and replaced by a more integrative approach. Correlative to this comprehensive reevaluation, Baverstock (2021) declares that though genes had been the touchstones of biology for decades, they are not the leading elements in cellular functions, phenotypic inheritance, or evolutionary development.

Brown and Hullender (2022) archly insist that evolutionary biology must "face the known deficiencies" and itself evolve. Macroevolution via survival of the fittest cannot be salvaged through arguments for random genetic action or any other Neodarwinian-centered mechanisms (Brown and Hullender 2022). Corning (2020) contends that the major problem is that genuine progress in characterizing evolutionary development can only be made once the obdurate tendency for biologists to portray natural selection as an evolutionary driver is uprooted. It is time to accept that natural selection does not do anything in an active sense. Selection is merely a metaphor for organismal-environmental complementarity. Further yet, that filtering process is not even focused on genes. As Marijuán and Navarro (2022) maintain, the standard Neodarwinian paradigm overreaches its natural limitations. Purporting that evolution is driven by random mutations and natural selection to the exclusion of other mechanisms is one of the "great blunders of science." Consequently, CBE is no minor extension of that narrative. It is a radical separation from the past.

Every generation of scientists over the last century and a half has offered critical adjustments to Darwinism. Not unexpectedly, in an evolving complex field of inquiry involving overlapping disciplines, errors are unavoidable but can still become ingrained, ultimately sourcing reflexive and axiomatic biases. Such was the last century's unwavering commitment to an evolutionary narrative of the power of natural selection to gradually shape random genetic mutations. From our contemporary informed vantage, it is surprising how long this belief system sustained such resolute allegiance despite the accumulating evidence to the contrary. Any can observe the inherent creativity invested in this planet's striking creatures. Each can independently judge how likely it might be that these stunning outcomes could materialize from mere random jostling. Fortunately, 21st century research has overcome that disquieting disconnect, revealing a single coherent, seamless biological continuum. The centrality of cognition represents the true well-spring of that prodigious biological creativity, materializing from intelligent cells seeking cellular solutions to environmental stresses. Once this cardinal principle is fathomed, the paths by which cellular creativity channels into robust biological outcomes become readily apparent.

Critical to the differentiation, and no matter its origin, self-referential cognition is the enabling biological force and co-incident with the inception of life as embodied in the cellular form (Miller 2016). Given this, it is likely that the instantiation of self-reference represents ultimate causation (Miller et al. 2020a). From this incontrovertible foundation, scale-free basal cellular intelligence emerged. That instantiation of self-referential cognition with its co-existing embedded liberties and constraints inaugurated permanent 'optimal' cellular rules across evolutionary space-time. Within this obligatory context, an expansive range of cellular problem-solving tools arose. Once self-reference supervened, life began and was thereafter explicitly defined by it. All living things share this essential co-equality. Consequently, no privileged level of causation can be assigned within the networks that enact levels of higher biological complexity (Noble 2012).

In those networks, genes are not a linear program. Instead, they form an informational tool within an overarching cell-wide information management system that governs cell-cell interchanges and responses to both internal and external environmental stresses. Accordingly, the information content of biology clearly resides within the dynamics of the system and not within any material participle of the information content itself (Miller 2018). As the cell is a highly integrated functional entity, all its living elements must participate in the reception, measured assessment, and ultimate deployment of information, constituting its defining self-referential observer/participant status (Miller 2018, Miller et al. 2019, 2020a,b).

The requisites of self-referential cellular cognition and entailing dependence on conjoining measurements for survival necessitate that each cellular entity attaches to a conforming information-space referencing platform (Miller et al. 2020b). How else might cells with highly divergent genetic complements and degrees of subspecialization sufficiently share and reference information to permit unifying cellular predictions? How might tens of trillions of participants assess information with sufficient concordance to willingly exchange resources and enable holobionts?

Since all cells dwell in ambiguity leading to superpositions of choices, N-space should be considered the locus where biological Bohmian implicates co-exist (Miller et al. 2019, 2020b). When so considered, the elusive character of cellular self-reference sharpens. Self-reference is the embodied cellular ability to conduct the measured assessment of imprecise information by sensing its uncertainties. A competent sensory apparatus must be capable of receiving ambiguous environmental cues and concurrently sense that any received information is imperfect. Consequently, self-reference is the inception of biological observer/participant status in the context of its superimposed implications. Accordingly, self-referential awareness can be regarded as the birth of the biological implicate realm (Miller 2016, 2018, Miller et al. 2019, Torday and Baluška 2019). Thus, the crux of biology is settling those implicates into living expression. Biology is about making choices and problem-solving. Crucially, self-awareness permits the postponement of biological action to a sufficient extent that it can be subjected to measuring assessment prior to any contingent molecular deployments. Cellular 'doubt' is its capacity to sustain some superpositions of implicates before explicate action. Therefore, cells collapse the implicate realm into a focused bandwidth of choice contingencies before explicit biological expression. Fortuitously, this line of reasoning offers potential insight into 'higher' intelligence, which can now be construed as a function of the species' capacity to juggle greater numbers of implicates before explicate biological direction.

The need for a conjoint information referencing platform relies on a specific constitutive cellular reality. Each cell creates its own reality. All of the information that any cell has at its disposable is self-produced as an exclusive internally measured interpretation of environmental stimuli. Necessarily, interpretive meaning is muffled and distorted by physical barriers and thermodynamic requirements. These limitations include the membranes that permit the relative negentropic status of cell and its successful maintenance of homeorhetic balance.

Notably, a quantum framework focuses the essence of cellular problem-solving. Self-referential measurement drives biological and evolutionary development through entangling cellular quantum coherences, resonances, criticalities, and instabilities (Torday and Miller 2020). From among these living uncertainties, the problem-solving collapse of cellular implicates represents predictions as least uncertainties. Necessarily then, concordant measurements in N-space are a precondition of any resolution of cellular problems to least uncertainties. When all the information that any cell has is self-generated, shared measurements are the only cellular pathway toward higher levels of effective information content as having achieved greater measuring value. Incontestably, mutualizing reference points must be available from which measurements can achieve meaning in those collective terms.

Once cells are correctly understood as intelligent and measuring, all subsequent multicellular life must be regarded as forms of strategic information management. Consequently, the origin of the cellular impulse for collaborative engineering clarifies. Collective engineering is strategic information management at any scale, whether human, termite, or cell. Furthermore, for any co-engineering, there must be a conforming informational architecture for both measurement and communication as predictions. A standard engineering definition is "the action of working artfully to bring something about". From this, cellular creativity comes into focus. Since cellular engineering is problem-solving in the face of environmental stresses, and because the imposed patterns of environmental cues are not always predictable, that problem-solving engineering must be considered an act of cellular creativity. Biological novelty springs from within this empowering interplay.

The same coherent narrative better explains genetic co-option and convergent evolution beyond the dominant notion of selective pressures acting on random genetic variations. For example, cephalopods and vertebrates independently evolved camera-lens-style eyes with closely matching features even though their last known common ancestor was more than 500 million years ago. Similarly, across Metazoa, limb patterning exhibits homologous attributes. A recent research report identifies a network of genes essential in squid eye development that also play a crucial role in limb development across animals, including *Drosophila* (Neal et al. 2022). As these researchers conclude, this co-option represents the skillful 'hijacking' of genetic tools to problem-solve creatively. Placing this genetic co-option into a framework of cellular co-engineering and problem-solving in which genes are tools is undoubtedly a more robust explanation than random genetic shifts and selection-producing convergent outcomes. Thus, a scrutable pathway to new forms now exists. Alternative engineering pathways emerge from fresh tools and new partnerships, leading to novel forms as distinct solutions to cellular problems.

Consequently, phenotype must be reappraised. Phenotypes are conjoint measured environmental predictions made by intelligent cells in the continuous strategic defense of individual cellular self-identity. Therefore, evolution is the succession of those various multicellular strategies to meet the planetary environment. This advantaged outcome is achieved through the perpetual internalization of the external environment in defense of cellular homeorhesis, thereby sustaining its critical self-identity.

The same initiating forces that enabled the first competent cell have perpetuated in reiterating and inventive forms for billions of years. Consequently, the basic cell should be considered the first instance of successful niche construction (Torday 2016). That cellular niche construction with its further internal niches of the nuclear genome, organelles, plasma membrane, ribosomes, endoplasmic reticulum, and other internal components is a highly competent integration of biological structural materiality and entrained energy. However, it is also a manifestation of natural informational engineering as self-referential self-organization in the guise of biological form (Miller 2018). The previously mysterious origin of living self-organization focuses. Self-reference is self-organization. All life spills forward from this elemental equality. Notably, though, even as self-reference licenses flexible living confrontation with unbending physical forces, deliberate freedom of action can only thrive through the imposition of countervailing multilevel constraints (Torday and Miller 2020). Both must entwine to sustain the rich biological variations that populate the planet.

Scholars will continue to discuss and dissect what consciousness means, and scientists will seek to identify those structures whose intracellular resonances yield conscious perception. Even when many of those are uncovered, we may not be closer to the fundamental element that licenses 'knowing'. However, it is possible to put these essential questions aside and still make useful progress in biology. Evolution is the product of cognitive awareness as an instantiated state function, embodied within the cellular form, and scale-free as a basal aliquot across biology as its energizing force (Miller 2016). By accepting this immanent base, Cognition-Based Evolution offers its novel reassessment of biological and evolutionary development based on this earthly gift in all its replete ramifications.

Any perusal of the scientific literature on consciousness leads to the rather disquieting discovery that there is no common agreement about its definition. For CBE, the meaning of effective cellular consciousness can be clearly articulated.

Cellular cognition equates with knowing a state of internal preference and competently maintaining it through flexible continent responses to environmental stresses. This definitional stance regards consciousness/cognition as action linked to experiences. There is no 'hard problem' of consciousness with cells, since all information inputs interact with a highly integrated senomic architecture and all information assessment is self-produced. Necessarily, everything that impacts a cell by crossing its intelligent plasma membrane is an experience since cellular preferences derive from that linkage as homeorhetic equipoise. It is this 'preference' that distinctly distinguishes living processes from computers or thermostats. And further, 'preference' includes the cellular appreciation of an advantaged self-referential state linked to 'knowing' that others are self-similar similar. Within this cell-based dynamic of conjoining cognitive awareness, multicellular forms meet environmental stresses through continuous conscious adaptation. Through those partnerships, life spills forward.

From this ground state, biology undergoes a crucial shift in perspective, permanently extending beyond its observational and descriptive roots and its decades-long obeisance to the primacy of genes. Genes are an imperative aspect of cellular life as an essential repository of memory, serving as a flexible engineering tool. Yet, genes do not drive anything. Instead, a modern genomic idiom regards genes as pliant participants within a sensitive feedback loop between genome and phenotype as a critical element in cellular problem-solving (Auboeuf 2021). Crucially, this reciprocating interplay enmeshes within a network of complex symbiotic relationships in holobionts. Consequently, evolution must be squarely centered within a framework of systemic variations in informational architectures supporting organismal self-sufficiency, differential self-construction, and reproduction (Marijuán and Navarro 2022). Consequently, a genuine informational approach to evolutionary development is required (Marijuán and Navarro 2021).

CBE proposes and fully articulates this alternative framework. Biology can now be firmly based on the impact of the measured assessment of information on cellular life and the strategic information management capacities of networking cells. Science is measurement, and, at last, biology becomes the science of biological measurement (Miller et al. 2020a). To prosper, it must identify what to measure and how to measure it accurately. That appraisal requires understanding those dynamic processes requiring particular scrutiny, which are readily identifiable in CBE. The dynamics of cell-cell communication and the cellular parameters that enable natural cellular and viral co-engineering to sustain collaborative living are the definitive objects of this interrogation. Notably, the exact crux of future biological studies focuses when placed in this frame. The cardinal determinants in biology and its perpetual evolution are the measuring capacity of cells and the continuous flow of downstream skills that flow from the empowerment of intelligent cells. Genes, as memory, are the cellular record of their collective solutions to environmental problems writ in biological form. Rather than agency, they form a record of collective environmentally-validated cellular predictions as cellular solutions to environmental challenges. Therefore, every cell is both the embodiment of 'self' and also a repository of information space-time memory (Miller et al. 2019, 2020b).

The Four Domains endure because the planet consistently filters for the integrity of cellular measurement as the perpetual focus of information space-time memory

(Miller et al. 2020a). The obligatory recapitulation of the unicellular eukaryotic zygote is an explicit manifestation of a constant recentering of the measuring capacity of the cellular form to accord with initiating principles. Why is this measuring capacity so scrupulously sustained? Because the measuring capability of cells is the essence of cognition, and cognition defines the living state. Selection acts to preserve the measuring capacities of cells since that equates with self-referential integrity.

From within this reanimating frame, a productive congruity emerges. Evolutionary biology transforms into the science of why, how, what, and with whom cells measure and communicate under differing conditions of environmental stress. Unshackled from Neodarwinism, biology can now be articulated as a verb, cast from lively processes and essential harmonies, rather than the passive default of random conformities and grim confrontations.

In 1970, the great evolutionary biologist Ernst Mayer said: "The theory of evolution is quite rightly called the greatest unifying theory in biology" (Ernst Mayr 1970). Yet, cellular intelligence is not a theory. It is measurable and has been proven. From our contemporary vantage, we can offer a productive amendment: *cellular cognition is the great unifying force in biology.* Talking about evolution without placing intelligent cells at the center is equivalent to subtracting life from the study of biology.

The ambiguity of biological information yields a living imperative. Unlike machines, biology and its evolution are irresolution. Biological information is imprecise, and therefore, cells must measure. They measure individually to sustain their specific states of homeorhetic preference. They measure collectively to improve the validity of that available information in context, driving cell-cell communication. Because they can measure together and communicate, cells can engineer to create cellular niche constructions as their collective expression of cellular preferences. Clearly then, life is the still inexplicable coalescence of the cellular ability to sense information, discern its ambiguities, and act deliberately to render self-referential biological solutions to environmental stresses.

Thus, all biology and its evolution is a continuous narrative of the intelligent, measuring cell. Life began as the self-referential attachment to information space-time. Cognition is its privilege, ambiguity its context, communication its means, genes are its tools, and evolution its yield. A further illuminating reduction crystallizes from these foundations. Biology is self-referential cellular problem-solving, and evolution is the planetary narrative of their solutions despite all intervening uncertainties.

References

Auboeuf, D. 2021. The Physics–Biology continuum challenges darwinism: Evolution is directed by the homeostasis-dependent bidirectional relation between genome and phenotype. Progress in Biophysics and Molecular Biology 167: 121–139.

Baverstock, K. 2021. The gene: An appraisal. Progress in Biophysics and Molecular Biology 164: 46–62.

Bongard, J. and Levin, M. 2021. Living Things Are Not (20th Century) Machines: Updating Mechanism Metaphors in Light of the Modern Science of Machine Behavior. Frontiers in Ecology and Evolution 9, 650726.

Brown, O.R. and Hullender, D.A. 2022. Neo-Darwinism must Mutate to survive. Progress in Biophysics and Molecular Biology 172: 24–38.

Corning, P.A. 2020. Beyond the modern synthesis: A framework for a more inclusive biological synthesis. Progress in Biophysics and Molecular Biology 153: 5–12.

Crkvenjakov, R. and Heng, H.H. 2022. Further illusions: On key evolutionary mechanisms that could never fit with Modern Synthesis. Progress in Biophysics and Molecular Biology 169–170: 3–11.

Goldenfeld, N. and Woese, C. 2011. Life is Physics: Evolution as a Collective Phenomenon Far From Equilibrium. Annual Review of Condensed Matter Physics 2(1): 375–399.

Marijuán, P.C. and Navarro, J. 2021. From Molecular Recognition to the "Vehicles" of Evolutionary Complexity: An Informational Approach. International Journal of Molecular Sciences 22(21): 11965.

Marijuán, P.C. and Navarro, J. 2022. The biological information flow: From cell theory to a new evolutionary synthesis. Biosystems 213: 104631.

Mayr, E. 1970. Populations, species, and evolution; an abridgment of Animal species and evolution. Cambridge, Massachusetts: Harvard University Press.

Miller, W.B. 2016. Cognition, Information Fields and Hologenomic Entanglement: Evolution in Light and Shadow. Biology 5(2): 21.

Miller, W.B. 2018. Biological information systems: Evolution as cognition-based information management. Progress in Biophysics and Molecular Biology 134: 1–26.

Miller, W.B., Baluška, F. and Torday, J.S. 2020a. Cellular Senomic Measurements in Cognition-Based Evolution. Progress in Biophysics and Molecular Biology 156: 20–33.

Miller, W.B., Torday, J.S. and Baluška, F. 2019. Biological evolution as defense of "self". Progress in Biophysics and Molecular Biology 142: 54–74.

Miller, W.B., Torday, J.S. and Baluška, F. 2020b. The N-space Episenome unifies cellular information space-time within cognition-based evolution. Progress in Biophysics and Molecular Biology 150: 112–139.

Neal, S., McCulloch, K.J., Napoli, F.R., Daly, C.M., Coleman, J.H. and Koenig, K.M. 2022. Co-option of the limb patterning program in cephalopod eye development. BMC Biology 20(1): 1.

Noble, D. 2012. A theory of biological relativity: no privileged level of causation. Interface Focus 2(1): 55–64.

Noble, D. 2021. The Illusions of the Modern Synthesis. Biosemiotics 14(1): 5–24.

Torday, J. 2016. The Cell as the First Niche Construction. Biology 5(2): 19.

Torday, J.S. and Baluška, F. 2019. Why control an experiment? EMBO Reports 20(10).

Torday, J. and Miller, W. 2020. Cellular-Molecular Mechanisms in Epigenetic Evolutionary Biology. Cham, Switzerland: Springer.

Woese, C.R. 2002. On the evolution of cells. Proceedings of the National Academy of Sciences 99(13): 8742–8747.

CHAPTER 21

Appendix
Differential Features Between Cognition-Based Evolution and the Neodarwinian Modern Synthesis

Cognition Based Evolution	Neodarwinian Modern Synthesis
Self-referential cellular cognition propels all biology and evolution	Selection-determined gene frequencies drive evolution
Evolution is the narrative of the intelligent measuring cell and its co-existent virome	Random mutation/selection
Overarching evolutionary framework is viral-cellular information assessment, communication, and information management	Random genetic mutations as replication errors/differential survival
Living state as an information management system/informational interactome based on cellular measurement	Selection-biased replication
Biology depends on a universal relational information space-time matrix (N-space)	Space-time is non-relational in a biological sense
Every cell attaches to N-space as a Pervasive Information Field	No equivalent concept
Multicellularity depends on an conjoining attachment to an N-space Episenome for concordant measurement	No equivalent concept

Evolution as symbiogenesis/ collaboration/mutualism at all scales	Evolution as competitive differential fitness
Collective engineering is strategic information management	Selection
Evolution is largely non-random (Intelligent cells harness stochastic phenomena, directing them into productive biological co-engineering outputs)	Evolution is primarily random as an epiphenomenon of replication errors
Symbiosis and collaboration dominate evolution	Selection/differential fitness/competition
Crick's Central dogma is invalid	The Central Dogma remains a useful concept
Natural viral-cellular engineering/ niche constructions enact forms	Forms arise from random mutations/ selection
Viruses as intermediaries of communication and co-engineering participants with cells	Host/pathogen model
Viruses as a primary driver of evolution through viral symbiogenesis	Host/pathogen model
Horizontal genetic transfers as expressions of infectious disease dynamics drive epigenetic phenomena as a significant evolutionary driver	Host/pathogen model
Integrated proteomics/transcriptomics as part of self-referential cellular problem-solving	Transcription is selection dependent
Self-referential cellular homeorhetic preferences (equipoise) guides resource utilization	Selection, random mutations
Selection, random mutations information underpins biology in a self-referential	Blind stochasticity
Triadic energy-information-communication underpins biology in a self-referential information cycle	Random genetic mutations/ rare directed mutagens/Selection determined gene frequencies

Biologic forms as informational architectures and collective informational motifs through informational engineering	Forms based on genetic mutations and survivorship bias
The senome as a whole-cell sensory apparatus serving as the cognitive gateway for the cellular appraisal of information.	Cells as foci of molecular/energetic reactions
The cellular senome is a critical component of self-referential cellular information management	No equivalent coordinating concept
Multicellularity is secondary to shared information assessment and energy resourcing as part of an informational interactome	Multicellularity a product of a competitive fitness landscape
The genome is a flexible read-write informational repository in continuous reciprocation with the cell	The genome is largely fixed, changing mostly through random replication errors
Biological variations result from coordinated natural cellular engineering and niche construction	Biological variation results from random genetic errors subjected to selection
Deployed mutations are primarily non-random	Random genetic errors drive variations
Genes as tools of a self-referential cell-wide information matrix as a repository of cellular solutions to environmental problems	Genes control biological processes
Phenotype as self-referential environmental exploration	Phenotype is selection-dependent
Phenotypes as cumulative Terminal Additions through adaptive cellular-molecular responses	Variation/Selection
Phenotypes are produced by natural viral-cellular co-engineering, representing collaborative predictions	Phenotypes result from selection, leading to genetic frequencies
Biological variation is due to changes in cell-cell signaling/natural viral-cellular engineering	Variation secondary to blind mutations

Natural selection is a post-facto filter of precedent cellular problem-solving	Natural selection actively shapes evolution
Selection for collaboration to effect continuous organismal-environmental complementarity	Selection for differential fitness
The primary object of filtering selection is the measuring cell	Selection targets macroorganisms
Evolution serves the four perpetual domains (Prokaryota, Archaea, Eukaryota, Virome); Macro-organic evolution is an epiphenomenon	Macroorganic reproduction/selection drives evolution
Biological novelty via Epigenetic impacts vary viral-cellular co-engineering as problem-solving	Random genetic variations
Deterministic cellular creativity solves problems relating to ambiguous information	Fitness/selection; stochastic variables
Holobionts as deeply integrated multispecies eukaryotic-microbial organisms as a consenual 'we'	Microbes as colonizers-Host/pathogen/ symbiont
Holobionts as collective cellular environmental problem-solving	Holobionts are selection-dependent outcomes
Holobionts as environmental predictions	Selection
Primacy of immunology to sustain self-referential integrity	Immunology as a selection driven secondary phenomenon
Meiosis maintains species integrity, and reinforces collective adaptive immunity as powerful force to protect average species	Meiosis assists genetic variation
The eukaryotic zygotic recapitulation is information management, re-centering cellular processes, as a crucial participant in epigenetic modulation	Zygote is part of the selection-driven macroorganic reproductive cycle.
Macro-organisms remain permanently anchored to unicellular roots; evolution has depended on continuous exaptations of unicellular tools	Macro-organisms are largely independent of unicellular requirements

Speciation is an immunological epiphenomenon, largely driven by infectious disease dynamics	Isolating physical or ecological barriers; disruptive or sexual selection; genetic drift
A species is the biological expression of a conforming information set as its N-space Episenome; coordinate life requires concordant measurements	Genetic and phenotypic resemblances and common traits
Extinctions are primarily the result of infectious disease dynamics, expressed in term of Multi-generational Attritional Loss of Reproductive Fitness (MALF)	Little considered in Neodarwinism, loosely attributed to changes in climate
Speciation and extinction are two separable expressions of infectious disease dynamics	Distinct phenomena with different causes
Evolution depends on collective cellular measurement to improve its validity requiring an attachment to information space-time	Selection/mutation
In holobionts, all development is co-development; all evolution is co-evolution	Evolution and development in holobionts proceeds largely separately between eukaryotic body cells and the microbiome with separate selective influences
Self-reference is self-organization	Self-organization is an emergent phenomenon
Life is self-referential cognition	No formal definition
A theory of mind is an extension of cell theory	No consensus about mind or its definition
Being is doubt. Evolution is self-referential irresolution	Random genetic variations and selection

Index

For Product Safety Concerns and Information please contact our EU
representative GPSR@taylorandfrancis.com
Taylor & Francis Verlag GmbH, Kaufingerstraße 24, 80331 München, Germany

www.ingramcontent.com/pod-product-compliance
Lightning Source LLC
Chambersburg PA
CBHW060355220326
41598CB00023B/2930